WATER

WATER

THE FATE OF OUR MOST PRECIOUS RESOURCE

MARQ DE VILLIERS

First published by Stoddart 1999
Revised (paperback) edition published by Stoddart 2000
McClelland & Stewart trade paperback (second) edition published 2003

National Library of Canada Cataloguing in Publication

De Villiers, Marq, 1940–
Water : the fate of our most precious resource / Marq de Villiers. – 2nd ed.

Includes bibliographical references and index.
ISBN 0-7710-2641-2

1. Water-supply. 2. Water-supply – Canada. I. Title.

HD1691.D48 2003   333.91
C2003-902380-D

We acknowledge the financial support of the Government of Canada through the Book Publishing Industry Development Program and that of the Government of Ontario through the Ontario Media Development Corporation's Ontario Book Initiative. We further acknowledge the support of the Canada Council for the Arts and the Ontario Arts Council for our publishing program.

Designed by Cindy Reichle
Typeset in Bembo by M&S, Toronto
Printed and bound in Canada

This book is printed on acid-free paper that is 100% recycled ancient-forest friendly (100% post-consumer recycled).

McClelland & Stewart Ltd.
The Canadian Publishers
481 University Avenue
Toronto, Ontario
M5G 2E9
www.mcclelland.com

1 2 3 4 5   07 06 05 04 03

For my grandfather, Johannes Jacobus de Villiers,
who farmed in a hard land and knew the value of water

CONTENTS

WATER

Why repeat mistakes when there are so many new ones to make?
— René Descartes

We're all downstream.
— Ecologists' motto,
adopted by Margaret and Jim Drescher,
Windhorse Farm, New Germany, Nova Scotia

INTRODUCTION

~~~~~~

The opening sentences of the first edition of this book described a nice little bucolic scene, a small spring that seeped out behind the barn on our farm near Maynooth, in the deciduous forest belt of Ontario. The spring bubbled sleepily from the ground, I wrote, and if you were really quiet and there was no wind in the trees, you could hear it making little burping noises, like a baby content at the breast. The water trickled through the grass and into a stone runnel, left there by a farmer long gone, and then formed a pool, ducked underground for a while, resurfaced in a small wetland, and disappeared into a ravine. Lower down it became a creek and, joined by others, a stream, a river, a lake, and then . . . well, it fetched up in the sea, where it lived a while.

As a first approximation of the hydrological cycle, this was accurate enough, as far as it went.

If I were starting now, I think, the passage would have darker tones, because in the interim, water matters have become much more dire. In many – too many – parts of this world of ours, a small spring wouldn't be left to chuckle cheerfully away; it would have to serve as the primary water source for twenty, fifty, a hundred people, who would squabble over it and in all likelihood exhaust it or pollute it. Or, if the spring were left alone, some farmer or industrialist further downstream would intercept it, and abstract the water, effectively stealing it from those even further down. Or someone would pour into it effluents containing toxins, and it would become unfit to drink, noisome, and deadly, and pour into the ocean as a diluted chemical stew.

When I first started looking at the business of water, I assumed I knew quite a bit. Most of us do, I think. I mean, how complicated can it get? Rain, rivers, lakes, and the sea . . . the endless cycle . . . the endlessly renewable resource . . . Alas, I was wrong. I knew, for example, where the water for my small spring came from, and how the forest cover sustained the water table that made it possible, but I had no real grasp of the intricate interconnectedness of the global hydrological system. I knew that the trees on my farm protected the groundwater, but I never knew the scale of the clearing of forests and wetlands that was occurring around the globe, for example in China, and how that clearing led to alternating bouts of drought and flood. I could see for myself how industrial effluents were pouring into, say, Russia's Volga, but I didn't understand the wholesale assault our species is making on the system that sustains us. I always knew how scarce water was in many parts of the world, but I didn't really comprehend the calamity facing many regions of our increasingly overpopulated planet. These things we have had to learn, and they are still going unreported – or at least under-reported.

And yet, and yet . . . The state of the world's water *is* dire, but I think constant jeremiads are as misguided as innocent ignorance. The hydrosphere is complicated, but it can be untangled and understood. There are problems with water, but they *are* solvable.

In the years since I first published this book, some readers called and wrote to say they had found an incongruity between what they saw as my gloomy recounting of the world's water woes, and my "relatively optimistic" view of the future. Yet I don't believe the two views are mutually exclusive. I continue to believe that, serious as the world's two overlapping water crises are, they are not without solutions. On the first crisis, the crisis of supply, there is enough water for everyone, if we manage it correctly; we just need the political will and therefore the money to do what we already know needs to be done. On the second crisis, the crisis of contamination, apparently intractable problems have shown themselves to yield to focused action – the Rhine River is the most obvious example, and the Germans the people to copy. Here, too, we know what needs to be done, and solutions can be found in existing knowledge, applied through politics, driven by citizen will. Proper use of market forces and the application of true cost-accounting to polluters (so that they can therefore pay their real share of the necessary cleanup) are not pie-in-the-sky economics but sensible business practice, of benefit to us all.

I also continue to think that engineering today ranks among the most creative of all endeavours, and that engineers will be part of whatever solutions we derive.

My environmentalist readers have indicated a mistrust of my reliance on technology; among many of them there is an almost Luddite suspicion of technological solutions. I, on the other hand, think sophisticated engineering can be usefully combined with a conservationist ethic to provide solutions that can be arrived at in no other way. It's true, as some have pointed out, that, were we to provide cheap and energy-efficient desalination of sea water, to take a popular example, one of the net results would be to encourage even more growth in places where, ecologically speaking, it shouldn't happen, such as in the deserts of the southwestern United States. This is the same argument that has brought home to traffic engineers that adding a lane to a freeway never alleviates congestion, but only encourages more cars onto the roads. So, in a way, I agree. But I never said that water woes can be solved in isolation to other human-caused problems. It all, in the end, comes back to the engine that drives all environmental problems – the astonishing growth in human populations. Meanwhile, people are dying of contaminated water, and in other places the water is simply running out. If engineers can save the resource, they should be enabled to do so.

As before, I would like to point interested readers to others in the field whose voices are worth hearing. The urgency of the world's water problems has encouraged academicians and scientists all over the world to study water, its availability and quality, its uses and abuses. Published sources range from the intricately technical (the permeability of schist formations) to the ecological and political (transborder water and resource conflicts). There are hundreds of books on the subject, and quite literally thousands of papers. But in the flood, some names stand out, a helping hand for those of us floating by.

Among them is Sandra Postel, the nearest thing in the water world to a philosopher – always thoughtful, cogent, intelligent, with a point of view as limpid as a forest pool. Her *Last Oasis* is the call to arms most often cited by others in the field, and her latest book, *Pillar of Sand*, is an examination of the intricate problems of irrigation. More recently, as director of the Global Water Policy Project in Amherst, Massachusetts, she has turned her abundant energy to entrepreneurship, seeing to the installation of cheap drip-irrigation schemes across the developing world. Peter Gleick is the organization man of the water world: his books and papers are meticulous and precise, a fund of data for those who need it. My views on Russia's Igor Shiklomanov are expressed near the end of my final chapter. I came to see him, in the years I spent in the water world, as a central figure, authoritative and magisterial. Leif Ohlsson's book *Hydropolitics*, which covers much the same ground as this volume, though aimed at the development community, has drawn together a stellar list of academics, and I have cited some of them in the text. Malin Falkenmark, also a Swede, is the person most quoted by other hydrologists; her work is at the core of modern water research. For cogent summaries of technical hydrology and the hydrosphere, the *Encyclopaedia Britannica,* both in its print and online forms, remains essential for the amateur. It is often neglected by scholars, who sometimes forget the meticulousness of its science.

Of the journalistic accounts, Marc Reisner's classic *Cadillac Desert* stands out. It deals primarily with the American southwest, but it is almost compulsively readable, and folded into the enjoyable polemic is a huge amount of information. There are many other passionate and readable books on water, among them Joyce Shira Starr's *Covenant over Middle Eastern Waters: Key to World Survival.* Lester Brown's Worldwatch Institute (from which Sandra Postel emerged) has produced other writers of note, among them Janet Abramovitz, Brian Halweil, and, of course, Brown himself, always a thorn in the side of the complacent. The Washington NGO Population Action International, while not primarily focused on water, has produced a good deal of meticulous research. On the skeptic's side, I was drawn to Eugene Stakhiv, of the U.S. Army Corps of Engineers, an engineer who writes (and talks) in clear, jargon-free, persuasive prose; and to the Spanish hydrologist, Ramón Llamas Madurga, who has the temerity to question received wisdom.

We *can* know what's going on, and we *can* fix what's gone wrong. That's the point of it all.

I | WATER IN PERIL

Is the crisis looming, or has it already loomed?

I was in the Kenyan town of Narok the night a group of Maasai *morans*, warriors going through their rites of passage to tribal elder, clashed with the thuggish national police of President Daniel arap Moi. The cause of the ferocious riot that followed is of no consequence – warrior exuberance had gotten out of hand, and the police had overreacted – but, to be safe, I left them to rattle their spears and truncheons at each other and took refuge in a nearby village. There I was invited in by a family of Gabbra, who lived in a tiny four-hut complex three kilometres from the nearest well.

The senior woman of the household, Manya, invited me to stay and pressed on me unwanted gifts of food she couldn't afford. It was, I knew, a typical African welcome.

In return, I picked up one of the four yellow plastic drums piled in front of the hut – it had started its life as a bulk container for vegetable oil in some far-off industrial city – and offered to help her fetch water. The family laughed, politely, but it was obvious what they were really thinking: white people, *mzungu*, are so inept. Fetching water, after all, was woman's work. So I stayed behind with the men, who were smoking and gossiping on a wooden bench outside one of the huts.

Later that evening Manya and her daughters returned, each with a fifteen-litre pail balanced on her head. They swayed down the trail, singing one of their working songs to pass the hours, as they had done that morning, and as they would do on the days that followed, and as they expected to do, if they thought about it at all, forever.

A little later we ate corn mash and fried banana and sucked on mangos. I declined the water, partly out of politeness and partly out of fear. The well was an old one and had originally been used by fifty families. Now, four times as many drew water from it, and they had to dig down further every year. A few months earlier, Manya said, two workers had descended into the well and had passed up the muck in buckets, deepening the well by the height of a man. The water was muddy and smelled unclean.

All over East Africa — indeed, all over Africa — it is normal for people to walk a kilometre or two or six for water. In the more arid areas, they walk even greater distances, and sometimes all they find at the end is a pond slimy with overuse. More than 90 per cent of Africans still dig for their water, and waterborne diseases like typhoid, dysentery, bilharzia, and cholera are common. Many Africans are a stew of parasites. In some areas the wells are so far below the earth's surface that chains of people are required to pass the water up.

In Mali a few months later I stayed in a village in which an American NGO had installed a solar-energy pump and a galvanized storage tank; it was still working perfectly five years after it had been put into place. In Niger, across the border, a similar pump had broken, and one night a child had opened the stopcock on the tank and the water had all run out, soaking into the parched earth. The child was beaten, but it was too late. The water was gone and the villagers all moved. They never returned.

A year after that, I visited a family of walnut growers in California's Central Valley. They had a drilled well out back, but had recently had to refurbish it because the water had run dry. They were down to 230 metres before they struck water again. They didn't mind. Their trees and gardens were irrigated by water brought in by the California Aqueduct from the Colorado River and supplied to them at ten cents a cubic metre, far below the cost of either gathering or transporting it. They were careful water managers, however, and meticulously metered the amount of water given to each tree. Yet they were members of the local golf club, whose fairways and greens were watered and fertilized all summer to preserve the lushness that golfers demand. They saw nothing incongruous in their behaviour.

The rainfall in that part of the Central Valley was only fifteen to eighteen centimetres a year, the same as the Kenyan plains. The water table was even lower – no one in Kenya could afford to install the kind of pumps that would deplete a subterranean aquifer, or "draw it down" at unsustainable rates, faster than it could be naturally replenished. Same climate, same rainfall, similar families. But when Bonnie, the walnut grower, wanted to fill her swimming pool, she turned the faucet without a second thought. Manya had never seen a swimming pool. Lazing in the water had never figured in her dreams.

<hr/>

The trouble with water – and there *is* trouble with water – is that they're not making any more of it. They're not making any less, mind, but no more either – there is the same amount of water on the planet now as there was in prehistoric times. People, however, they're making more of – many more, far more than are ecologically sensible – and all those people are utterly dependent on water for their lives (humans consist mostly of water), for their livelihoods, their food, and, increasingly, their industry. Humans can live for a month without food, but will die in less than a week without water. Humans consume water, discard it, poison it, waste it, and heedlessly change the hydrological cycles, indifferent to the consequences: too many people, too little water, water in the wrong places at the wrong times and in the wrong amounts. The human population is burgeoning, but water demand is increasing twice as fast.

There are, therefore, not one but two overlapping water crises, the crisis of supply and the crisis of quality. Or, put another way, there is a sufficiency of water on the planet if we manage the resource correctly; the real problem is providing consumers with water that is fit to drink.

The environmental movement, accustomed by now to fits of gloomy Malthusian soothsaying, has forecast increasingly common collisions between demand and supply. Even officials of so sober an institution as the World Bank have joined the chorus. Ismail Serageldin, the bank's vice-president for environmental affairs and chairman of the World Water Commission, stated bluntly a few years ago that "the wars of the twenty-first century will be fought over water." Although he was roundly criticized for this opinion, he refused to disavow it and has frequently asserted that water is the most critical issue facing human development. The former UN secretary-general Boutros Boutros Ghali said something similar about water wars. So did the late King Hussein of Jordan, who had obvious cause to mean it. Egypt has more than once threatened to go to war over diversions of the Nile. In the United States, the intelligence community has turned its suddenly alarmed attention to resource conflicts that could further destabilize peace and encourage terrorism. The National Intelligence Council warned in 2001 that "as countries press against the limits of available water between now and 2015, the possibility of conflict will increase." Water is in crisis in China, in Southeast Asia, in southwest America, in North Africa – indeed, in much of Africa except the Congo, Niger, and Zambezi basins. Even in Europe there are shortages – *drought* is no longer a word alien to England, where water tables were dropping throughout the early 1990s. In many parts of Europe, downstream towns and cities are beginning to feel the consequences of the careless alteration in age-old hydrological ecosystems, as rivers suddenly rage out of control, wetlands dry up, and contaminants enter the groundwater. Yes, even in Europe there is a crisis in water supply and management, as groundwater tables sink and rivers are reduced to a trickle or increased to a destructive flood.

And in all these places, the quality of the water that is available is bad, and getting steadily worse. The UN-sponsored Third World Water Congress, which met in Stockholm in August 2001, concluded that "a looming water crisis could threaten one in three people by 2025, generating as much conflict in this century as oil did in the last." The report was written off as typical envirohyperbole, but its data, as opposed to its conclusions, proved harder to rationalize away.

In the winter of 2000 I travelled north from Agadez, in Niger, towards the Algerian town of Tamanrassat, which nestles at the foot of the Ahaggar Mountains. At an oasis along the way, the small pond of water, the only source in the community, because the well had temporarily filled in, was contaminated by the corpse of a dead camel. It was half in and half out of the water, and had clearly been there for some time, because it was decomposing, and the water around it was a slimy green in colour and contained bits of reddish camel dung. No one had taken the camel out because no one knew whose it was, and to mess with another man's camel in the Sahara is a risky thing. Prudent travellers to the oasis that week dug shallow wells near the pond, hoping the sand would filter out the detritus. Less prudent but still cautious travellers dipped their goatskin bags into the pond as far from the camel as they could. Others, inured to the sight and stench, simply filled their skins where they could and went on their way.

When I got back home after these travels, I would tell groups of people the anecdote of the dead camel. Their reactions were uniformly appalled – *if the water was making them sick, why didn't they do something about it, what was wrong with those people?*

In the end, as I listened to the reactions, the incident transformed itself in my mind from a story to a parable, the Parable of the Dead Camel. So I would tell the audiences of some other things I had found on my travels, dead camels of their sort, only much larger and much more deadly – our own planetary dead camels. For everywhere you looked, what we were doing to our water resources more and more resembled the carelessness and contempt for safety exhibited by those Tuareg nomads of the deep desert.

Oh, there were and still are skeptics, just as there are those who still don't believe in the notion that each generation is simply the earth's steward, holding it in trust for generations to come. These skeptics believe the problem is overblown, and, even if it isn't, it will surely be solved through human ingenuity and technological advances in the future. But these people are a constantly shrinking minority. Everywhere you look, there are signs that the water supply is in peril, dead camels of our own making:

- Estimates from the U.S. government and the United Nations – in a rare display of unanimity – maintain that, by 2015, at least 40 per cent of the global population, or about three billion people, will live in countries where it is difficult or impossible to get enough water to satisfy basic needs.
- UNESCO director-general Koichiro Matsuura maintained in March 2003 that the average supply of water available per person will drop by one-third within twenty years.[1]

- About 250 million people inhabited the earth two thousand years ago. By 2020 there will be 400 million along the North African shores and in the Middle East alone. And the water supply is shrinking as fossil aquifers are used up.
- An extensive survey of the world's rivers found only two fit for drinking in their "wild" state; the survey was widely derided, and politicians in a score of jurisdictions could be seen manufacturing photo-ops by scooping river water into their mouths. The authors of the survey stuck by their findings. Not a single major American river is any longer safe to drink from without chemical treatment.
- In the southwest states of India there were riots in 2001 as many towns simply ran out of water in an exceptionally dry summer. A dozen people were killed. The army was deployed to deliver tanker loads of water to desperate villagers. Water looters were shot.
- In the Punjab and in Bangladesh, where there is flooding almost every year, the rate of drop in the water table is more than a metre a year.
- Bangladeshi water for almost eighty million people is contaminated with varying degrees of natural arsenic.
- Drought ravaged much of North America in 2002; on the contrary, massive floods were the problem in Europe.
- The beaches along North America's Great Lakes – the source of drinking water for many millions in Canada and the United States – routinely carry warning notices in summer: "Unfit for swimming."
- Intestinal infection in Bolivia caused by poor sanitation and contaminated water became the number-one child killer; when the government, through its private surrogates, increased prices in a vain attempt to pay for a system upgrade, the people took to the streets in a series of major water riots. Two people were killed, but the item was ignored by the Canadian press: that

same day a white Zimbabwean farmer had been killed, and his death dominated the news.

- A tributary of the Danube was poisoned by cyanide runoff from mine tailings when a retaining dam collapsed. All the fish in the tributary died; that the Danube itself wasn't worse affected was ascribed to the fact that its fish were already hazardous to eat, and the river already unfit even for swimming.

- In Turkey, dam-building went on apace. The Ilısu dam, the second-largest in the country after Atatürk, exiled five thousand Kurds and swamped scores of villages; Kurdish campaigners say the dam was a plot to erase all traces of Kurdish culture, and have raised the spectre of ethnic cleansing. Syria and Iraq, which consider the Tigris's water as much theirs as Turkey's, protested vigorously, to no avail. On the other hand, the Turks are equal-opportunity flooders: the Birecik dam is nowhere near Kurdistan, but when it began filling in May 2000 it buried eighty-two priceless archaeological sites in nine villages, some of them dating back to paleolithic times.

- The water level in Russia's once-pristine Lake Baikal, the deepest freshwater lake in the world, is sinking steadily. At the same time, the quality of its water deteriorates as effluent from unregulated factories and pulp mills pours into it.

- In millions of hectares of northern China, the water table is dropping at a rate of one metre a year. Irrigation – and its wasteful runoff – is blamed. Beijing can now supply itself only by diverting water from farmers, who then give up farming and retreat to the cities – adding to the water demand there. Huge diversion schemes are afoot to bring in water from the water-rich and flood-prone south, but this may not be enough, or may not be in time to match need to supply.

- In 2001, a bloom of "red tide" algae off the south China coast was a thousand kilometres long and more than a metre thick; it was blamed on runoff from China's notoriously polluted rivers.

- In the same country, a hundred tons of lettuce[2] were withdrawn from the market as unfit for human consumption; they had been irrigated with water from the Yellow River.
- Bangkok, which had been pumping about 2.5 million cubic metres of groundwater a day from beneath the city to supply its growing population, began to sink. Hydrologists estimated that the amount removed was twice the safe level, but the government paid no attention. The city sank five centimetres last year, and the rate is accelerating. The city will be below sea level by 2050.
- The level of the Dead Sea has plummeted more than ten metres this century. The relentless sun is one culprit. Another is the agreement in 1981 between Israel and Jordan to increase the volume of water they could take from the River Jordan, which has been reduced to little more than a drainage ditch. In northern Israel the Sea of Galilee, which gives much of the south its water, is shrinking and threatening to turn saline.
- In Gaza, overpumping is reducing the hydrological pressure, which is letting the sea water in, and the wells are producing water that is less and less potable. Already Jordan, Israel, the West Bank, Gaza, Cyprus, Malta, and the Arabian Peninsula are at the point where all surface and ground freshwater resources are fully used. Morocco, Algeria, Tunisia, and Egypt will be in the same position within a decade.
- Lake Chad – once, it was supposed, one of the sources of the Nile – is shrinking at a rate of nearly one hundred metres a year. Already, in dry years, humans can wade across it safely, if they are wary of crocodiles and hippos.
- Water supplies in the Nile Valley itself – the cradle of civilization – are in peril. Egypt is an efficient user of water, but Egyptians are consuming virtually all the available supply, and the population is growing at more than 3 per cent per year. There are a million new Egyptians every nine months.

The human "need" for water depends on definitions. The crisis, real though it is, is to some degree a management problem, a matter of allocation and distribution, and not just a pure problem of supply – although in some places, such as North Africa and the Middle East, it is that, too. Dr. Peter Gleick, a water guru who runs his own think-tank in California, defines water needs as "access to basic drinking water and water for sanitation," which seems straightforward enough. By this test and according to the latest data, most of Africa, most of Asia, and western South America fail. More starkly, over a billion people have no access to clean drinking water, and more than 2.9 billion have no access to sanitation services. When I published the first edition of this book in 1999, I wrote that "the reality is that a child dies every eight seconds from drinking contaminated water," a figure that was widely disbelieved. But since then the sanitation trend has been getting sharply worse, mostly because of the worldwide drift of the rural peasantry to urban slums, and the real figure is now probably closer to a dead child every six seconds. Of course, Gleick's measure is personal and humanitarian, and doesn't factor in agriculture or industry. Other water experts define needs differently. Population Action International, for example, maintains that the number of people living in "water stress and water scarcity" was 436 million in 1997, and projects that the percentage of the world's population without enough water will increase fivefold by 2050. Figures from the UN show it will be worse than that. The Danish scientist Per Pinstrup-Andersen, director of the International Food Policy Research Institute, says that one in every five countries is likely to experience a severe shortage within twenty-five years. Hydrologist Malin Falkenmark suggests that any nation with less than 1,000 cubic metres per person per year is water-scarce. And what she calls "water stress" occurs in any country with less than 1,700 cubic metres per person per year. Most hydrologists have adopted this measure to denote severe water shortages. (It takes 1,100 cubic metres to grow the food needed for one person's nutritious but low-meat diet for a year.)

But these figures don't necessarily tell you much about the world's flashpoints. How should they be calculated? There are a number of criteria: where the water supply is static or falling; where there is a dependency on water supplies from outside national boundaries; where rainfall is unsteady or meagre; where populations are increasing; and where there are incompatible demands for water from competing internal sources (agriculture, basic population needs, industry). In Africa alone, by these measures, 300 million people, one-third of the continent's population, already live under conditions of scarcity, and this number will likely increase to more than a billion by 2025. Nine of the fourteen nations of the Middle East already face water-scarce conditions, and populations in six of them are projected to double in twenty-five years. India could join the list of water-scarce countries by 2025, almost entirely because of population increase. China, with 22 per cent of the world's population and only 6 per cent of its fresh water, is in serious trouble already: one-third of the wells in the northwest have gone dry, and more than three hundred cities have suffered water shortages.[3]

Sandra Postel has calculated that "if 40 per cent of the water required to produce an acceptable diet for the 2.4 billion people expected to be added to the planet over the next thirty years has to come from irrigation, agricultural water supplies would have to expand by more than 1750 cubic kilometres per year – an amount equivalent to roughly 20 Nile rivers, or to 97 Colorado rivers. It is not at all clear where this water is to come from." Most projections, in fact, indicate that more than 90 per cent of the extra demand expected between now and 2025 will be in developing countries, with less and less water available for irrigation as developing economies shift to higher-GDP-yielding activities.

Worldwide, more than three hundred river systems cross national boundaries. Hardly any of the world's major rivers are contained within the borders of only one state, and even fewer now that the world's last great empire, the Soviet Union, has broken up. Watersheds seldom acknowledge humankind's political conceits and pay little attention to frontiers. Downstream problems are not always solvable if upstream is in another country. Were Ethiopia to divert or use substantial portions of the Blue Nile, Egypt, entirely dependent on the Nile for its moisture, would be starved of water, and Egyptian politicians have always made it clear they would have no option but war, were that to happen.

Wars, or threats of wars, have been made in several riparian systems. The water resources of the Golan Heights and Gaza have figured largely in the military minds of Israel and its neighbours. The Jordan, Yarmük, and Litani rivers have all been subject to military planners, and Israel has always treated water as a matter of national security. Water, and the Indus in particular, has poisoned relations between India and Pakistan; India and Bangladesh squabbled for decades over the Ganges, and, though both these disputes have been tentatively resolved, there are several unsettling internal water issues that have frequently threatened to end in violence and have several times spilled over into riot, murder, and assassination: militants from Tamil Nadu state (formerly Madras) have threatened guerrilla warfare on neighbouring Karnataka (what used to be Mysore), and Sikh separatists have manipulated water issues to their gain. Iraq, Syria, and Turkey have each mobilized troops in defence of water rights on the Euphrates and Tigris. In Europe, upstream "grooming" of the Rhine and the draining of its safety net, the Rhine wetlands, have caused downstream flooding; industrial pollution is another irritant. The United States has essentially "stolen" the Colorado from Mexico, much of it to irrigate the deserts of Arizona and California, but a good deal of it to fill swimming pools in Los Angeles and fountains in Las Vegas. In return, the Mexicans "owe" Texan farmers 1.5 million acre-feet (nearly 1.9 trillion litres) from the Rio Grande, and in 2002 there were demonstrations along the international border. The Paraná, dammed and flooded, has caused friction between Argentina and Brazil.

Only one-third of the water that annually runs to the sea is accessible to humans. Of this, more than half is already being appropriated and used. This proportion might not seem so much, but demand will double in thirty years. And much of what is available is degraded by eroded silt, sewage, industrial pollution, chemicals, excess nutrients, and plagues of algae. Per capita availability of good, potable water is going down in all developed and developing countries. In the gloomy forecast of an eminent food bureaucrat, "worldwide use [of water] has become so excessive that the implications for irrigated food production are considerable."[4] Mostafa Tolba, a grand old man of the environmental movement and a former director-general of UNEP, the United Nations Environment Program, put it this way: "Just to match demand, major water projects will have to be started within the next ten years, or global supply will be overtaken." But most of the "easy" sources for water have already been exploited, and much of the water is in places it isn't needed. Demand, it seems, will inevitably intersect with supply. And then what?

2 | THE NATURAL DISPENSATION

How much water is there, who has it, and who doesn't?

The icebergs drift past the harbour entrance at St. John's, Newfoundland, massive, beautiful, and deadly, but only to vessels blinded by hubris, incompetence, or sheer rotten luck. In the aftermath of the dizzying success of the movie *Titanic*, Concorde-loads of affluent Europeans jetted in to the Newfoundland capital to gawk. They would go out in Zodiacs, circling the bergs, some of them as large as small islands drifting in the Labrador Current, stately as dowagers. The Europeans would snap away on their Nikons, but the photographs, when they got home, never did the icebergs justice – they never captured the awesome scale, the pristine beauty, the blue glow in the sunshine. Even when the photographs showed the icebergs drifting past the harbour entrance, with human habitation in the background for comparison of scale, they never looked as they did at sea, pale, frigid sapphires, the breeze in the summer sun frosty from their passing.

Icebergs smaller than houses, the Newfoundlanders contemptuously refer to as "bergy bits." One such, no bigger than a pickup truck, was seen dashing itself against the rocks in the fishing village of Upper Island Cove on Conception Bay. We'd been talking about icebergs that afternoon.

"You really should taste the water," said Ron Whynacht, who is from Lunenburg, Nova Scotia. "Twelve-thousand-year-old water. The cleanest, purest water you'll ever have. Drink it clean, by itself. Or twelve-thousand-year-old water mixed with twelve-year-old Scotch."

Indeed, local companies have tried marketing "iceberg water," but the products failed. They were a victim of the pervasive cynicism about advertising: no one believed the claims were true.

The floes that drift past Newfoundland have broken off from Greenland glaciers. They began there as snow in prehistoric times; the cold and the dry air inhibited evaporation, and there they stayed. Snow fell year after year and, in the centuries that followed, the snow was compacted to crystals of ice, each crystal incorporating some of the air of the original snow as bubbles, which disappear only through compaction at depths exceeding a thousand metres. The Greenland glaciers, like most of the world's ice rivers, have been in a fairly steady state since the last ice age, gaining as much through new accumulation as they lose through breakages, although this is changing in worrying ways. The ice moves gradually towards the ocean, not just because of basal sliding and the pressure of new ice forming but because of the peculiar nature of the ice itself, a crystalline solid close to its melting point, which can "flow" just as water does, only more ponderously, as consistently as any river but infinitely more slowly. The glaciers can be hundreds of metres thick and might grind the bedrock as they move. When they retreated last time they left behind moraines, the ground-up debris of their passing, and for centuries farmers have planted there, little thinking of how the soil came to be.

Glaciers may well retreat further, if the dire warnings about planetary warming are true. Already certain glaciers, like the Muldrow and Variegated glaciers in Alaska, sometimes surge rapidly and occasionally catastrophically, for no apparent or known reason. A study released in 2002 by the University of Alaska at Fairbanks measured sixty-seven Alaskan glaciers and found that the melt rate was three times faster than predicted, even given the rising temperatures over the past three decades. Keith Echelmeyer and his colleagues found that, since the mid-1990s, Alaska's melting glaciers had been dumping enough water into the oceans to raise sea levels by .2 millimetres a year – which sounds risibly tiny, but represents a very great deal of water.[1] British Columbia's Athabasca Glacier has retreated two kilometres in a hundred years, and the rate is increasing. In Alaska's interior, the thawing of the permafrost has caused what the locals call "drunken forests," the trees tilting and leaning as the ground subsides. And more recently and alarmingly, some of the Antarctic's major ice shelves have suddenly shrunk. In March 1998, for example, the Larsen B ice shelf abruptly lost a 200-square-kilometre block that collapsed into the sea, leading scientists at the U.S. National Snow and Ice Data Center to say, gloomily, that "this is the beginning of the end." This forecast could seem overwrought had not the Larsen A shelf, all 1,300 square kilometres of it, not disintegrated entirely in 1995.[2]

In Upper Island Cove, Ron was still talking about iceberg ice. "It will fizz and crack when it melts, or when you drop it into a drink," he said. "Crystals under pressure."

We were staying at a small inn on Conception Bay run by Barbara and John Mercer. John was a man who knew how to take a hint; he looked at the bottle of Scotch on the table, and went to fetch a bucket and an axe. Ron and John scrambled down the fifty-metre cliff and balanced precariously on the sea-washed rocks to grapple a bergy-bit fragment, about twice the size of a human head. They heaved it into the yellow plastic pail and took it back to the inn. While they changed their sodden boots and water-soaked pants, another guest, Jim Lockhart, went at the block with a kitchen knife and a cleaver and split it into ice cubes.

Radio isotope dating had put the age of a similar chunk at somewhere between eight and ten thousand years old. I looked it up later. It had a pH balance of 5.4. I knew pH, the negative logarithm of the hydrogen ion concentration in moles per litre, ranged from 0 to 14, with the lower numbers denoting increasing acidity, which made this ancient chunk slightly acidic, but not as high as, say, acid rain. There were minute traces of ions of potassium, sodium, magnesium, calcium, chloride, sulphate, and bicarbonate. There was also a tiny amount of dissolved silica – about .30 parts per million. Hardly any water is truly pure. Only distilled water is just H_2O, and the hydrologic cycle is not as simple as I had first supposed. How did minerals, even in minute amounts, get picked up by evaporation?

Jim dropped chunks of ice into tumblers. They were clear, as clear as, well, spring water. The blue of the bergs was no illusion, but it was the refraction of light that drew the colour of the sea into the crystals of ice. Jim added a sprinkling of white wine to one of the glasses. The ice crackled a little, but no more than ice made from tap water. It didn't mean Ron had been wrong about the crackling. It more likely meant the ice was older than he'd thought, and from deeper down, possibly deeper than a thousand metres.

I put a small piece into my mouth and crunched. The water slowly melted. I tried a bit of tap water on the side. After the iceberg water, it tasted heavy, pregnant with additives. Chlorine traces, probably – concerns for health will beat those for taste any time, and the public-utility people have other things to worry about than pristine clarity.

In a good many parts of the world, the worry is finding any water at all.

~

It's already a cliché in hydrological and Green circles that water is at once our most precious and most abundant asset. Many years ago, Adam Smith pointed out that water, which is vital for life, costs nothing, whereas diamonds, useless for life, are valued highly. Water is everywhere. Humans, if rendered down, are 70 per cent water, and Frank Herbert, in his series of ecological novels about the desert planet Dune, was psychologically astute when he had the inhabitants, the Fremen, "taking back the tribe's water" in the rendering tanks when one of their number died. All life depends on water – indeed, life probably began in water. Water's curious heat-retaining properties steady the climate and make life on our planet sustainable. Without clean water, disease and misery take their toll. Without water, we die.

It is also a cliché to say that the world is not short of water, but rapidly running short of usable water.

~

Where did water come from? The most common assumption is that the earth itself is around 4.6 billion years old, formed by gravity from cosmic junk, clouds of ionized particles around the sun, debris left over from the somewhere-sometime explosion called the Big Bang. This cosmic tip-heap coalesced to form a protoplanet, which grew by gravitational attraction of even more junk (what the cosmologists call "particulates"). This was the Hadean Eon: a sort-of earth existed, but there was no atmosphere, no ozone layer, no continents, and no oceans – and most definitely no life. Around half a billion years later, give or take an eon or two, things had settled down enough to precipitate rocks; the oldest known rocks are in Greenland, and have just celebrated their 3.9 billionth birthday. The earth was still aflame with volcanoes and bombarded by asteroids, meteorites, and whatever else was floating in the interstellar void and intersecting with our nascent planet, but those oldest rocks show signs of having been deposited in an environment already containing water. There is no direct evidence for water for the period between 4.6 and 3.8 billion years ago. But suddenly, there it is.[3]

The prevailing theory is that the atmosphere was created from the release of gases from volcanic eruptions. As the eons passed, the first lightweight silica and aluminum rocks, which are typical of continental land masses, formed. The surface of the earth cooled, and water vapour in this spanking new atmosphere condensed to form the water of the oceans.[4] Which raises the question: Where did the water vapour in the atmosphere come from in the first place? What was it in the volcanism that was our first "weather" that produced water? Or perhaps it was already present in all that cosmic junk – comets, after all, are sometimes little more than frozen lakes of water – and it came from space, an alien and infinitely curious little molecule. For water is curious, much more curious than it might at first sight appear, and is actually little understood. Why is it, for instance, that water is the only substance whose solid form is less dense than its liquid one (a phenomenon that has profound implications for aquatic life)? There are versions of ice, formed under high pressure and in intense cold, that are amorphous – non-crystalline ice that can flow and shift shape. Why this is so, no one yet knows. Scientists have discovered that water is made up of hexameters of hydrogen atoms arranged in what is called a "cage" structure, and that the smallest theoretical drop of water is made up of six molecules arranged as a cage. Which means what? No one knows this either.

As a liquid, water has special thermal features that minimize temperature fluctuations. First among these features is its high specific heat – that is, a relatively large amount of heat is required to raise the temperature of water. The quantity of heat required to convert water from a liquid to a gaseous state (latent heat of evaporation) or from a solid to a liquid state (latent heat of fusion) is also high. This capacity to absorb heat has several important consequences for the biosphere, including the ability of inland waters to moderate seasonal and daily temperature differences, both within aquatic ecosystems and, to a lesser extent, beyond them.[5]

In any case, water appeared, precipitated from what were not yet called the heavens. Then, around 2.5 billion years ago, life on earth began. And it began, almost certainly, in water.

Darwin and the early evolutionists imagined life evolving in a pool of soupy, chemical- and nutrient-rich water, an idea still pretty much accepted today, although there is a small but influential scientific subset that believes that life, too, might have come to us from space, ready-formed, cosmic nuggets among the infinite dross. "Runoff collected in a small volume is the most likely means of achieving the necessary concentration of ingredients," says Gustaf Arrhenius, a geochemist at Scripps Institution of Oceanography. In the dry language of chemists, "ponds may have further concentrated compounds on the internal surfaces of sheet-like minerals, which attract certain molecules and act as a catalyst in the subsequent reactions. Two aldehyde phosphate molecules thus united form a sugar phosphate, a possible precursor to organic life" – as though that explained anything at all.[6]

~

How much water is there?

In the last few years there has been increasing evidence that water is far from scarce in the universe. In March 1998, for example, a meteorite that fell to earth in Texas (breaking up a basketball game in the process) was found to contain water inside crystals of rock salt. We now know that Europa, a pretty little moon of Jupiter, holds more water than all the oceans of earth, a "shell" of liquid and frozen water 160 kilometres thick. "Water is very common in the outer solar system," says Ronald Greeley, a planetary geologist from Arizona State University. Water traces have been found in the sun, on the moon, the planets and their moons, asteroids, and even distant stars. "Water molecules must have been a major constituent of the solar nebula from which the planets formed," according to Robert Clayton, a physicist at the University of Chicago. A huge cloud of oxygen, hydrogen, and other gases in the constellation Orion pumps out enough water molecules to fill the earth's oceans sixty times a day.

On earth, the total amount of water has almost certainly not changed since geological times: what we had then we have still. Water can be polluted, abused, and misused, but it is neither created nor destroyed, it only migrates. There is no evidence that water vapour escapes into space – a popular theory for those speculating about the fate of, say, Mars, but out-of-fashion since the moon, with no atmosphere of its own, was found to contain significant amounts of water in certain deep crevasses, "somewhere between 2.6 and 80 billion gallons" (a nice, tidy 10 billion gallons translates into a shower for one person lasting about 3,800 years, or 12,500 Olympic-sized swimming pools, or one rather small lake – a tiny amount by earthly standards but ample for a small moon colony). And in late 1998 NASA announced a new mission – to "map" the water on Mars.[7] A small amount of water, called juvenile water in hydrological circles, flows to the surface through "outgassing" of the mantle in geologically active zones (a.k.a. volcanoes), but it is thought to be balanced by water consumed in the hydration of minerals.

Water exists, then, in a closed system called the hydrosphere, and contemplating the hydrosphere and the hydrologic cycle is almost enough to make a skeptic believe in the omni-existent Gaia, the notion of the biosphere as a single, self-regulating organism. The system is so intricate, so complex, so interdependent, so all-pervading, and so astonishingly stable that it seems purpose-built for regulating life.

The hydrologic cycle is the way water circulates through the earth's systems, from a height of fifteen kilometres above the ground to a depth of some five kilometres; it is a self-regulating, quasi-steady-state chemical system that transfers water from one "reservoir" to another in complex cycles. These reservoirs include atmospheric moisture (cloud and rain); the oceans, rivers, and lakes; groundwater and subterranean aquifers; the polar icecaps; and saturated soil (tundra or wetlands). The cycle is the process of transferring water from one state or reservoir to another through gravity or the application of solar energy, over periods that range from hours to thousands of years.[8] The whole system works for terrestrial life only because more water evaporates from the oceans than returns to it directly in rain or snow. The balance falls on land, and it is this balance that makes our lives possible, for when the rain falls, it falls as fresh water. There is not only quantitative but also qualitative renewal: the process scrubs the water of its impurities and delivers potable, usable water to the biota, which includes us.

For obvious reasons, most evaporation is from the ocean. The notorious gangster Willie Sutton robbed banks "because that's where the money is," and the oceans yield up most evaporation because that's where most of the water is. But evaporation also takes place from lakes and rivers, soil, and even snow and ice, in which case it's called, for reasons now obscure, sublimation. Plants also exhale water; the evaporation of water through minute pores, or stomata, in the leaves of plants is called transpiration. Most hydrologists simply lump transpiration, sublimation, and evaporation together and call it evapotranspiration.[9]

The best estimate among the many educated guesses – the estimate of Igor Shiklomanov and his State Hydrological Institute in St. Petersburg – is that there are some 1.4 billion cubic kilometres of water on earth, in liquid and frozen form, in the oceans, lakes, streams, glaciers, and groundwaters. And even Shiklomanov, a formidable figure in the water world, and the man the United Nations selected to do its world inventory of water resources, suggests that this is a crude guess – no one really knows how much water is stored in underground ice in permafrost regions, for instance, or in bogs and marshes. His own estimates on these matters, he admitted to a UNESCO conference, were based on numbers arrived at "computationally, under fairly crude assumptions." The amount of ice on earth is, after all, constantly changing. About eighteen thousand years ago, ice covered one-third of the earth's surface. Now, only about 12 per cent remains icebound, and that proportion is shrinking. If global warming is real, it will shrink more and faster.

Some 1.4 billion cubic kilometres. How much water is that? The number is so huge it's hard to grasp. The total weight (mass) of the world's water is the same as the mass of half the rocks on earth – about 5 per cent of the mass of the earth's entire crust.[10] How to put this into any kind of human perspective? A common metaphor, useless but often trotted out by water statisticians, is that "if you smoothed out the earth's crust, water would cover the earth to a depth of 2.7 kilometres." All of which means what? How many cubic metres is the average iceberg? The average lake? The average river?

Hardly any of this 1.4 billion cubic kilometres is useful for human consumption. More than 97 per cent is ocean water, too salty to drink or to use for irrigation. Freshwater stocks are only 2.5 per cent of the total, and spreading that amount evenly over the globe would make a skin a mere seventy metres deep.

But even this 2.5 per cent of the water supply isn't all usable. A trivial amount is in the air at any one time in the form of rain, fog, or clouds, about .001 per cent of the total. An even more trivial amount is in the "biosphere" – in us and other living things, including plants – about .00004 per cent. But really significant amounts, slightly more than two-thirds of the total, or 24 million cubic kilometres, are locked into polar icecaps and permanent snow cover. And a large percentage of the remaining 16 million cubic kilometres lies too far underground to exploit, imprisoned in the pores of sedimentary rock.

Freshwater lakes and rivers, which are where humans get their usable water, contain only about ninety thousand cubic kilometres, or .26 per cent of the world's total supply of fresh water, which is itself less than 3 per cent of global water supplies. Put another way, if all the earth's water were stored in a 5-litre container, available fresh water would not quite fill a teaspoon. Or, to use the metaphor of spreading the stuff evenly over the globe, available fresh water would make a layer only 1.82 metres deep – a really tall person who couldn't swim, say a basketball player, could stand upright in that and not drown.

Fresh water is renewable, at least in the sense that the hydrologic cycle evaporates water from the oceans and returns a good deal of it to the land. This water eventually makes its way back to the oceans through rivers, streams, lakes, and underground aquifers. An enormous amount of water evaporates from the land and oceans annually, using up about half the solar radiation reaching earth. Shiklomanov, as usual, has found his computer equal to the task and has calculated the amount at about 505,000 cubic kilometres. Other estimates put it higher, at nearly 575,000 cubic kilometres. Looked at another way, the top 1.4 metres of the sea evaporates every year.

The time the water stays put in any one place is called its "residence time." Residence times vary tremendously, from ten days for the atmosphere to somewhere around thirty-seven thousand years for the sea. Lakes, rivers, ice, and groundwaters have residence times somewhere in the middle. Most rivers completely renew themselves quite rapidly, in about sixteen days. Groundwater and the largest lakes and glaciers can take hundreds or even thousands of years. If the Upper Island Cove bergy bit that Ron and John grappled into their yellow bucket really was twelve thousand years old, that would put it near the middle of the range.

Almost all the half-million cubic kilometres of evaporated water, around 458,000 cubic kilometres, or 90 per cent by Shiklomanov's measure, falls back into the sea as rain or snow. But since evaporation exceeds precipitation on the seas, the net difference, somewhere around 45,000 or 50,000 cubic kilometres, falls on otherwise dry land. Another 60,000 cubic kilometres is rain and snow of purely local, or non-ocean, origins. About one-third of all the precipitation falling on land – somewhere around 34,000 cubic kilometres – goes back to the oceans in rivers and groundwater runoff.

"Runoff" is the renewable aspect of water resources, the dynamic part of long-term water reserves, an index, if you like, of viable water supply. And so at last we get to the human community's usable fresh-water supply: about thirty-four thousand cubic kilometres a year.

Humans are already using more than half of that – 35 per cent for irrigation, industry, and households, and another 19 per cent to meet instream needs. It sounds as if there is capacity to spare, but the remaining half is the hardest – and most expensive – to acquire. All the easiest aquifers have been tapped, the easiest rivers dammed. The population is still increasing at alarming rates. The ecological costs of using all the water in any system have become only too apparent. Water demand tripled between 1950 and 1990, and is expected to double again in thirty-five years. Where is this water to come from?[11] Shortages are closer than they might appear.

Twenty per cent of global runoff comes from the Amazon Basin alone, while some parts of South America are the driest on earth: the rain gauge of Arica, in Chile, routinely records zero annual precipitation, and in the early part of the twentieth century did so for forty years straight. In the Sahara, the greatest desert on earth, there are virtually never clouds. On the wet end of the scale,

Mount Waialeale in Hawaii has recorded more than 11.5 metres of rainfall in a single year.

Antarctica is frozen and Australia arid, and together they contribute very little to global runoff. The runoff from Europe and Asia is pretty close to the world average; it is lower in Africa and North America, and very much higher in South America – again because of the Amazon. The twenty-eight largest freshwater lakes in the world, overwhelmingly concentrated in northern regions where glaciation has scored deep holes in the earth's crust, account for 85 per cent of the volume of all lakes worldwide. Lake Baikal in Russia alone accounts for one-quarter of all the world's lake-held fresh water (23,000 cubic kilometres). Africa's Lake Tanganyika is second in volume (19,000 cubic kilometres), and Lake Superior, on the U.S.–Canadian border, is third at 12,000 cubic kilometres. The North American Great Lakes, the world's largest lake system, account for 27 per cent of global lake volumes.

Of the twenty-five largest rivers of the world, three are in Africa (the Congo, Niger, and Nile, with a combined runoff of 1,982 cubic kilometres); four are in South America (the Amazon, Paraná, Orinoco, and Magdalena, with a combined runoff of 8,829 cubic kilometres); eleven are in Asia (the Ganges, Yangt'ze, Yenisei, Lena, Mekong, Irrawaddy, Ob, Chutsyan (Quijang), Amur, Indus, and Salween, with a flow of 5,722 cubic kilometres); five are in North America (the Mississippi, St. Lawrence, Mackenzie, Columbia, and Yukon, with a combined flow of 1,843 cubic kilometres); and only two are in Europe (the Danube and the Volga, with a combined flow of 468 cubic kilometres).

In national terms, Brazil has the most water, containing one-fifth of all global resources. The various countries of the former Soviet Union are collectively second, at 10.6 per cent of global fresh water. China (5.7 per cent) and Canada (5.6 per cent) are third and fourth.[12]

China and Canada have virtually identical resources, but China's population is thirty times greater. Water cannot be counted in isolation to human need and numbers. There were only one billion people on earth in 1850; today there are that many in China alone, and the earth's population crossed the six billion mark at the start of this millennium. The Finnish ecologist Pertti Vakkilainen put it this way to a water conference in Paris: "A billion people will be born in the next ten years, so time is critical. We must measure time now in human beings, no longer in seconds, hours, or years."

Nor can water be assessed in isolation to its other, non-human, purposes. Even if we could, we wouldn't shift Brazil's 20 per cent to, say, the Sahara. Doing so would put an end to the planet's greatest rain forest, which we have come to understand is second only to the oceans as its respiratory system. To do so would be like placing a giant vise around the earth's lungs.

Still, that 34,000 cubic kilometres of available water would be enough to supply every human on the planet with about 8,000 cubic metres a year – an ample allocation, if you remember that Malin Falkenmark has suggested that 1,700 cubic metres per person per year is the cutoff between a country being water-stressed and reasonably comfortable.

But that eight thousand cubic metres is a misleading figure, because global water is not distributed evenly. There are places that don't need it that have too much, and places that desperately need it that haven't nearly enough. Water is often in the wrong places at the wrong times in the wrong amounts. And the distribution patterns are constantly changing, which means that stress is appearing in places that had never imagined it before. Even in places like . . . England.

～

"There is far less good water to drink in England and Wales than was previously supposed, the Environment Agency claims in a report issued today." In these words, the British media, in March 1998, reported that a supposed buffer of about one billion litres of water didn't in fact exist. The study, which looked closely at rainfall patterns over recent years, says that the northwest, Thames Valley, and Wales and the west of England are the hardest-hit areas, though there is an increasing incidence of droughts in general, and certain critical aquifers are shrinking. In the fourth year of the major drought of the early 1990s, the *Economist* reported, the margin between supply and demand in England had shrunk alarmingly. In the two most critical areas, the Thames Water District and the Southern Water District, the margins were only 4 per cent and 3 per cent, respectively.

The BBC, in its Water Week coverage in the summer of 1998, trotted out an expert from the Institute of Hydrology who said the increasingly common droughts in Britain – and in Western Europe – were due to changing weather patterns over the previous ten years. Less rain had been falling in the summer, and less was falling over the parts of Britain with the densest population. "In the U.K. as a whole," he said, "rainfall over the last ten years has been very close to average – it's just that the distribution has been very unusual." He said that similar changes had also taken place in Britain in the 1850s and 1930s. Groundwater levels in parts of the Thames Valley were the lowest in the twentieth century.

These "shortages" should probably be viewed with some degree of skepticism. The British water companies had been privatized in 1989, and were no doubt seeking justification for the average 44-per-cent increase in prices to consumers they had subsequently imposed. They had also been stung by a report that 4.5 million litres went missing every day from leaky pipes. And in the summer of 2002, the problem was flooding, not drought. But still . . . water shortages? In England?

Europe has, on the whole, plenty of water, about 4,000 cubic metres per person per year, based on a 2000 population of 510 million. In only isolated instances are Europeans without access to safe water, and those instances are generally caused by civil war or by temporary pollution problems. (For hydrological purposes, which are slow to react to political realities, the countries of the former U.S.S.R. are calculated separately. For those countries, there is relatively plenty of water, except for the Central Asian republics, which are largely desert: 15,220 cubic metres of water per capita for a population of around 290 million.) Even without Russia, the averages are skewed by the Scandinavian countries, which have water to spare (90,000-plus cubic metres per person in Norway, 22,000 in Finland, 20,400 in Sweden, and a whopping 624,000 in Iceland). The situation in Spain, however, is much more dire, at an average of 2,800 cubic metres, and much lower in the east and south, where consumption is passing critical levels. The Balearic Islands, Majorca especially, were essentially out of water in 1998 and were reduced to tankering it in from the Spanish mainland; they set about to build desalination facilities as fast as possible. The south of France is not much better off, and the situation is compounded by the serious industrial pollution pouring down major rivers into the Mediterranean. The French, with two of the largest private water companies in the world, are beginning to charge realistic delivery costs to the affluent, with their swimming pools and Jacuzzis in the hills of Provence.

The Northern African countries south of the Mediterranean are generally worse off than those on the European shores. The Mediterranean Basin as a whole used about 280 cubic kilometres of water in 1990, a demand that has doubled in this century, and increased by 60 per cent in the last twenty-five years. "By the year 2025, practically no southern Mediterranean country will have resources exceeding an average of 500 cubic metres per person per year."[13]

The major water problem in Europe is pollution, apart from instances of expanding deserts in southeastern Spain, generally thought to be caused by unsustainable farming methods and groundwater depletion, and salination of aquifers in many parts of Spain, in Hungary, and on the Yugoslav–Croatian border. Europeans are still using their rivers as convenient sewers, and there is hardly a metre of European coastline without some level of pollution. The worst is the Mediterranean coast from Barcelona around to the toe of Italy, but there are many other areas almost as bad: from Venice to Trieste; around the city of Athens; the Danube delta; the Volga delta; the east coast of the Gulf of Bosnia; all of the Gulf of Finland; all of the Gulf of Riga; the coast on the southern Baltic from Gdansk to the Sound at Denmark; the whole area around Copenhagen, Gothenburg, Oslo, the mouth of the Elbe, and all Friesland; the coast from Amsterdam to Le Havre; the Thames Estuary and the Strait of Dover; the eastern British Isles from the Wash to Edinburgh; and the Bristol Channel. Less polluted, but still causing concern, are the Aegean, the Dardanelles, the eastern Black Sea, and parts of the Caspian. In 1998, of the 472 British beaches designated by the European Union as bathing beaches, only forty-five were found free of pollution, and the commissioners' report was full of horror stories – raw sewage lapping the sands, children gaily topping sand castles with used condoms, and swimmers catching gastrointestinal illnesses and, on a few occasions, viral hepatitis.

Eastern Europe – the former Soviet Union and its satellites – suffers from ghastly levels of pollution. In parts of the Czech Republic and Poland, the landscape is still black with emissions, and the waters run a bilious shade of yellow.

Most of the major lakes of Europe, however, are still relatively pollution-free, perhaps because a majority are in Scandinavian countries with sparse populations. The largest lakes in Europe are in Russia: Ladoga (17,679 square kilometres) and Onega (9,720 square kilometres. Baikal is in the Asian part of Russia). Many of the major rivers in Europe have been – and to some degree still are – heavily polluted. But there is progress: the Rhine, the sewer of Europe, is once again clean enough for salmon.

There are no true natural deserts in Europe.

North and Central America have, at first sight, water to spare. For a population of 427 million, there is an available and renewable water supply of 6,945,000 cubic kilometres of water, or 16,260 cubic metres per person per year. But, again, the figures are crude. Canada has more water than the United States, by about half a million cubic kilometres, with a tenth of the population. Many parts of the United States have plenty of water – or would have, if people weren't polluting so much of it. But in other parts they are draining aquifers by recklessly mining them dry, the depredation compounded by a snarl of laws and regulations designed for a simpler era, when natural resources seemed to be limitless. The most notorious is the "use it or lose it" rule, by which landowners are forced to find a way to use their full allocation of water, or risk losing it to someone who will.

The United States has a theoretical availability of over nine thousand cubic metres per person per year, more than five times the stress level. Yet there are water shortages. Virtually all the available rivers have been dammed, and already more water is being shifted from one place to another than in any other country on earth, and major wetlands have been thoughtlessly drained. Still, there are positive signs: the overall demand for water has been falling steadily in the United States and consumption in certain places like Boston has dropped substantially. The Americans are also giving more thought not only to the natural functions rivers perform but to the restoration of wetlands, most notably in Florida.

The most deprived part of the region is the Caribbean. Many of the islands are too small to have real rivers, so reservoirs are not possible. Worst off is Barbados, with a paltry 190 cubic metres per person. Haiti is next, with a potential resource of 1,540 cubic metres per person, closely followed by Jamaica, Cuba, the Dominican Republic, and El Salvador. In Haiti, Honduras, and Nicaragua, less than half the population has access to safe drinking water.

Mexico is relatively parched, with a potential supply of a little less than four thousand metres per person. Parts of Mexico were always desert, but these areas are spreading throughout the northern part of the country, largely because of misuse. Other human-caused deserts are extending in many parts of the Americas, including southwestern Utah and Oklahoma, parts of southern California, the southern half of Arizona, most of New Mexico, western Texas, and southern Nevada. The remaining soil in many of the same regions is rapidly becoming saline though evaporation caused by repeated irrigation, and impossible to cultivate even if the water were available. The salinity belt extends from parts of the Canadian prairies (mostly Manitoba, Saskatchewan, and southern Alberta) all the way through the High Plains to the mouth of the Rio Grande.

Industrial pollution is largely an American and Canadian problem, except for the Havana area of Cuba, where the beaches are a lot less safe than the capital's sunny tourism posters will tell you. The Great Lakes are cleaner than they used to be – by a small margin – but there are still reservoirs of agricultural and chemical runoff, and heavy metals in health-hazardous concentrations. Among the worst-polluted places are Lake Michigan, southern Lake Huron, southern Georgian Bay, Lake Erie, Lake Ontario, the lower St. Lawrence River, the northern coast of Maine, the coast from Georgia to Boston, Puget Sound, San Francisco Bay, and Los Angeles to Ensenada, Mexico.

Dozens of cities are still pouring raw sewage into waterways. Sewage outfall pipes still dump stormwater mixed with sewage

into New York's Hudson River, despite the city fathers' loony plan
to build swimming beaches in lower Manhattan. Both Halifax,
Nova Scotia, and Victoria, British Columbia, Canadian provincial
capitals on opposite coasts, have no sewage treatment at all.[14]

South America averages a hefty 34,960 cubic metres of water per
person per year for a population of 296 million, but the figures are
skewed by the Amazon Basin, the greatest reservoir and rain forest
on earth and the greatest source of the planet's terrestrial bio-
diversity. The Amazon is shrinking, but fortunately not as fast as
some recent dire predictions suggested. In 1994, Al Gore, then
the U.S. vice-president, asserted that 20 per cent of the Amazon
had already been deforested and that deforestation was continuing
at the rate of 80 million hectares a year. The real figures seem to
be that 9 per cent has vanished, and the attrition is 21 million
hectares a year at its worst, in the 1980s, and only around 10
million by 1998. This revision does not mollify the Greens – nor
should it. In their view, any deforestation of this essential global
resource is irresponsible.

Paraguay is the only American country where less than 50 per
cent of the population has access to safe water. Peru is worst-off in
South America, with a mere 1,700 cubic metres per person of
potential availability, and little Surinam best-off. Surinam's admittedly tiny population of 420,000 people is awash in 468,000 cubic
metres of water each.

The greatest lakes in the continent are Maracaibo in Venezuela,
at 13,338 square kilometres; Titicaca in Peru and Bolivia, 9,288
square kilometres; Poopó in Bolivia, a salty 2,590 square kilometres; and Buenos Aires, shared by Chile and Argentina, at 2,240
square kilometres. The driest parts, and places where desertification
is proceeding apace, occur in a sweep from Peru, around Lake
Titicaca, to Lake Poopó in Bolivia, and in another large area in

northwestern Argentina, east of Mendoza. Numerous spots in the Peruvian Andes and in northern Chile are suffering from human-induced salination of the soil. The greatest deserts on the continent are Patagonia, in southern Argentina, and the Atacama, in northern Chile.

There is severe pollution in Lake Maracaibo and the Gulf of Venezuela; at the mouth of Río Magdalena, Colombia; in the Gulf of Guayaquil, Ecuador; at Blanca Bay, Argentina; at the mouth of the River Plate (Plata) between Argentina and Uruguay; on the coast south of São Paulo and around Rio; and along the coast at Recife. But the worst polluter on the continent is Brazil. The cowboy capitalism of the Brazilian interior, and a robust Brazilian disinclination to follow the rules of officials they consider corrupt, have led to an outpouring of chemical and industrial pollution exceeded only in Eastern Europe and parts of China. More than 130 tons of mercury are still washed onto the banks of the Tapajós River every year from the gold-mining industry. It is probably no coincidence that a 1994 study revealed an epidemic of birth disorders as well as severe bouts of chemical poisoning among adults.[15]

Africa has a disturbingly low water-resource potential of 6,460 cubic metres for each of its 650 million people, and even this amount is inflated by the Congo River and the moist tropics. Africa also has the greatest desert on earth, the Sahara, which covers 8.6 million square kilometres, and three other major deserts, the Libyan (really an extension of the Sahara), the Kalahari, and the Namib. And these deserts are growing, especially in the Sahel, or south shore of the Sahara, but also at its northern fringes in Morocco, Algeria, and Libya. There is also desertification at the mouth of the Senegal River, in Somalia around Mogadishu, and in a worrying strip in Southern Africa – from northwestern Mozambique through much of Zimbabwe, the northwestern parts of South Africa, along the western Orange Free State, and through the Karoo to the west coast north of Cape Town.

As many as twenty-two African countries fail to provide safe water for at least half their population: Guinea-Bissau, Guinea, Sierra Leone, São Tomé and Principe, Mali, Niger, Nigeria, Cameroon, Congo, the Democratic Republic of Congo, Angola, Lesotho, Swaziland, Burundi, Mozambique, Madagascar, Uganda, Kenya, Ethiopia, Somalia, Djibouti, and Eritrea. The worst pollution in Africa is along the Gulf of Suez and the mouth of the Nile, near Maputo in Mozambique, and around Tunis, but there is troublesome pollution also at the mouth of the Congo River, along most of the coast of South Africa, in all of Tanzania north to Ungama (Formosa) Bay in Kenya, at Djibouti, in the Gulf of Aqaba, and, as we have seen, along the Mediterranean coast.

Africa has some of the greatest lakes in the world, among them Victoria, shared by Kenya, Tanzania, and Uganda, at 69,484 square kilometres; Tanganyika, shared by Burundi, Tanzania, Congo, and Zambia, at 32,893 square kilometres; and Nyasa, shared by Mozambique, Malawi, and Tanzania, at 29,600 square kilometres. Lake Chad, shared by Cameroon, Chad, Niger, and Nigeria, was once more than 20,000 square kilometres, but by the early 1980s it had reduced to 17,806 square kilometres, and is still shrinking rapidly.

～

An average for Asia (with a population of around three billion and available water resources of 10,114 cubic kilometres) represents another crude measure. Some countries, such as Laos, have more than 55,300 cubic metres per person per year. But others, such as heavily industrialized Japan, are dependent on only 4,400 cubic metres per person. The critical countries are China, India, and Pakistan, which together account for more than two billion of the total population, and there the picture is much more dire: China has no more than 2,295 cubic metres per person, and that mostly in the south; India has even less, at 2,240, and is heavily dependent on the Ganges and the Indus rivers in the north; Pakistan, with fewer than two hundred million people, is better off, at 3,435 cubic metres per person. Less than half the population has access to safe water in Afghanistan, Bangladesh, Bhutan, Cambodia, India, Indonesia, Laos, the Maldives, Myanmar (Burma), Nepal, Pakistan, Sri Lanka, Thailand, and Vietnam.

There are many deserts in the region, including the greatest and one of the world's oldest, the Gobi, 1,295,000 square kilometres in extent, which is shared between Mongolia and China. Other significant deserts are the Takla Makan (Talimupendi), in northwestern China; the Thar, shared by India and Pakistan; and the Mu Us, on the border of Shaanxi Province, China, and Inner Mongolia. The evidence for their spread is ambiguous, though it is thought that desertification is occurring in a huge sweep, starting in Iran in the west and cutting an arc through Pakistan, Afghanistan, Turkmenistan, Uzbekistan, Kazakhstan, Tajikistan, Kyrgyzstan, and into western China, including a massive stretch north of Beijing. There is another bad section on the eastern edge of the Great Indian Desert, south and west of Delhi to the sea, and in two spots north of Lahore.

The main water problems in Asia are China's looming shortage, with all its implications for the world's food supply; the squabbles over the Ganges and Indus rivers to India's north; and contending claims to the Mekong, the greatest river of Southeast Asia. The most polluted waters in Asia are the Chinese coast, the coast at Bombay and around Malabar, the coast at Madras, the coast at Calcutta, the Ganges Delta, the Gulf of Martaban, Myanmar (Burma), all the waters around Singapore, Jakarta, Manila, Hong Kong, and Shanghai, the southern tip of Korea, and most of Japan, from Tokyo south to Kitakyushu and much of the western coast.

Most hydrologists lump the Middle East (the Levant, the Arabian Peninsula, Iran and Iraq, and west to Turkey) in with Asia. But the politics of the region are unique, and so are its water problems, the most obvious of which are the fractiousness of Israel and its neighbours over shared water resources, and Turkey's role as upstream provider to Iraq and Syria. The region holds about 190 million people, 60 million in Turkey alone, and has a shared water availability of 370,000 cubic kilometres. Bahrain and Kuwait have no water of their own; at the other end of the scale, Iraq has 5,430 cubic metres per capita per year, Iran 1,719 cubic metres, Israel 389 cubic metres, Jordan 318 cubic metres, Lebanon 1,854 cubic metres, Syria 3,780 cubic metres, and Turkey 3,174 cubic metres. Three of the region's largest lakes, the Van, the Tuz, and the Beyşehir, are in Turkey. The Dead Sea, the largest lake in the Levant, is more saline than the oceans.

The region is arid: the Syrian, Arabian, and Rub' al-Khali deserts between them cover nearly 3.25 million square kilometres. There is clear evidence that the deserts are spreading, most obviously through south-central Turkey and down the Euphrates River system in Iraq to the Persian Gulf, and in a sweep from Jordan through Syria.

The worst pollution is in the Red Sea near Mecca, the whole coast of Israel, the Bosphorus, the eastern Black Sea, the mouth of the Kura River in Azerbaijan, and the mouth of the Euphrates.

What of Oceania – Australia, New Zealand, Fiji, Papua–New Guinea, and the Solomon Islands? The Papuans are awash in water, nearly 186,000 cubic metres for each of the country's 320,000 people. New Zealand is not much worse off, at 91,800 cubic metres per person. Australia has slightly more than 19,000 cubic metres per person.

Australia, at 4.7 million square kilometres, is the sixth-largest country in the world. It is also the driest inhabited land mass on earth. It has the least river water, the lowest runoff, and the smallest area of permanent wetlands on the planet. Australia is not as barren as the Sahara, nor as arid, for even in the heart of it, in the region around Lake Eyre, there is still an average of twenty centimetres of rainfall a year. There is also a relative abundance of trees and shrubs – drought-resistant acacia and eucalyptus – and even in the most parched regions something grows, usually the grasses called spinifex. But there are hardly any permanent freshwater lakes, and groundwater is the only practical source of water for most Australians. So it is not surprising that Australia stores more water per capita in reservoirs and dams than any other country. Agriculture, the largest user of water, accounts for more than 70 per cent of total consumption. All agriculture in Australia is irrigated agriculture.[16]

Until the 1990s, the emphasis was on capturing as much water as possible, damn the ecological consequences. That other national activities, such as cutting forests, damaged the soil's water retention properties has only recently come into public consciousness (the Aussies have been cutting down their forests at a rate second only to Brazil, and in the more fertile regions, such as northern New South Wales and in the wheat belt of southwest Western Australia, the forests have disappeared altogether). Desertification is proceeding apace in southern Queensland and into northern New South Wales, as well as on the eastern edge of the Simpson Desert, in southwestern Queensland, and into northeastern South Australia.

There is still minimal pollution of inland waters in the country, but considerable coastal pollution around Melbourne and Sydney. Other coastal areas, including those near Perth and Brisbane, are less polluted.

In the intensively cultivated Murray–Darling Basin in southeast Australia, excessive withdrawals in the 1980s caused increasing water salinity and worrying algae blooms from the rivers, and in 1993 the Murray–Darling Ministerial Council cranked itself into a state of high anxiety. After an audit of resources that year by its commissioners, who found a decline in native fish populations, shrinking wetlands, and ever-increasing salinity, the council put a stop to further river diversions in 1995. This regulation was an interesting political achievement, since the basin spans five Australian states and territories with varying water policies, and is knotted up in such historical regulations as the post-colonial agreement that New South Wales and Victoria guarantee a minimum of water deliveries to South Australia.

So much for the "natural" supply, but what of its utility for humans? In 1998 at a UNESCO water conference in Paris, I listened to Peter Gleick, a short, neatly bearded, precise fellow who exudes a justified air of authority and competence, who was up on the platform arguing passionately for different ways of looking at water. The crude global or continental measures are all very well, he was telling the assembled hydrologists, but they tell you nothing about the human costs. Even national figures can be misleading, as we know – parts of Bangladesh may be flooded, others parched. Or a country can be under water one month and stricken with water shortages a few months later. Water availability is one measure, certainly. Actual water use – withdrawals from the system – is another, and it tells you different things, some of them difficult to interpret. A further way of looking at water is to see how we can provide for the basic water needs of all citizens. "The actual amount you give to each individual – the number of litres – is not that important," he said, "[In most cases it is] only that we move from zero to something."

Gleick paused and leaned forward on the lectern. "The cost of not providing basic water for drinking and sanitation will far outweigh the cost of doing so," he said. These costs were currently running between $100 and $200 billion a year, for health care and social welfare alone. "About half the modern world doesn't have the same basic amenities the ancient Romans took for granted."

Gleick recommended that UNESCO adopt a "human entitlement" of 50 litres per person per day. "Drinking water, 5 litres; sanitation water, 20 litres; bathing water, 15 litres; food preparation, 10 litres. Total, 50 litres." These figures, he pointed out, are far below even the minimal average withdrawals per capita in the most water-poor of countries. "This is not a technological issue," he continued. "The technology is easily available. It is a political and organizational issue. Water is a social good – we all agree on that. People should pay for its use, to encourage efficiency and as a recognition of its value. But perhaps a universal 'lifeline rate' should be established, and anything above that should be priced much higher. To water a lawn, for example, should be truly expensive."

I was only half-listening, because this was familiar stuff. I was still puzzling over something else he had said earlier about the amount of water withdrawals on a continent-wide basis. It has become a hydrologist's truism that 1,700 cubic metres per person per year was the cutoff between a country being water-stressed and reasonably comfortable, yet when I looked at Gleick's figures for actual water use, as opposed to available water, no one was really using 1,700 cubic metres. Even the Americans and Canadians, the greatest water hogs, were using only 1,693 cubic metres a year, out of a resource much greater than that. In descending levels of use, Oceania (the Pacific Islands, Australia, New Zealand, Papua–New Guinea, and Fiji) used 907 cubic metres, Europe 726, Asia 526, South America 376, and Africa a puny 244.

And I had just that morning finished reading a pamphlet put out by Population Action International which had dealt with revised UN population estimates for 2050, projections made in 1996 that were sharply lower than those the United Nations had been gloomily forecasting only two years earlier. Even with the

downward revision of population forecasts, the "medium projection" numbers showed that in a world of 9.4 billion people in 2050,
a billion people would be in "water scarcity" and 970 million in
"water stress." Another report, published by Johns Hopkins School
of Public Health, had said that "caught between the growing
demand for fresh water supplies on the one hand and limited and
increasingly polluted supplies on the other, many countries are
making difficult choices." How to square these circles?

When Gleick's lecture was over and he had made his way to the
coffee machine in the lobby, trailed by a wake of hydrologists from
a dozen countries, I asked him about the slipperiness of the
numbers. "Your book is called *Water in Crisis*," I said. "This conference is called *Water – A Looming Crisis*. How do you match this
1,700 cubic metres a year with actual consumption and come up
with a crisis?"

"First of all," he said, "you have to be careful with these
numbers. To some degree, water statistics are a technocratic illusion. A thousand cubic metres a year doesn't necessarily mean water
stress. Israel is well below that, at about 300 cubic metres per person
per year, and while they're on the edge, no one at present suffers
from a lack of basic amenities. In Nigeria, which has lots of water,
more than half the population goes without safe drinking water." It
doesn't mean, he suggested, that if a country draws less than its
available resources, it is living thriftily. It might simply mean that
the infrastructure is a shambles. He'd just been approached, he said,
by *National Geographic* magazine, which was planning a major
feature on water and wanted to use some of his data. "They made
the same point to me. 'No one uses 1,700 cubic metres,' they said.
But that's only one way of looking at it. They were ignoring, for
instance, the use of rain-fed agriculture. When that's added in,
many, many places approach 1,700 metres." He shrugged. "You
have to look at total resources, renewable resources, usable renewable resources, the ability to transfer water from water-rich to
water-poor places, the development level of the economy, the

annual consumption, and the deprivation level, all matched against population trends and economic resources. When you do that, you'll see that there are crises in many places."

He sipped from his coffee and grimaced. Machine-made French "cappuccino" was not going down well; he was a Californian, after all. "And another point. Forecasts should be scrutinized carefully and used cautiously. In the 1960s, water consumption in America was forecast to increase by up to 150 per cent by the year 2000. This has not happened for a number of reasons. A fundamental change in attitude, for one thing, a shift from profligacy to conservation, has stabilized demand, and withdrawals are in fact shrinking slightly."

Just before he turned to walk away, he repeated a point he'd made on the podium: "It is difficult," he said, "to see the abyss before you fall into it. Are we really on the edge? Take Mexico City, for example. They're providing basic water needs for their citizens. But we already know that unsustainable pumping of the local groundwater has caused parts of the city to subside by nearly twenty metres. They're now bringing water in from three hundred kilometres away, pumping it uphill a substantial distance. Population is still growing explosively. How close to the edge are they? How close to catastrophe? That's why we should be looking at these things. In many places, unpleasant surprises are inevitable."

3 | WATER IN HISTORY

How humans have always discovered, diverted, accumulated, regulated, hoarded, and misused water.

The ancients might have seen God everywhere, but they also saw opportunity. Superstition was often accompanied by a politics both cynical and wise, and by engineering skills of a very high order. Water has been used to nurture cultures, but also as a weapon of war, and as a convenient cover for crimes best hidden. Of course, nothing in that prevented ancient cultures from withering away, victims of hubris, or drought, or, most likely, of both.

I have this passage in my notes, cobbled together from various texts over the years. The line about the Sadd-el-Kafara Dam comes from the ever-reliable *Encyclopaedia Britannica*:

Cultivation terraces, some of them more than 3,000 years old, can be found in many parts of Asia Minor, Europe, Africa, Asia and Andean America. The earliest urban societies all depended on systems of dikes, canals, and aqueducts, many of them designed and built to exploit and control seasonal flooding. Some of these were significant in size: a dam built below what is today Cairo somewhere in the third millennium

B.C., at Sadd-el-Kafara, was estimated to contain a hundred thousand tons of earth and rock. There are qanats [horizontal wells] in the Iranian desert three thousand years old, some of them still functioning.

I once ran into an archaeologist in Tanzania who wrote prose like that – workmanlike, sturdy, but essentially dull. He knew it, too, and despaired of it, because what he really wanted was to communicate his passion for the ancient lives he found in the runes and ruins of his field trips, and all he could produce was prose that would make graduate students' eyes droop. I found him in his tent, perched on the stony floor of the Great Rift Valley, near the Ngorongoro Crater, and close to the Olduvai Gorge, where the famous Leakey family began untangling the story of the human species. This is one of the most eerily evocative places on earth. Nearby, impressed in the mud many millions of years ago, are the Laetoli Footprints, fossilized imprints of two adults and a child, our earliest ancestors, left there for the contemplation of their distant descendants. The archaeologist, whose name was Witman, had been in the area for more than seven years, funded by his university in Germany and by the grants he could scrounge from American foundations. When I came across him, his head was in his hands and he was staring gloomily at a pad of paper in front of him. The page was blank. I started to laugh. A blank page is something all writers know only too well.

He said nothing for a while, then he suddenly sprang to his feet and dashed outside. "Come with me," he said, and strode off into the bush.

I looked at the man who had brought me here, a Swahili-speaker of unknown tribe from the coast near Dar es Salaam, but he just shrugged and raised his eyebrows. "Might as well," he said, and set off in pursuit.

We caught up with Witman at the foot of a stony hillside. He was perched on a boulder, teetering a little.

"What is it?" I asked. "What do you see?"

He pointed to the left, where the hill flattened into the plain. "Over there," he said, "about fifty miles, that's Lake Manyara. Up from there is the Mosquito River. It's new, I think. It didn't exist centuries ago. There were no rivers here, but there was water up on the plateau, there." He shifted a few degrees on his boulder, chopped his arm upward, pointing at a crest we couldn't see. "So what were people doing, farming here?"

"They were?"

He scrambled up the hillside a hundred metres or so, calling for us to follow. Near the top, he stopped. "Look. See . . ." He pointed to his feet. He was standing in a shallow gully, which ran arrow-straight across the face of the hillside, with a gentle downslope. "These are irrigation trenches. At the bottom of the hillside there were small fields with stone borders where sorghum was grown. The neatness! The precision! Some of those fields are immense, but some of them are less than a metre square! Think of the kinship patterns and the needs that drove them to carve up the land in ever-smaller parcels. This is an irrigation system. There are better-preserved ones further up, but these are far, far older, we haven't yet worked out how old."

"So what's the problem?" I asked.

"The problem," he said, "is that there were cultivators here, pos-sibly a millennium ago or more, in this arid place, which was arid even then, a culture with villages, laws, government, chiefs. . . . What I mean is, it's so exciting, imagining these people, these van-ished and long-gone people, and putting yourself into their heads and thoughts, imagining their lives. . . . And then you write it down, and it looks so dull. . . . 'So there are stone channels in the hills?' people say; 'big deal, these are no pyramids.'"

"So you're writing for students and scholars, not for a screen-play," I said, a little unsympathetic. Indeed, I admired him for his imagination, but not, when I saw it later, for his prose.

I went back to look at those stone channels built by that long-ago people. Not pyramids, no, but the engineering skills were far from insignificant. I tried to imagine those men – and women, this being Africa, where the women did most of the work – toiling in the long-ago, rolling these stones up the hill, trenching, terracing. Who had been the overseer, the person with the vision, the surveyor-in-chief? What kind of person could imagine taking water to a place where no water was possible? Further up the Rift, another archaeologist, John Sutton, had found many more of these channels, engineered responses to water needs emerging spontaneously where they were needed. Sutton was working in similar countryside. There, too, the surroundings were too dry for agriculture, but a few rivers rose on the lusher Crater Highlands, and an unknown number of centuries ago these rivers were also harnessed and channelled into even longer and better-preserved stone-built canals, cunningly embanked and levelled and meticulously maintained, some of them more than five kilometres long, ending at tiny stone-enclosed fields covering more than two thousand hectares.[1] Like Witman's, this unknown culture, here and at other sites by lakes Eyasi, Manyara, and Natron, has vanished, abandoned for reasons unknown – perhaps because the nomadic Maasai were moving into the area, perhaps because increasing population had overstrained resources, and perhaps because deforestation for firewood had critically wounded the streams, causing them to dry up.

I thought back to my notions of who had built these channels. I admitted that I was prone to my culture's failure of imagination. Somehow, we have lost the real sense of history. We have come to hold the quaint notion that, because people lived long ago, they must perforce be more "primitive," more driven by superstition, than we are, and we are constantly surprised at evidence of planning, forethought, ingenuity, and skilfulness. I thought then that it ill behooves a culture in which large numbers of citizens believe in abduction by aliens (and in which a woman was recently charged with attempted murder in a U.S. court for hiring a fortune teller to put a spell on an enemy) to pour scorn on the arcane beliefs of ancient cultures, and even less so when it is clear that folklore cohabited comfortably with engineering and management skills of a very high order. More than 2,400 years ago, Darius the Great of Persia, a ruler noted for his monumental building projects and great administrative skills, encouraged the construction of new water sources by exempting from taxation "for five genera- tions" the revenue derived from the building of new *qanats*, an early instance of government pump-priming of the economy by subsidizing entrepreneurs. *Plus ça change* and all that − a primary solution for modern water problems is to treat water as an eco- nomic good; to Darius, this was already plain good sense.

Still, it is interesting, the folklore that has grown up around water. Water is so necessary, so central to all life, that it isn't sur- prising it took on a significance beyond the intrinsic. In a world where gods inhabited everything, gods most certainly inhabited water. And it is always interesting, I thought, how politics and reli- gion intersected where it came to water.

> *Spring up, O well,*
> *Sing ye to it*
> *Thou well dug by princes*
> *Sunk by the nobles of the people*
> *With the sceptre, with their staves*
> *Out of the desert, a gift . . .*

This song, from ancient Jewish Apocrypha, referred to a custom in which a new well is lightly covered over and then miraculously uncovered in the presence of the tribe's leaders, thereby assigning it as the property of the whole clan. In modern times we'd be more cynical. We'd call this ceremony a photo op and impute base- ness to the politicians who orchestrated it. The song served as a kind of ribbon-cutting ceremony. In either case, the community ended up with a benefit it hadn't had before.

The peoples of Mesopotamia and the Middle East, water-stressed even in biblical times, have water tales running all through their sacred writings. In the Book of the Prophet, believers are enjoined to share water with whoever needs it, a basic part of the human obligation towards strangers. Did not the world, after all, result when God divided the waters of the deep from the waters of the air, and so created the earth? In the Koran the Lord says, "We clave the heavens and earth asunder, and by means of water we gave life to everything." God is the fountain of living water; the bean-counters of exegesis have said there are more than two hundred references to water or wells or oases in the Bible, and I see no reason to doubt it. Doesn't the Bible say that Moses means "drawn from the water" in Hebrew?

The Gnostics regarded water as the primary element, and their influence was felt among the Jews. Judah ben Pazi, passing on the wisdom of Rabbi Ismael, had the wise man saying: "In the beginning the world consisted of water within water; the water was then changed into ice and again transformed by God into earth. The earth itself however rests upon the waters, and the waters on the clouds."

The Well of Miriam, the source of life for the Jewish wanderers in the Sinai, is said to be located in the deeps of the Sea of Galilee (which might, in truth, once have stretched through Eilat to the Dead Sea). The mystic of the Kabbalah, Isaac Luria, felt that the water of the Well of Miriam was one of the centres of cosmic energy; anyone who immersed himself in these waters would understand the mysteries, and his memory would never fail.

Muslims believe that the Dome of the Rock in Jerusalem is closer to God than any place on earth, and that beneath it originate all the sweet waters of the planet. The ancient cultures had the same notions. The Egyptian god Hapi, a male, was often shown with two full breasts, from one of which flowed the northern Nile,

the other, the southern. Nun (or Nu) is Chaos, the primordial
ocean, the germ of all things and all beings. He was the "father of
the gods" and is sometimes found represented as a personage
plunged up to his waist in water, holding up his arms to support the
gods who have issued from him.[2] The Sumerian god Aspu was
the governor of the Sweet Waters; Ea of Mesopotamia, the god of
water, set up his palace in the Fertile Crescent where the Euphrates
and the Tigris converge; Gilgamesh rooted the Tree of Life in the
same place. For the Babylonians, the god Nun personified the idea
that water was the source of all life, that historically the earth came
forth from the water, and that water was the quickening element of
all creation. "I shall shut up the heavens, and no rain will fall" –
always the curse of a vindictive and cruel god.

The power of the myth persists. There is a fountain called the
Pure One in Egypt. It is said that, if a menstruating woman touches
the water, it begins immediately to stink, and the fountain has to be
emptied and cleaned before the water becomes drinkable again.
Gods live in wells, streams, rivers, lakes, and the oceans. Water gods
govern the rains; many old cultures believed that, if the rains failed,
the one way to get them going again was to deal directly with water
gods in the palaces where they lived. Perversely, some thought
the best way to get the gods cranked up was to irritate them. Throw
some defiling thing into the water of the well – that'll get 'em going!

Around Accra in Western Africa, the pond spirits were locally
powerful. They must be particularly irritable these days, since the
modern Ghanaian capital doesn't treat its sewage but dumps it into
dry wells to seep away; in the rainy season the wells overflow and
pollute the city and the groundwater.

Among the Dogon of Mali, whose farmers are famously
creative with the meagre water they receive, water was a practical
matter but also integral to the creation myth, in which water is a
divine green seed that impregnates the earth so that it brings forth
twin green beings, half-man, half-serpent. Humans originated

when the supreme celestial god, Amma, created in his likeness Nummo, the god of "wet" water. Water, the life principle, is closely involved in the rites accompanying childbirth. When the afterbirth is expelled, proving that the child has indeed been born, one of the midwives takes a mouthful of water and sprays it gently over the baby. The cool touch of the water makes it cry out – by which the child has "officially" received the gift of speech.[3]

Lest we be tempted to believe that these were just quaint primitive customs, well, they were – but they were just as common among the quaint primitive Europeans as anywhere else. Water worship and holy wells survived in England long after the Druids were a distant folk memory. Pope Gregory, deliberating on the English in a letter to the Abbot Mellitus in the year 601, recommended countering the water superstitions of the locals not by destroying their places of worship but by subsuming them by sprinkling "holy water upon said temples . . . that they might be converted from the worship of demons to the worship of the true God." But the old ways often persisted. An old folklore book records the supplication of a Scottish peasant: "O lord, Thou knowest that well would it be for me this day an I had stoopit my knees and my heart before Thee in spirit and in truth as I have often stoopit them after this well. But we maun keep the customs of our fathers."[4] In 1893, R.C. Hope published *The Legendary Lore of the Holy Wells of England*, in which he catalogued the names of 129 saints to whom anciently venerated wells had been transferred, the Holy Virgin being the most common, closely followed by the ever popular Saint Helen, the tabloid heroine of her time (she was the English-born mother of Constantine, the first Christian emperor).

The most famous holy well in Britain is at Holywell in Wales. Winefride, the beautiful daughter of the chieftain, Thewith, resisted Prince Caradoc's advances after she had dedicated her life to virginity and God. He caught up with her at the top of a hill, cut off her head, and, after it rolled to the foot of the hill, a copious stream burst forth, forming a well that was lined with fragrant moss, the stones at its foot "tinctured with the blood of the youthful martyr." Winefride was subsequently miraculously reunited with her head and became a nun, dying in 1660.

The indefatigable Brothers Grimm recorded many instances of well worship along the Rhine Basin. "Sacred wells and fountains," they wrote in *Teutonic Mythology*, "were rechristened after saints, to whom their sanctity was transferred." Ancient customs and rituals survived, among them oaths and ordeals, consecrations and processions, but clothed in Christian forms. In some customs there was little to change. "The heathen practice of sprinkling a new born babe with water closely resembled Christian baptism."

As in other endeavours, religion blends with pseudo-science, itself developed as an explanation for the unexplainable. In arid regions in ancient times, the mere finding of water was high art. Water divination was a valued skill, and water diviners ("dowsers") have existed in all cultures at all times. They are still active in Europe and America, and rural people in search of a well commonly call on such a one, trolling across a field holding a forked piece of wood, or even a coat hanger bent into the right shape. Like many urbanites, I've watched them work, not knowing enough even to be skeptical – maybe there is some principle at work here, ill-understood but efficacious. Knowing where to drill for water is a serious business in South Africa's Karoo region, for example, because water is scarce there and the drills must go very deep. I once watched an old Boer patrolling a hectare or two waiting for the twinge that showed him where the aquifer was to be found. He was a grand old fellow, with a beard like Abraham the Prophet, but he always wore baggy khaki shorts and knee socks, which made him look like a wrinkled old Boy Scout. He used a yew fork for his work, imported from England; he claimed that yew knew water better than any other wood, and no one dared to argue. It took him most of the day, criss-crossing the homestead a hundred times before he was satisfied, but when

the drill went in it found water. No one was ever surprised; everyone knew that water-dowsing worked. It was a natural law, like
the monsoons and man's venality.

One of the most famous diviners of this century was Sapper
Stephen Kelly, of the Third Australian Light Horse Brigade, who
found water in the desert for an entire army in the Gallipoli campaign in 1915. His exploits were meticulously recorded by army
bureaucrats: challenged by a skeptical Brigadier General Frederic
Hughes to tell him where under the endless sands he should have
his men dig, Kelly didn't hesitate, found water at the first go, "and
in little time had thirty wells going, with sufficient water to supply
every man with a gallon a day and every mule with its six gallons,
and this of pure cold spring water instead of the lukewarm liquor
from kerosene tins off the transport."[5]

Superstition – and engineering. Didn't Solomon say, "I made
pools of water to irrigate a forest springing up with trees . . . thus
I gained more wealth than anyone else before me in Jerusalem"? It
was in Babylon, of course, that Nebuchadrezzar II built his
Hanging Gardens to please his Median wife, Amytis, who was
pining for her verdant homeland. What she got was on everyone's
Seven Wonders of the World list. They were described by the
ancients with suitable awe: a massive series of terraced roof
gardens, each lined with layered reeds, bitumen, and lead to
prevent seepage to the levels below. The gardens were irrigated
with pumps from the Euphrates River and from a deep well, the
water hauled to the top with a chain pump.

The digging of shallow wells was the earliest organized change
from the communal waterhole. Wells went deeper as the need (i.e.
the population) grew and as the tools were developed. Well-
drilling went astonishingly deep astonishingly early, as early as
neolithic times. The Egyptians had perfected core drilling in stone

quarries by 3000 B.C., although the wells of the time, of large diameter, lined with stone and constructed with human and donkey labour, rarely went much below fifty metres. In China, however, a churn drill was developed two millennia ago, and some of the wells these devices produced went down more than one kilometre into the soil and rock, though they were merely made of wood and powered by humans, and sometimes took years, and even decades, of unremitting labour to finish.

Europe, cut off from the civilizations to the east and their engineering skills, came late to well-drilling, perhaps because water was relatively easy to find in the temperate climate. In the early twelfth century, "flowing wells" were discovered in Flanders and England. In the same century, percussion drills were invented; monks made them, as they had introduced so many other innovations, such as flow-through toilets and in-house sanitation. Carthusian monks from Lillers, in the province of Artois, France, drilled an early well in 1126, a thing of much marvel. A few years later other religious in nearby Gonnehem drilled four wells one hundred metres into the fractured chalk below; they gushed so strongly that they drove a water mill nearly four metres above ground, and flowing wells have been called "artesian" after these Artois wells ever since.[6]

By 1340, Bruges in Flanders had a municipal water-distribution system, a central cistern from which water was pumped by buckets on a chain underground to public faucets. Two centuries later, London set up an awkward contraption under London Bridge, a series of five waterwheel-driven pumps, which supplied the whole city with water.

The earliest irrigation schemes were called *qanats*, which were both well and aqueduct. *Qanats* made the great urban civilizations of Mesopotamia possible, and they are still widely used, from Afghanistan through Iraq and Iran all the way west to Egypt. A *qanat* is essentially a horizontal well. Aquifers, groundwater close to the surface, are found on uplands, often the foothills of mountains. A shaft is bored horizontally into this alluvial fan, which is usually fed by mountain streams. The tunnel is constructed to slope gradually downward, so the water flows entirely by gravity to its destination, which in some of the more remarkable old *qanats* can be fifty kilometres away.

If you fly over the desert in southern Iran you can easily track the course of the longest *qanats*. Every few hundreds metres, stretching for kilometres across the stony plains, you can clearly see a mound of earth surrounding a hole bored into the soil, each hole a vertical shaft down to the *qanat* tunnel itself. These rows of manholes, for that's what they are, have multiple functions. Their primary purpose is a means of extracting the rubble and soil from the excavation of the tunnel itself. They are also access points for repairs, breathers to aid circulation and water purity, and sources of irrigation water – at the lip of many a manhole there is still a rickety contraption of poles and rope, homemade winches. The *qanat* itself can be many metres below the surface, the water protected from evaporation in the desert sun. Many of these *qanats* have been operating for thousands of years; until the building of catchment reservoirs in the 1930s, Teheran got all its water from a dozen *qanats* that put out something like eight hundred litres a second. Because most Iranian towns of any consequence are supplied by *qanats*, they have been built close to mountains and have a canal running down a main street.

Throughout the Middle East and ancient Mesopotamia (Iran and Iraq) are interconnected tunnels and systems of cisterns of great complexity and huge scale. King Herod's fortifications on the waterless stones of Masada, for instance, had bathing pools and fountains. The technique spread from the Fertile Crescent eastward to Afghanistan and westward into Egypt; a functioning *qanat* was irrigating many hectares of land west of the Nile in 500 B.C. *Qanats* are still used in other parts of North Africa. Most of the so-called oases in the pre-Saharan regions of Morocco are created artificially, some by diverting water from mountain streams, others by underground *qanats*. A few of them stretch almost fifty kilometres to the Atlas or Anti-Atlas mountains.

Another theory has the idea coming in from China through early traders, or through the educational efforts of early Buddhist monks who travelled back and forth over the Karakoram Range. *Qanats* are still common on the Balochistan Plateau of Pakistan, and shafts sunk into fields are a characteristic part of the landscape.

The ancients were also shrewd enough to know that water was not just water. It had to be clean water. Texts dating from as early as 2000 B.C. outline techniques for cleaning water through sand filtration and boiling. A guide for pilgrims to the supposed tomb of the apostle St. James in Compostela, Spain, made sure that pilgrims of the eleventh and twelfth centuries knew "which waters were fair, and which foul, which were fit to drink, and which not fit for asses."

Many ancient cultures also understood that waste should be separated from potable water. Mesopotamia was the first civilization to formally address sanitation problems by connecting many homes to a storm-water system that carried waste away. Poorer homes had latrines connected to deep vertical shafts in the ground lined with perforated pipe to provide dispersal cesspools. By 1700 B.C. at King Minos's palace in Crete, drainage systems carried waste through terracotta pipes into stone sewers. Rain-fed cisterns and aqueducts provided constant-flush toilets, so recently reinvented in ever-wasteful Nevada. Shortly before the Christian era, Rome completed its central sewage system, which conveyed sewage and storm water into the Tiber; on the streets, large terracotta jars were provided at intervals for the use of passing strangers, and street vendors would rent passersby a "privacy cape," so they could do their business in peace.

In China, where ancient Chinese scholars invented toilet paper (or so the Chinese say), a two-thousand-year-old toilet has been found, complete with running water, a stone seat, and comfortable armrests. It was buried in the tomb of a Han-dynasty emperor, who evidently had thought through all his needs for the afterlife. So much for Thomas Crapper's place in the history of civilization . . .[7]

The Romans were the greatest builders of water-distribution systems in the ancient world, but they learned their techniques from the *qanat*s of the Fertile Crescent and from the Assyrians, who three millennia ago used more than two million huge blocks of limestone to construct an arched aqueduct to bring fresh water to the city of Nineveh. The Assyrians were inventive engineers; the water was carried across a river valley in a limestone channel ten metres above grade and one kilometre long. Even more brilliantly, they managed to construct an inverted siphon to increase the flow into the aqueduct, a feat not duplicated until the nineteenth century. Less positively, the Assyrian king Sennacherib burnt Babylon to the ground in 695 B.C. and diverted a major irrigation canal so that its waters would cover the ruins, hiding the crime. The Assyrians themselves were said to have learned their techniques from their predecessors, the Sumerians.

Most of the Roman aqueducts were built in a spurt of construction between A.D. 312 and 455 throughout the empire, and their arches remain impressive even today. Remnants can still be seen in Italy, Greece, France, Spain, Syria, Morocco, Libya, and the old Persian Empire, and some of them are still in use. The Roman system, described in detail by Sextus Julius Frontinus, the supervisor of the empire's waterworks, used eleven major aqueducts to bring water more than forty kilometres to the city in sinuous, curving channels that were themselves almost a hundred kilometres long, mostly in underground tunnels made of stone, terracotta, and a variety of other materials, including brass, wood, and, notoriously, lead – an early instance of inadvertent industrial pollution. The system was gravity-fed and flow-through, like the latrines of Rievaulx Abbey in Yorkshire and the trendy toilets of California; the surplus water was used to power the city's fountains and to flush its sewers into the Tiber.

No one improved on this Roman delivery network until the industrial era, when the Americans invented pipes that could sustain pumping under pressure. These pipes were made of asphalt-treated wood, banded with iron. For those interested in industrial archaeology, pipes like these still take water to the ocean-side villages of Witless Bay in Newfoundland.

All these developments pale, of course, before what the Californians have done, which is to spend billions of dollars diverting entire rivers, sometimes pumping them over a water basin and a continental divide, to bring a cultivated civilization to a place where it should not be – in the middle of a desert. Their network is still not finished, and perhaps never will be (for it grows ever more grandiose), but it already brings in water some nine hundred kilometres, with an annual yield of over four million acre-feet (there are 1,234 cubic metres, or 1,234 metric tons, per acre-foot). The Californians have made the desert bloom – but at what cost in money, resources, and impact on the environment, they are only just beginning to find out.

Irrigation underpinned most of the ancient cultures whose imprint has been left to us. As we shall see, irrigation carries its own dangers, and many irrigating cultures perished through their very success, expanding until they exhausted the resources on which their success was based. One of the interesting things about ancient water-dependent civilizations – cultures that relied on irrigated agriculture – is the reason for their collapse. Why did the cultivators of Tanzania who so tormented archaeologist Witman vanish? Why did the Sumerian culture disappear, to be replaced by the Assyrian and then by the Babylonian cultures, which disintegrated in their turn? Why did Persopolis disintegrate? What happened to the ancient civilization of Fatehpur Sikri in northern India? Why did the Hohokam culture of the Arizona desert vanish?

What put an end to the empires of ancient Ghana and Zimbabwe? In many cases, of course, the answer is war – and a more powerful adversary. But as Marc Reisner put it in *The Cadillac Desert*, "explaining the collapse of ancient civilizations is a cottage industry within the anthropological and archeological professions, like the riddle of the dinosaurs." Reisner himself favours a simple solution: increasing salinity brought on by poorly drained irrigation.

The real calculus of decadence is almost certainly more complicated than that. A single explanation doesn't account for the cultural exhaustion that allowed the "barbarians" to take Rome – lead pipes just won't do as an answer. A single explanation doesn't account for the way that small European cultures flourished in sequence, each briefly taking its place among the empires of the powerful, only to be replaced by another similar culture – first Aquitaine and Navarre, then Holland, then Portugal, then France, then Britain.

Anthropogenic degradation of the environment – disruption of the essential hydrological cycle – must play some role in the disappearance of ancient cultures. Catastrophic droughts no doubt contributed, aggravated by desertification. Arid lands can be made desert by human action, and deserts can be extended by careless destruction of the soil; deserts where man has been active longest are more arid, more barren, than "natural" deserts. There is some certainty, for instance, that the great migrations of the Golden Horde across the steppes into mother Russia were set off not by power rivalries or wealthy empires, but by aridity and the hunger for water. Human-caused salinity no doubt has its place among the villains. Nevertheless, cultures can survive anything but an outright shortage of water.

If superstition and engineering have been entangled in ancient cultures, so have superstition and politics. The Talmud says no man may sell water from a public cistern; and, while it is required that the laws of hospitality be followed, the needs of the townsfolk and their thirsting animals must prevail over the needs of foreigners.

Even where outsiders' need is great, the person who controls the water has the right to partake of it first. A man in the desert who comes upon a thirsting stranger is entitled to satisfy himself before he shares his water bottle. Charity is a high virtue, but the preservation of your own life is a higher one.

Joyce Shira Starr, whose book *Covenant over Middle Eastern Waters* rooted diligently among the Jewish, biblical, and Koranic water sayings, found an apt quotation from Maimonides' *Book of Acquisition*: "When people have fields along a river they water them in the order [of their proximity]. But if one of them wants to dam up the flow of the river so that his field may be watered first, and then reopen it, and another wants to water his field first, the stronger prevails. The cistern nearest to a water channel is filled first in the interests of peace."

The Koran says no man may abuse a well. If the owner has a surplus of water, he must provide it to strangers and their cattle, but not for irrigating their crops. Desert waters are the source of all real property. Thus power politics: the right of first use, the right of might, the seeds of conflict where powers are uneven and needs great, the water politics of the Middle East and North Africa today, and the genesis of water wars, as we shall see.

4 | CLIMATE, WEATHER, AND WATER

Are we changing the first, and will changes in the other two necessarily follow?

That climate governs weather, that weather dictates water distribution, and that water distribution controls life is obvious enough. Climate is linked to the earth's orbit and its axial tilt, to the rotation of the earth and the consequent wind patterns, to oceans and ocean currents, and to the amount of solar radiation reaching the earth, and therefore to evaporation and precipitation patterns.

Climates change? The El Niño effect, which brings periodic flooding to otherwise arid regions and can abruptly change food chains and ecological systems, is the best-known popular example, but there are many others, of varying timelines. About one-third of the planet, for example, has been covered by ice in fairly recent geological time. New anxiety over the instability of the Antarctic ice sheet has led to closer studies of the region, and these have turned up evidence that the ice sheet did indeed collapse at least once, during a recent (by geological standards) interglacial period less than two million years ago. One reason the Antarctic sheet could be a worry is its sheer size. Its complete collapse would increase mean sea levels by somewhere between three and five metres, with obvious catastrophic effects. Another is its composition. The coastal ice sheets are fed by "rivers" of ice flowing from the interior. If the coastal sheets collapse, the flow of these rivers would likely accelerate, to what timetable is unknown, but the change could be measured in centuries rather than millennia.[1]

We are, of course, now in another interglacial period. And the critical question is no longer whether global warming is real, only whether it is being intensified by industrial waste gases like carbon dioxide. Are we humans changing the climate in such a way as to affect both weather and water, and therefore our own lives and futures?

Much attention has been focused on so-called "extreme weather events" said to be caused by global warming – more and more severe hurricanes and typhoons, tornadoes, and floods. The evidence for the assertion remains, in fact, ambiguous. Severe weather events are certainly causing more damage than hitherto, but largely because vulnerable areas have been colonized by ever-expanding populations. It's true, however, that the average temperature of the continental United States increased by .6 Celsius in the twentieth century, while rainfall went up by somewhere between 5 to 10 per cent, mainly in intense rainstorms. And if the projections are correct, temperatures are likely to rise another three degrees during the twenty-first century, and will be accompanied by much more precipitation and much faster evaporation, leading to a greater frequency of both very wet and very dry conditions.[2] Worryingly, however, especially for Canada and the northern United States, global warming may have other, less immediate but more long-lasting effects: some of the newer models (GCMs, or General Circulation Models) show much of the current Canadian and northern American agricultural lands turning arid and unproductive – that is, Canada may get warmer, but also a lot drier. The evidence is contentious and disputed (some models show, on the contrary, only small changes, and there are also theories that say eastern Canada and Western Europe will get a lot colder as the planet warms up, as the Gulf Stream is deflected from its course). But there is an undercurrent of worry – even occasionally panic – in the climatological journals.

~

One morning not long ago, I walked north from the centre of Timbuktu through the dun-coloured streets, passed by the old mud-built university that had made the city a centre of Islamic learning when it was founded in 1340, skirted a couple of shabby mosques with their distinctive prickling of structural poles protruding into the air, and eventually came to a great depression in the desert floor, maybe fifty metres deep. It was early in the morning. I had left the Prefecture of Police, where I had collected one of the obligatory passport stamps. The area was peaceful, again, though Tuareg bands were still said to be harassing travellers in the northern oases, and the police were watchful and suspicious. A block further I found a police van stuck in the sand, its wheels spinning impotently. Crowds went by, paying it no mind and never offering to help – Fulani women in startling scarlet and gold robes; Tuareg men, veiled and hidden, in their distinctive indigo. The desert sand was twenty centimetres deep, much thicker in places, and it was like walking across a beach; the houses were nearly half a metre below street level. Soon, I guessed, they would be overtaken and buried as the Sahara continued its relentless march. The previous day I'd had a gloomy conversation with the *chef de mission culturelle* of Timbuktu, whose losing job it was to protect the decaying-but-historic city from poverty, neglect, rapaciousness, and the Sahara.

"Money," he said. "Money's the real problem. Without money, you can do nothing. You cannot sweep back the Sahara with a broom."

UNESCO had declared Timbuktu a World Heritage Site, a part of
the cultural patrimony of the human species and therefore worth
protecting. But it had provided no money.

"Are things getting worse?" I asked.

"Oh yes, of course," he said. "In the old days the traders came
up the Niger River and stayed here before attempting the crossing
to Cairo or Marrakech. They had gold with them, and ivory, and
they bought salt. The town was full of rich things —"

"No," I said, "I meant the desert."

"Oh, of course. Once there were gardens to the south, to the
banks of the Niger, when canals brought the river water here.
Now, there is only sand."

"The Niger is at least ten kilometres away from here, isn't it?"

"Yes. It used to be much nearer. But the desert is closing in."

The poor people were cutting down the few scraggly trees for
firewood. They had no other source of fuel and no option; you
can't eat goat meat raw.

In the old days, he said, the Tuareg would never cut down trees.
They needed the shade when they grazed their stock, and in the
wadis the trees grew unmolested. But life was changing. Even
the Tuareg needed money, and the only way they could get
money was to come to town. There, they settled down, and
instead of maintaining their lives as poor but healthy nomads they
became slum dwellers. It is a planetwide problem, and the *chef de
mission* knew he had no answers. Once, Timbuktu had been a
prosperous city of one hundred thousand inhabitants. Slowly it
shrank, until barely twenty thousand were left. Now, the nomads
are swelling it again, but it's no longer prosperous. All these new
citizens need water.

Young boys were playing soccer in the depression in the north
end of the city, scampering about in the forty-three-degree Celsius
heat. Beyond them a group of camels snickered and grumbled. At
the far end of this depression a pathway became a stairway that
spiralled downward. I went to the edge of this deep hole and

looked in. Another twenty metres down a ramshackle construction of poles held a creaking pulley and a massive rope. This was one of the few public wells in Timbuktu.

Ahmed, a Tuareg who had taken me out into the desert a few days earlier, was with me. He had pointed out the communal ovens on street corners; each family was allotted an hour with the oven to prepare food. It was a way of saving firewood. "You have to share," Ahmed said, "you can't waste anything. A boy I know spilled a bucket of water last week and they beat him for it. Two buckets a day and a family can survive. We must teach care with water."

"Is there still water in this well?"

"Oh yes. But soon the men will have to go down and dig another ten or twenty metres."

"It's sinking?"

"The water comes more slowly now."

We went back to the crest of the depression and I stared northward. Footprints led off across the plain. I knew there were still nomads out there – Ahmed's family had an encampment a few kilometres away, their tents and lean-tos anchored in the lee of a dune, their cattle grazing a hundred kilometres further. Beyond that, there was nothing between Timbuktu and the dusty little Moroccan town of M'hamid but fifty days' journey by camel, through sand, sere scrub, stones, and secretive caves in the Ahaggar Mountains in which early man had left memories of better days. And oases, of course. It was the occasional oases that made the desert traffic possible. Oases meant water. Deep water, but precious.

Once, and not that long ago in geological time, there was plenty of water in the Sahara, which was covered in verdant grasslands. The evidence is everywhere: not just the cave paintings or the occasional hippo fossils, but middens of cattle bones and signs that grazing animals once roved here in large numbers. So did elephants, giraffes, rhinoceroses, gazelles, and ostriches. Crocodiles and fish stirred the waters of lakes and rivers. A mere 10,000 years ago, cypress, oak, lime, alder, and olive trees once grew on the grassy uplands. There is persuasive evidence, in fact, that the Sahara has alternately been a desert and prairie in long, 10,000-year cycles – it was a desert 40,000 years ago, 30,000, 20,000 . . . Now, there is only gravel, boulders, mountains, sand, and salt, dry lakes, and wadis.[3] Interestingly, current theories maintain that the most-recent desiccation of the Sahara reinforced the emergence of a settled culture on the Nile, as the early Nilo-Saharans, grassland herdsmen and proto-cultivators, were driven from the centre of the desert to its fringes, to find what water they could. By this reckoning, the flowering of ancient Egypt was a by-product of climate change. Of course, it is an Egyptologist's truism that climate's effects, the annual flooding of the Nile, sustained the culture that did emerge.

Even in recent memory, the Tuareg watered their goats and cattle at shallow lakes. No more.

~

Although humans are making deserts worse, they didn't cause them.

At the other end of Africa, around Swakopmund on the fringes of Namibia's Skeleton Coast, is the Namib Desert, which is supposed to be one of the oldest on the planet. As evidence, scientists point to the survival there of peculiar plants, such as the Welwitschia, a primitive tree that is largely subterranean, with only two leathery leaves visible above ground. This curious plant very likely existed before the evolution of flowering plants, to which most terrestrial vegetation now belongs. There are also animals in the Namib Desert marvellously adapted to life with almost no water and at extremely high temperatures, including a little beetle that stands quivering in the sand, its fanlike tail spread in the air to collect what moisture there is, looking for all the world like a miniature maquette for a choreographer of alien ballets.

Most of the earth's deserts are not as old as the Namib. The majority are thought to be of fairly recent geologic origin, probably dating from the steady cooling and consequent aridification of earthly climates that has been the primary feature of the present era (the Cenozoic era, in technospeak), starting about sixty-six million years ago. Desert plants likely evolved even more recently, maybe only ten million years ago, most of them in what used to be the Tethys Sea, a larger ocean along the Mediterranean and Caspian axis.

Most deserts, also, are in the subtropical zones to each side of the saturated equatorial belt. Their existence there is due to something the climatologists call the Hadley Cell, another example of a closed circle in nature. At the equator, where the solar energy reaching the earth is at its greatest, evaporation is obviously fastest, and the air near the ground is heated more quickly. It rises, expands, and cools, and when it cools it condenses, and the rains fall in tropical storms, great drenching sheets of water. The now drier air continues to rise, and at high levels moves away from the equator, pushed outwards by the wall of continually rising hot air. Over the subtropics it cools and starts to descend. It has already lost most of its moisture near the equator, and as it descends it is compressed and becomes warmer, its relative humidity declining even further. By the time it gets to the ground, it is hot and dry; it then moves back towards the equator to complete the cycle as hot, arid wind.[4]

Desert temperatures are extreme. The highest air temperature so far recorded was in Libya, 58°C (136.4°F). The soil can get even hotter, with 78°C (172° F) having been recorded in the Sahara. Night temperatures can be dramatically colder. The lack of cloud cover allows the heat to radiate rapidly outwards at night, and in winter, desert night-time temperatures are routinely below the freezing point.

By definition, rainfall in deserts is meagre, ranging from almost zero in Libya and some South American coastal deserts to about

250 millimetres in the marginal lands of Southern Africa. At Cochones, Chile, no rain fell at all in forty-five consecutive years between 1919 and 1964. Usually, however, rain falls in deserts for at least a few days each year – typically fifteen to twenty days. When precipitation occurs, it may be very heavy for short periods. For instance, fourteen millimetres fell at Mash'abe Sade, in the Negev Desert in Israel, in only seven minutes on October 5, 1979.

Occasionally, the rain that does fall fails to reach the ground because of the heat and the aridity. The potential evaporation rates in deserts can be extraordinarily high, typically 3,000 millimetres a year or more. Death Valley in Nevada has recorded rates of over 4,000 millimetres. Even where rain does fall, it is erratic, and in some years may fail altogether. The reasons are not well understood. Regions where the mean annual rainfall is between 250 and 400 millimetres are called semi-deserts; they, too, are barely arable, and humans can survive only by using them as grazing lands for livestock. Or, of course, through irrigation.

Deserts, while entirely natural, can also be of anthropogenic origin – a consequence of human actions. It's not altogether certain whether the Sahara itself is an early consequence of human action; the evidence is very thin, though it hasn't stopped extravagant speculation. But even if humans had nothing to do with "causing" the Sahara, they certainly contributed to its spread and the degradation of the surrounding land in the Sahel. As vegetation is stripped from the land, the surface dries out and reflects more of the sun's heat. This condition in turn alters the thermal dynamics of the atmosphere in ways that suppress rainfall. Other experts suspect that increased dust or other atmospheric pollutants could be causing changes in the climate.[5] This is what is meant by desertification.

The Israelis, who know as much about it as any people on earth, define it as "the degradation of natural resources in an ecosystem such as soil, water, vegetation and wild life. Global climate changes as well as human intervention influence the rate at which desertification occurs or is averted. The dry land sub-regions in the East Mediterranean are characterized by intense sun radiation and the poor quality of water resources."[6] Desertification means declining water tables, and the salination of topsoil and the remaining water. Increased erosion and the extinction of vegetation, however caused, make the climate drier. Even within natural deserts, human activities can magnify the effects. The Sonora Desert of the American southwest and Mexican northwest, which for centuries had been in precarious ecological balance, with a meagre rainfall keeping a complex ecology of plant and animal life alive, has become almost entirely barren through human causes, and its natural wildlife has all but vanished.

Even development that is beneficial on balance can affect the process. Constructed lakes can increase precipitation in some areas, sometimes a considerable distance away – Lake Kariba on the Zambezi River is perhaps the best-known example. But the increased rainfall in some areas is accompanied by reduced rainfall elsewhere. This doesn't mean there's a fixed amount of moisture to evaporate or rain to fall. Weather is not a zero-sum game. It just means that local consequences of large-scale environmental engineering are often unforeseen. The draining of Sudan's Sudd Marshes, or at least the creation of the Jonglei Canal that will push the Nile's water more quickly through the region, will benefit downstream Egypt by ensuring that a larger proportion of the Nile waters reaches its territory. But a lessening of the evaporation from the marshes risks changing the microclimate that makes Sudan's grain-growing areas possible – another source of potential conflict in a situation already fraught with tension. As the Israelis put it:

Economic development can accelerate the process of desertification if left unchecked. Construction of railroads, highways and airports can interfere with ground water sources and natural replenishment and adversely affect the existing agriculture upon which it depends. Intensification of agricultural production without proper attention to water allocation, preservation of water quality, and environmental pollution resulting from use of pesticides can harm both plant and animal life and limit the potential for long-term demographic development. Similarly, pollution from industry and mining activities may pose a threat to the ecosystem in the future. Sparse pastures may disappear entirely as a result of over-grazing.[7]

Population growth, the engine of all increased resource demands, is at the heart of the desertification problem. More people mean more animals, and more vegetation cut for fuel wood and construction. The additional herds needed by greater numbers of humans trample and compact the soil, reducing the infiltration of the little water that exists, causing erosion and damaged soil. In the Sahel, humans have learned to use cattle dung as a source of fuel, but even there what sparse vegetation remains is being hacked back for kindling, a violation that leads to increased erosion and wind damage. I remember standing in N'Djamena (Fort Lamy), the capital of Chad, in the harmattan, a windblown sandstorm, in which the visibility was less than a dozen metres, and watching the soil of the Sahel blowing overhead, to be deposited sometimes hundreds and even thousands of kilometres from its source, saline and leached of any nutrients it might once have contained. It is the classic vicious circle: the desert causes ever more desperate measures by humans to survive, and those human activities increase the desert. Only a break in the population upswing can stop the desert's relentless progress.

The increasing aridity of the Saharan fringes has imposed itself on international politics several times in the last few decades.

In 1984, to take an obvious example, the rains had failed, again, and the ecological refugees were on the move in the Sahel, again, only this time tracked by the cameras of the BBC and PBS, taking their dusty and desiccated lives into the living rooms of the affluent West. It wasn't the first time thirst had killed, and, alas, it hasn't been the last. The United Nations estimated that some two hundred million people in twenty African countries were severely affected by the 1984 drought and its related famine. The survival of thirty million people was at imminent risk.

Drought is a natural phenomenon, and so are deserts. Famine, by contrast, is a political catastrophe that occasionally happens as a consequence of drought. There is no doubt that this Sahelian famine was to a considerable degree political. The Derg, the fanatical rulers of Ethiopia, had used food shortages as a political device. In the Sudan, the rule of Colonel Jaffar el-Nemeiry made things worse than they would otherwise have been, and it was only the heroic efforts of the UN relief effort, under the energetic and inspired direction of Canada's Maurice Strong, that prevented millions from perishing. It must have been a wonderful sight to see Nemeiry, a dictator of the brutalist school of politics, bullied in his own office and threatened with international opprobrium if he didn't give in, which he did. Many felt at the time that Strong should have been a shoo-in for a Nobel Peace Prize, but the Swedish committee inexplicably passed him over.

Part of the cause of the Sahelian famine was political venality, corrupt and brutal regimes whose priorities were, in the almost literal sense, unhuman. Part was also due to poverty, to the degradation of the environment that poverty causes, and to too many humans living in the wrong place. Some small part was due to cycles in weather patterns: Gaia, if she exists, doesn't take much notice of where humans choose to live. And part of it was due, almost certainly, to climate change.

Climates can and do change, sometimes very rapidly, even without man's intervention. They have changed in historical times – as we have seen, the Sahara became desert at about the same time that the Nile cultures were evolving, and with astonishing, and worrying, rapidity. A study by the Potsdam Institute for Climate Impact, reported in the journal *Geophysical Research Letters*,[8] suggests that the transition from savannah to desert took place in two discrete "episodes," the first between 6,700 and 5,500 years ago, and the second in a 300-year burst sometime between 4,000 to 3,600 years ago. The first episode was dangerous, but not disastrous. The second, appallingly short in geological time, tipped the climate system completely, and life in the Sahara changed abruptly. Summer temperatures climbed dramatically, rainfall almost ceased (to less than five centimetres a year), rivers and lakes dried up, and water tables began dropping.

The most plausible explanation for this transformation combines a theory of axial tilt with new notions of positive-feedback loops in an intricate interplay of ocean, land, vegetation, and atmosphere. In this hypothesis, slight and gradual changes in the earth's orbit and axis accumulated until they caused abrupt changes in vegetation and climate, so abrupt that in that mere three hundred years or so the Sahara changed from grasslands to the desert that exists today.

Another clear example was documented a thousand years ago in a colourful saga about the Norse settlement of Iceland and their colonization of Greenland: a confidence trick by an Icelandic outlaw called Eric the Red and his son Leif the Lucky.

Eric was, from contemporary accounts, something of an Icelandic loose cannon, the Norse equivalent of a Wild West outlaw. He and his father had left Norway around 960 "because of some killings," suggesting that Eric had made it to his longboat just before the posse. As a refugee in Iceland, he seemed at first to prosper. He made a good marriage and settled down to farm in the gentle valleys of Haukadale, in the south of the island. But his bad temper got him into trouble again and there was more violence. With another posse of aggrieved relatives sworn to vengeance on his heels, he set sail for a land far to the west that had never been settled, but had been described a century earlier by a storm-swept sailor called Gunnbjørn Ulfsson, who had painted vivid verbal pictures of mountains of ice, towering seas, and ghastly storms – none of which seem to have deterred Eric. Sailing west on the sixty-fifth parallel, Eric caught sight of the massive 2,500-metre ice mountain now called the Ingolsfjeld Glacier. He turned south, rounded what is now Uummannarsuaq or Cape Farewell, and explored the much less savage west coast. Liking it well enough, he returned to Iceland for one more skirmish with his enemies and to round up a few colonists, ambitious outsiders like himself. To persuade them to depart for parts unknown, he was careful to counteract the earlier horrific notions of Ulfsson, and prudently called the new country "Greenland." A contemporary, under no illusions about Eric's character, left a record of his reason: "He said that people would be much more tempted to go there if he gave it an attractive name."

In the summer of 985 the colonists duly signed up, and a fleet of twenty-five ships set sail from Iceland to the newly named Greenland. Eric's con game notwithstanding, Greenland was clearly more temperate than it is now. There was sufficient grassland on the western coastal strip to permit the cultivation of grains and the raising of sheep, goats, cattle, and pigs. Modern excavations of the settlement show that it had been reasonably prosperous. Eric's own farm had four barns and room for forty head of cattle, and the remains of a stone cathedral thirty metres long have been uncovered. The Icelanders, and therefore the Greenlanders, were resolute Christians.

For the next few centuries, the settlements were stable. That the Greenlanders explored the Arctic regions is known from the discovery of a stone cairn, dating from the fourteenth century, at almost seventy-three degrees north latitude, bearing the runic description "Erling Sigvatsson and Bjarne Thordarsson and Endridi Oddsson built these cairns." They also found "Vinland," though the exact location of this land of flowing streams, majestic trees, and wild grapes is still a matter of acrimonious academic debate.

However, in the fourteenth century the climate began rapidly and terrifyingly to deteriorate. For a while, ships still made their way west along the sixty-fifth parallel, but by the middle of the century they had to skirt the increasing pack ice and make a difficult circuitous route along more southerly latitudes. By the turn of the century the Inuit, who had migrated northward after the previous ice age, began to be forced southward by the cold, and there were repeated clashes with the Norsemen. A decade into the fifteenth century, the Greenland colony was completely cut off from the rest of the civilized world. The last reference to Greenland was in the Iceland Annals for 1410, when an Icelander returning home left a distressingly laconic account of his sojourn, a last elegy for a hardy and energetic people now defeated by nature. His was the last ship out: winter had abruptly shut down Greenland, and the exhausted remnants of the colony presumably perished.[9]

Climate change, not of man's making. But it defeated him, nonetheless.

~

Regional climate change can profoundly alter local hydrological ecosystems. It can affect the distribution of water basins, shift groundwater tables, and cause desertification in some places, flooding in others. Global climate change, obviously, will do the same thing, only on a grander scale.

Is climate change, in the form of global warming, happening now, and if so, what will its effects be on the availability of the planet's most precious resource? Are the record high temperatures we have been experiencing over the past decade an indication of anything other than a short-term cycle, about whose beginning and end we know little? In other words, is it a problem, and if so, is it self-correcting?

Are there more "extreme weather events," to use the current jargon, than there used to be? Are hurricanes worse and more frequent? Typhoons and cyclones? Are tornadoes more common and more violent? Are ice storms more likely than they were? Floods? Droughts?

Is earth going to turn into Venus, a hothouse of methane and poisonous gases, or Mars, a sterile and airless place, deadly to life? And if any of this is happening, is it human-induced?

One theory of climate change should generate caution in extrapolating from short-term trends. In the 1930s, the Serbian mathematician Milutin Milanković suggested that basic variations in the earth's movement could affect global climate. He had detected, he said, three cycles: a 100,000-year cycle of the planet's orbit, a 41,000-year cycle in the tilt of earth's axis, and a 23,000-year cycle in the wobble of the axis. Alas, theory sometimes collides with fact: according to these cycles we should now be in the middle of a long period of cooling.

And to introduce another note of caution and put the debate into perspective, consider that a *Newsweek* cover story in 1975 asserted that "meteorologists disagree about the cause and extent of the cooling trend . . . But they are almost unanimous in the view that the trend will reduce agricultural productivity for the rest of the century." Compare that to vice-president Al Gore's assertion in 1992 that "scientists [have] concluded – almost unanimously – that global warming is real and that the time to act is now." So either the world's scientists, as synchronized as a school of fish, are darting hither and yon in overreacting to the latest ambiguous data, or they aren't nearly as unanimous as the media reports, politicians, and Greens would like us to believe.

Until 1998 there was, in fact, some disagreement about what the earth's temperature really was doing – and temperature, you would think, would be one measure that could be recorded with a degree of certainty. Now it is conceded that the global average temperature has risen by around half a degree centigrade this century. This is a small enough change, but not necessarily re-assuring. Between 1940 and the mid-1970s, temperatures appeared to be roughly stable, but there is evidence that in the last decade the change has been accelerating. There is still room for some ambiguity. The evidence for change was deduced from measure-ments taken on the ground, and there are still arguments as to whether the temperature of the atmosphere is rising (early studies said No; subsequent studies claimed to have found an error and that the answer was Yes; further studies said the error might be an error).[10] The Intergovernmental Panel on Climate Change, a UN-sponsored group of scientists, concluded in 1996 that there was discernible evidence that humans have already begun to alter the earth's climate. The IPCC based its conclusions on three "fingerprint" studies: a clear warming pattern in the mid-troposphere in the southern hemisphere; the decrease in variations in daytime temperatures; and the statistical increase in hurricanes, cyclones, tornadoes, and destructive floods in many countries in the previous half-dozen years. Recent studies on the Antarctic ice sheet have made pessimists of many scientists who had remained skeptical.

One of the key factors is the rise in TMINs (temperature minimums). Since 1950, night and winter minimum temperatures have been rising, while maximum temperatures are also rising, if marginally less quickly. The cycle is self-sustaining: warming intensifies the hydrologic cycle. A warmer atmosphere holds more water (6 per cent for every degree Celsius), and the increasing cloudiness reduces daytime warming and retards night-time cooling by blocking outgoing long-wave radiation. The increased evapotranspiration and cloudiness also help explain why there is now more heavy rain and flash flooding than there used to be.

Ocean warming has contributed to the drift northward, since the 1930s, of semitropical marine plants and animals along the California coast. More recently, deep ocean warming has been reported in the Atlantic, Pacific, and Indian oceans, not only in subtropical regions but also near the poles. There have been recent radical reductions in the number of zooplankton near the ocean surface, and studies have found sharp declines in mid-ocean nutrient levels (leading one scientist to speculate that the ocean was becoming a "dust bowl" – a peculiar metaphor under the circumstances, but revealing). Speculative data indicate that warming sea temperatures may be the global capacitor (thermal sink) that is absorbing the past century's global warming by "storing" the surplus carbon dioxide.[11] If this is true, and the "sink" stops absorbing and begins transfusing, the earth's climate could change much more rapidly than earlier predicted.

There is none of that famous "unanimity" on the issue, but a consensus is building that the temperature increases resulting from a doubling of atmospheric carbon dioxide will change the world's basic hydrologic cycle by increasing both evaporation and precipitation. Depending on which study you look at, the rises are predicted to be somewhere between 7 and 15 per cent. Rainfall will get heavier, though not everywhere. Rainfall patterns are predicted to shift, though the studies lack precision. Some areas will become wetter, others drier, with profound implications for national water policies. Some water-stressed countries will likely go critical, while temperate regions will get more floods and more "heavy rain events" (which have already increased in the United States by 20 per cent since 1900). As temperatures rise, disease-carrying insects such as malarial mosquitoes will become more widespread. Lakes that are now pleasant recreational areas might become disease vectors for bacteria. Already, algae blooms are increasing at an alarming rate.[12]

At a meeting in Japan in 1997 the assembled nations agreed to a treaty that would at least slow the onset of global warming. The Kyoto Protocol would require dozens of nations to cut back on greenhouse emissions, some of them drastically. A few nations will meet their targets, but many will not. The Americans, for example, won't even try, and though Canada ratified the treaty in 2002, the mandated remedies were minimal and widely derided. Even if every country ratified Kyoto, the treaty's effects would be modest; an estimate by Tom Wigley, of the National Center for Atmospheric Research in the United States, suggested that the treaty, fully implemented, would keep the global temperature only four-tenths of a degree lower than if there were no treaty at all.

Many geographers now argue that prevention isn't enough, and that nations should prepare themselves to adapt to the inevitable changes. Proposed measures include breeding food crops that can better withstand drought, discouraging people from living on flood plains, and making irrigation and water use more efficient. Joel Smith, who bills himself as a "climate impact expert" at a company called Status Consulting, points out that action now will be cheaper than action later. The bridge between Canada's Prince Edward Island and the mainland is a case in point. It was built on pilings one metre higher than originally planned, to ensure that ships could pass underneath it if sea levels rise.[13] So persuasive has the evidence become that a number of major American corporations, including the now-disgraced and bankrupt water giant Enron and the chemical firm DuPont, maintained that it warranted "immediate action," and announced they would press Congress to guarantee financial benefits for companies that act, global agreement or not. The steam went out of the initiative as a parade of former Enron executives appeared sheepishly before congressional committees, but at the same time a few major oil companies, among them BP and Shell, announced that they would cut their own emissions by 10 per cent from their 1990 levels by 2002, and succeeded in doing so.[14]

Curiously, studies show that some places will become much colder as the earth warms up. If melting Arctic ice, for example, interferes with the Gulf Stream and the flow patterns of the major North Atlantic ocean currents, New England, Nova Scotia, Newfoundland, Greenland, Iceland, and much of Western Europe will become dramatically colder. On the other hand, they may get warmer. Predictably, skeptics exist. Among them is Eugene Stakhiv of the U.S. Army Corps of Engineers, the army offshoot that has built dams and "groomed" rivers across the United States for more than a century. There are, Stakhiv says, three legitimate ways of looking at climate change: it doesn't matter, or we should wait until we have more evidence, or we need to take precautionary measures. "It's fair to say the Corps is in a wait-and-see mode," he says.

I caught up with him at a coffee break at the same UNESCO water conference where I had met Peter Gleick. I found Stakhiv's own paper a model of clarity and logic, and wanted to tell him so, since neither logic nor clarity were necessarily hallmarks of the papers thus far presented. I asked him what he thought was happening to the global climate.

"Have you read all the studies?" he demanded, peering at me owlishly. Before I could confess that no, I hadn't read them all, he patted the air at around waist height, about a metre off the ground. "Pile them on each other, you'd have this much paper," he said. "I've read 'em all, and I can tell you, many are useless, some are useful, the data are ambiguous, and some of them are simply outrageous." Since I'd identified myself as a Canadian, he bored in on what he perceived to be Canada's shortcomings. "The Canadian GCM [General Circulation Model] on climate change is the worst of the lot," he said happily, grinning at me. "Hopeless. Apocalyptic. Which is why so many people like it. But we don't agree with it at all. For instance, the study forecasts the emptying of the Great

Lakes through global warming. Our most pessimistic estimate, on the other hand, is that there might be a drop of a metre at most. They might even become fuller."

In the paper he delivered to the conference, he said that there is sufficient evidence from recent climate-change impact studies "to offer some degree of optimism that well-managed resources systems can withstand all but the most severe climate-change scenarios postulated by the leading GCMs. In other words, even as the number of countries and regions that are susceptible to water scarcity and variability climbs, there is a reasonable degree of confidence that vulnerability (the degree of harm or damage) can, at a minimum, be stabilized at current levels, or reduced in most cases."

Several Corps studies on specific river systems in the United States were consistent with a larger study of the Nile, he said. The Nile study showed that even doubling the carbon-dioxide levels would have manageable effects on the river, and only a marginal effect on the Egyptian economy (perhaps a loss of 1.6 per cent). In the United States, the Corps study of the Missouri River system showed the most negative changes of all U.S. rivers, but even there the data were inconclusive and contradictory, ranging from a decrease in the Missouri flow of 35 per cent to an increase of 2 per cent. "Water management, which is what we do on a daily basis, deals inherently with uncertainties in supply and demand and with managing rapid change; the key is therefore institutional reform (making managers better able to respond) rather than engineering solutions to problems that might not ever exist."

Later, at a second coffee break, Stakhiv added another point. "Technological advances can mitigate problems. This is something all the studies ignore. But if you're willing to project climate change into the future, why not also be willing to project techno-logical changes and inventions? And look at the data. Everyone is crying that a crisis is imminent. But it seems that crises are always imminent. They never actually seem to get here. We always seem to cope." He took comfort in the fact that "many of the water-use projections in the United States have been, to put it kindly, simply wrong. In fact, water withdrawals are actually declining," he said, and pointed to metropolitan Boston. There, proper pricing mechanisms had sharply reduced demand.

Still, Stakhiv conceded, even without climate change, "we will have to find more water. There is still some reason for optimism, but the task will be extraordinarily difficult. Despite the concerted efforts of international organizations to upgrade the water management capacity of the fifty-or-so developing nations in Africa and Asia, not all will succeed equally. And there is no way of determining in advance where the shortcomings will occur."[15]

A recent study by Dr. Robert Mendelsohn and his colleagues at Yale goes some way to supporting Stakhiv's skepticism. Mendelsohn is an experienced environmental economist, but his study has been controversial, partly because of its results and partly because it was financed by the research wing of America's power-producing lobby. His conclusions suggest that, while some countries and regions may suffer, the overall economic impact of global warming is likely to be zero, a notion that has kicked off considerable debate in climatological circles. For example, David Glover, director of the Economy and Environment Program for South East Asia (EEPSEA), carefully refrained from endorsing the study, but suggested it deserves serious examination. "If its findings are correct, then drastic measures to reduce the use of fossil fuels might not be in the interests of developing countries. Such measures could excessively restrict the energy consumption of poor people and divert resources from more immediate environmental problems like water supply and sanitation."[16]

5 | UNNATURAL SELECTION

Contamination, degradation, pollution, and other human gifts to the hydrosphere.

We drifted down the Volga River in a small boat, its engine throaty but rugged, chugging our way past Yaroslavl and Plyoss, on our way to Nizhni Novgorod (formerly Gorky) and eventually to the Caspian Sea, still several thousand kilometres to the south. I had wanted to travel the full length of the Volga, for in Russia this isn't just another river. It's the longest river in Europe, true, but that's not what binds it to Russian hearts. In Russian folklore, the Volga is mother and mistress, companion and teller of tall tales. *Matrushka Volga*, Little Mother Volga, is Russia itself, as the Nile is Egypt.

But past the placidly bucolic village of Plyoss, so beloved of Russian painters, the landscape along the river began to change. First, the trees disappeared. Rather, they were still there, but they were all dead. Along some stretches, nothing larger than scrub seemed to live. The guidebooks gave a clue, pointing proudly to the paper- and glue-making factories, phosphorous and fertilizer factories, and shoe factories that dotted the banks, set down in their twentieth-century ugliness among the monuments to the fourteenth-century battles against the Tartars at Mogiltsy, the ancient monastery at Navaloky, and various estates, monuments, and villages.

On the right bank we stopped at the little town of Chkalovsk, an exceedingly dreary place named after its only local celebrity, an aviator named B.P. Chkalov, who in 1937 had made a non-stop flight to the United States over the North Pole and became a Hero of the Soviet Union for his achievement, his picture appearing in *Pravda*, his arm draped over his little aircraft while a beaming Uncle Joe Stalin looked on. We bought supplies at the meagre market there and moved on down river to the small city of Gorodets, where the great Russian hero Alexandr Nevsky died in 1263 after a peace mission to the Tartar capital of Bolgari.

Gorodets was first mentioned in the Russian Chronicles in 1152, the year it was founded by Yuri Dolgoruski, but the modern city is a disappointment. There's nothing left of its long history except an earthen mound that is said to be the remains of the ramparts burned down by the Tartars. The factory workers there were on strike, again. They always seemed to be on strike, sometimes because they hadn't been paid for months, sometimes just because the working conditions became intolerable. At the riverside, a worker handed out pamphlets demanding a cleanup of the environment. "Can you smell the air?" he demanded. "We have to breathe this every day of our lives, and it's killing us. The very air we breathe is killing us."

It was true, the air was thick, with an acrid smell and a chemical taste. On the left bank, a huge chimney was pouring a plume of smoke into the atmosphere. By now, the sight of the thousands of dead trees along the banks had become oppressive. Our boat had travelled for several hours through an eerie stillness – no foliage, no birds, no life. My fellow travellers were depressed. The image the landscape conjured, of a Mother Russia poisoned at the hands of the Russians themselves, was grim.

All the towns along this stretch of the Volga were horrors. Pravdinsk, for example: a 1926 guidebook said the city had been famous for its lace. Now it seems to consist largely of a pulp-and-paper plant and a cellulose factory, until recently named – appropriately, I thought – after the dreadful Felix Dzerzhinsky, who ran the Cheka, the forerunner of the KGB, for Lenin. The face Pravdinsk showed to the river was profoundly discouraging: a railway line with a tottering trestle bridge led to the pulp factory; lumber, spilled off the trains, littered the ground and rotted where it fell; and trunks that had been delivered to their destination lay in untidy piles. The neglected factory itself was falling down; elevated pipes, strung between buildings, were sagging, and others had been patched with old tires or were rusting through.

The town of Balakhna, on the right bank about thirty kilometres from Nizhni Novgorod, was even worse. A mining and smelting centre, it was one of the first large consumers of Volga hydro-electricity. But its factories were almost a parody of Victorian industrial exploitation – not only grimy infernos, but falling to bits. Grass was growing from the masonry of the walls, storage tanks were crumbling, roofs were sagging, and smoke poured from windows on all sides as well as from the chimneys. Downwind the air was yellow, a cloud of pollution that rolled over everything in its path, as relentless as a line of Nazi tanks, killing all vegetation in the way, shrubs, bushes, grasses, and trees, all dying. Upwind, scrawny trees still lived. As we passed one factory, I noticed a gully leading to the river, a runoff into the Volga; it was steaming, a sulphurous yellow in colour. A man was fishing just a few metres downstream. I hoped for his family's sake he caught nothing.

~

Many of the ancient chronicles detail saline and "bitter waters"; sometimes they are listed for public edification, so they can be avoided – a kind of early Surgeon General's Health Hazard Warning; in others they are regarded as just punishment for transgressions against the gods. One of the signs of the imminent Apocalypse is "the bitterness of all waters"; and anyone travelling through Eastern Europe, the former Soviet Union and its satellites – everywhere that the command-economy model operated, with its callous disregard for anything but narrow-focused abstract principle – could be forgiven for thinking that the Apocalypse was no longer imminent but in full cry. There's hardly a river, stream, or brook that isn't contaminated with the runoff from human misuse, whether industrial effluents, agricultural pesticides and herbicides, or worse. (The "worse" could be bacterial contamination – the river as disease vector – or the dumping of radioactive wastes.) A recent Czech study commissioned by the government found that three-quarters of all surface water in the country was "severely polluted," and almost one-third was so badly contaminated it couldn't support any fish at all.

When the Danube – the Donau, the Duna, the Dunav, the Dunarea, the Dunay – passes Bratislava, it is several hundred metres wide and a muddy brown in colour, no longer the *schönen, blauen Donau* of Johann Strauss's romantic elegy. It has already travelled a considerable distance, rising in the Black Forest mountains of Germany and traversing Austria. From here, it bisects Buda and Pest, turns south to Belgrade, skirts the U-trap of the Carpathian and Transylvanian Alps, and passes south of Bucharest to its delta at Sulina, where it empties into the Black Sea. The river's collection basin touches on eighteen European countries over an area of 817,000 square kilometres, a territory inhabited by eighty-seven million people. Like the Rhine, the Danube is no longer pure river but a manufactured waterway. To improve navigation, the slow, meandering sections were long ago dredged, deepened, and straightened; numerous canals were dug; and hydro-power plants, barrages, dams, reservoirs, navigation canals, and locks were constructed. Water was extracted for agriculture, city uses, and industry; in return, sewage was dumped into the waterway.

The Russian researcher Irina Zaretskaya, who works with Igor Shiklomanov, has tracked the Danube in detail. She is clearly not impressed, and her distaste for what has been done is obvious, even in the formal academic prose of her technical reports. Of the Danube's engineering works, she writes: "These works [carelessly] changed cross sections, coast lines, slope, bottom and suspended sediment discharge, as well as water quality." Elsewhere she has written: "The quantitative and qualitative depletion of water resources in individual regions of the basin has resulted in a critical situation, especially during dry periods. An increasing water resource deficit in the region can become a brake on the economic development of the countries." Her figures show that, already, 31 per cent of the Danube is extracted for human use, which will rise to 37 per cent in a decade. There is still enough water, but not by much. Availability is already only 60 per cent of the European norm, and in the driest year on record, 1949, the available water dropped to 1,500 cubic metres a year per person – not dangerous, but disturbing, "reaching critical values," in Zaretskaya's words. And the population is now considerably greater.

Zaretskaya tracked the deteriorating quality of the water as well as its reduced flow. Dozens of cities and half-a-dozen countries allow huge amounts of insufficiently purified storm runoff, industrial wastes, and agricultural pesticides to enter the river, virtually unmonitored. No one has ever added up the total or catalogued the horrors. What is known, she says, is that the Danube pours about eighty million tons of contaminated sediments into the Black Sea every average water-flow year.

Matters are somewhat better in Western Europe, where regulation has been tougher and the Green movement earlier and angrier. Yet, the little Espierre River in northern France and Belgium was the dumping conduit for a fertilizer plant until 1983, when its potassium levels were more than ten thousand times the naturally occurring levels – about the same amount as the entire North Sea.[1] In 2002 the Flemish water service, VMW, reported that pollution in the Ijzer (Yser) River estuary was so heavy that drinking water was threatened in the region; the culprit was atrazine, a herbicide used in maize production.[2] And until the 1980s, a single potash mine in Alsace dumped 15,000 tons of sodium chloride into the Rhine every year for fifty years, the equivalent of the combined "natural" loads of the Congo River and the Mississippi.

The Rhine is Europe's most important waterway, rising in the Reichenau above Lake Constance in Switzerland and flowing for 1,320 kilometres to the Waddenzee in the Netherlands, most of it through Germany. Its catchment area of 185,000 square kilometres takes in Germany, France, Holland, Austria, Luxembourg, Liechtenstein, and Belgium, even Italy, and contains more than sixty million people. But by the 1980s, the river was known as "the Sewer of Europe," a conduit for a multiplicity of poisons.

On October 31, 1986, a fire broke out in an electrical switching box in a riverside warehouse in Basel, Switzerland, nearly five hundred kilometres upstream from the place where the Sieg enters the Rhine. It was no one's fault, a mechanical system had failed. But flammable material was stored nearby, and that was definitely someone's fault. The fire spread quickly, engulfing a couple of storage rooms before racing into the main warehouse itself. The heat was intense and began to blister steel drums stacked against one of the walls. Then one of the drums exploded.

By the time Basel's firefighters got to the warehouse, which was owned by Sandoz (now part of Novartis), one of Switzerland's largest chemical manufacturers, there was little they could do about the fire itself. After hearing from the owners what the building contained, however, they backed off, donned their gas masks, and worked feverishly to build a catchment wall to contain the runoff, to prevent it pouring into the Rhine. Too little, too late: the wall collapsed, and more than thirty tons of poisons poured into the river, an evil brew of herbicides, fungicides, pesticides, dyes, heavy metals, and two tons of mercury.

A few hours later the toxic stew reached Germany. Bertram Müller, a hydrologist who had been working with the International Commission for the Protection of the Rhine, the ICPR, saw it pass. "The river ran red. Just for a while," he said. "The dyes, I suppose. Otherwise, it looked no different. It was the Rhine I had always known. But I knew that, as I watched, its creatures were dying. It was the most terrible feeling. I was frozen, sickened. I couldn't even cry. It was the worst case of chemical contamination of a European river, ever, and maybe of a river anywhere, ever."

More than a million fish, it was estimated later, died in the catastrophe. Livestock along the banks were poisoned as they drank from the river. All the way to Amsterdam, eight hundred kilometres from Basel, engineers shut down the intake valves for drinking-water systems: they could cope, barely, with the "accustomed" level of Rhine pollution, but this was impossible. The reservoirs of a hundred towns ran on empty for weeks.

"We believed," says Müller, "that the river would remain biologically dead for a generation. It was a despairing thought." Yet even this ecological disaster, perversely, was leveraged into good news. Chernobyl was still a recent and terrifying memory. The people and the politicians were frightened into action. The Rhine cleanup, begun with little popular enthusiasm more than thirty years earlier, was galvanized. They would, at last, do what must be done.

As little as four years later, the first salmon appeared in the upper reaches of the Rhine. It was, as far as anyone knew, on its own, and on examination it was too toxic to eat, but its mere appearance was a triumph.

"The rebirth of the Rhine . . . has to count as one of the great environmental success stories of the century," said Angela Merkel, Germany's environment minister, with considerable hyperbole but some justification.[3] Gottfried Schmidt, a biologist who has been working the Rhine's Sieg tributary in the ICPR's Salmon 2000 program, points out that the Sieg's water, once among the most polluted in Europe, is now almost drinkable without treatment. Dozens of dams and other obstacles have been removed, wetlands are being restored wherever possible, and thousands of salmon fingerlings are being released. Schmidt figures that, by century's end, somewhere up to twenty thousand adult salmon will spawn in the Rhine system. This would be just a tiny fraction of the historical levels, but a self-sustaining population of any size must be counted a success.

The ICPR is not stopping there. Although PCBs have been banned in the Rhine area for years, there are persistent traces in the river's sediment and in fish: a study of eels caught in four of the Rhine's countries showed alarming concentrations. Nitrate pollution is still going up, despite best efforts, coming mostly from waste-water treatment plants, cars, and power plants, but most of all from farm fertilizer runoff. Phosphates have been banned from detergents for years, but phosphate levels remain too high, and the commission in 1997 mandated another 17 billion Deutschmarks (nearly 9 billion Euros) to impose stricter thresholds. Heavy-metal concentrations in the river are down, but not out. Cadmium is still found. Lead, copper, and zinc are present in concentrations five times higher than the ICPR's "target" values. Mercury in locally caught barbel fish exceed allowable levels, and the fish are still unsafe to eat. Serious floods in 1993 and 1995 meant that more banks will have to be restored and more wetlands initiated.

Much to do, yes, but what has already been done is impressive. Pollution can, it is clear, be fixed. All it takes considerable political will. And even more money.

∼

As we have seen, coastal pollution is regarded as "severe" in the Baltic, in most of the North Sea, and all along the Mediterranean littoral, from Spain to Italy. Starting in about 1998 – no one knows for sure – a species of "killer algae" was spreading through the Mediterranean at about four hectares a day and at the end of 2000 had infected more than seven thousand hectares. The rapid growth of *Caulerpa taxifolia*, accidentally spread by man, was described by marine scientists as a "major biological invasion," and was found to be threatening marine biota between Toulon in France and Genoa in Italy. Sea urchins were trying to eat plastic and their own waste instead of the indigestible algae.[4]

∼

In many parts of Asia, Africa, and South America, things are even worse than this. In China, 80 per cent of the country's fifty thousand kilometres of major rivers are so degraded they no longer support any fish. Seventy per cent of China's catch once came from the Yangt'ze, but it has declined by more than half since the 1960s. The rapid industrialization has made the idea of pollution control moot. In the Yellow River, discharge from paper mills, tanneries, oil refineries, and chemical plants has poured into the water, which is now laced with heavy metals and other toxins that in places make it unfit even for irrigation. Traces of lead, chromium, and cadmium have been found in vegetables sold in city markets, and so have concentrations of arsenic. Farm chemicals washing into the sea are being blamed for massive blooms of algae. Shanghai recently spent some $300 million moving its intake of water further away from the city because nearby river water was too polluted. The biggest cause of water pollution, however, is sewage. Many of China's rivers contain ten times as many bacteria from human waste as waterways in Western countries. A World Bank study in 1997 put the cost of air and water pollution in China at $54 billion a year, equivalent to an astonishing 8 per cent of the country's gross domestic product.

The litany of problems is relentless. In the Mekong River, inland fisheries have dropped by half in the last twenty-five years. In the Ganges and Brahmaputra systems, industrial runoff has increased dramatically and the fecal coliform counts have reached crisis levels. In Pakistan, an emergency was declared in Peshawar when one thousand people were admitted to hospital after drinking poisoned water caused by leaking pipelines. The Yamuna River, which passes through Delhi, receives nearly two hundred million litres of untreated sewage every day.

In August 2002, a three-kilometre-thick toxic cloud shrouded much of southern Asia and affected the lives of millions of people. The rain that fell onto the land and rivers was a witches' brew of ash, acids, and aerosols – the result of forest fires; a dramatic increase in the burning of fossil fuels in cars, industries, and power stations; and emissions from millions of inefficient cookers. Scientists estimated that 80 per cent of the cloud was man-made.

In the Americas, rivers were in better shape, but often not by much. Buenos Aires, for example, treats only 2 per cent of its sewage, and the rest is dumped into waterways. I've already pointed out that the cities of Halifax and Victoria in Canada, the capitals of Nova Scotia and British Columbia, respectively, dump all their sewage, untreated, into their harbours. The bucolic little province of Prince Edward Island, Canada's smallest, is seeing its gentle meadows and pastures converted by agribusinesses into massive industrial farms, with large areas poisoned by agricultural runoff from massive potato fields – to the ludicrous extent that at fertilizer time local residents have taken to keeping their kids indoors and taping their windows shut to keep the pesticides and herbicides out.

There are so many people that ordinary human actions are causing major pollution problems – suntan oil from millions of bathers is now a serious pollutant on Mediterranean beaches. Similarly, in North America, residues of what are referred to in enviro-shorthand as PPCPs (Pharmaceuticals and Personal Care Products) are contaminating surface water, groundwater, and even drinking water, so far at relatively low levels but nevertheless with unknown consequences because of their reactive nature and long life. These products include drugs such as antibiotics, steroids, anti-depressants, narcotics, painkillers, tranquilizers, and every other prescription or non-prescription drug. They also include oral contraceptives, antiseptics, fragrances, shampoos, sunscreens, insect repellants, food supplements, caffeine, and nicotine.

The totals are discouraging, probably ranging as high as thousands of tons a year. German scientists have found sixty different drugs in their water samples. Clofibric acid, a drug used to reduce cholesterol levels, has been found in the North Sea; researchers estimate its presence there at between 40 and 96 tons, increasing by 50 to 100 tons every year. The U.S. Geological Survey has measured thirty-one kinds of antibiotics and antibacterial chemicals as well as a variety of hormone and birth-control compounds in its thirty-state sampling program. Water treatment plants seldom remove PPCPs, so they run freely into the waterways.[5]

And all those people eating all the meat they do (all those cheap hamburgers) mean huge quantities of farm animals, which mean elevated levels of nitrogen and phosphorus in groundwater. There are farms in Alberta for example, that have 25,000 animals in a lot the size of a city block. That many cows produce about 50,000 tons of manure per year, equal to the waste produced by a city of

250,000 people. A large hog operation can produce more sewage waste than the city of Los Angeles. An EPA study in the U.S. determined that in 1992 about two trillion pounds of manure was produced in U.S. livestock operations. It contained not just nitrogen and phosphorus but organic material, antibiotics, hormones, heavy metals, and microorganisms such as *E. coli*, cryptosporidium, and giardia.[6]

Everywhere you look, rivers are in trouble.

In Africa, the Senegal and Niger rivers, among others, have seen cataclysmic drops in fishing catches. In South America, Colombia's Magdalena River has seen its fisheries decline by two-thirds in fifteen years because of pollution from oil production. In North America, the Illinois River supported two thousand commercial fishermen at the turn of the century, but the landings have dropped 98 per cent, mostly due to heavy sewage loads.[7] In Orlando, Florida, two young boys became critically ill with amoebic meningoencephalitis after swimming in local waters. In Sydney, Australia, tap water was tainted early in 1998 with a dangerous parasite that may have come from dead dogs or foxes, leading to panic-buying of bottled water. A Dublin conference on Water and the Environment in 1992 finished with the depressing conclusion that "after a generation or more of excessive water use and reckless discharge of municipal and industrial wastes, the situation of the world's major rivers is appalling and getting worse."

It is not hard to see why. For centuries, rivers have been used as depositories for human wastes – as natural, God-given sewers, out-of-sight, out-of-mind flushing mechanisms. If you have something you don't want, why, rivers will take it away for you. Until recently, this solution caused few problems, and the system maintained itself in balance. Rivers and lakes are naturally efficient at self-cleansing. Natural processes such as sedimentation, aeration, mixing, dilution, and bacterial processing all helped break down wastes and return them to the natural environment. But growing populations and accelerating industrialization have overwhelmed the natural recycling properties of waterways, resulting in gross pollution and increasing threats both to human health and to the water supply generally.

~

In the west of England, in one of the verdant valleys of that most gentle of landscapes, there is a country house with a wood behind. Not a wild wood any more, because the old oaks and beeches have long since been transmuted into joists and beams and posts, but not groomed either. It's a mixed wood of hundred-year-old growth – an unruly jumble of larches and imported cedars, of impudent young oaks and elms, of wild cherries and junk woods of no particular lineage. Through the centre of the wood, at least in the spring and early summer, runs a brook. The marsh on the other side of the village where the stream once rose has been drained and ploughed, and the stream dries up in high summer.

A grandfather was walking in this wood with his grandchildren, a boy and a girl. The children were skipping along the stream banks looking for water spiders, tadpoles, frogs, minnows, any-thing that lived. But there were no frogs any more – they had all gone. The grandfather had read somewhere that frogs were dying all over the world. Their damp, smooth skin was apparently too porous for their own good, rather like a lung worn on the outside. The diminishing ozone and increased ultraviolet light burned them, and they were transparent to poisons. Nature's early warning system, the grandfather thought, like canaries in a mine – except that they are measuring poisons instead.

The little girl kneeled and dipped her face into the cooling waters.

"Don't do that!" the grandfather shouted, alarmed.

"Why not?" she asked, startled.

"You don't know what's in the water," he said, lamely.

"But I'm thirsty!"

"I know. But you can't drink that. It might make you ill."

"How can it do that?"

"It could have almost anything in it," he said again. "You can't be sure. An animal might have died upstream. Someone might have put something in it that would harm you . . ."

"Why would he do that?" asked the little girl, making an early gender assessment.

The grandfather didn't answer. Why? Because that's what people do. Once, when he was a boy, you could walk through the woods and everything would be fresh and new, the streams clear, full of life. Now . . . Last week he'd found a magpie, its beak held tight by a ring tab from a pop can, drowned in the stream. Even here there was junk – plastic wrappers, cigarette Cellophane, a discarded boot. He'd been in the Lake District the previous year, and there the windswept farms and gorse meadows still smelled clean and fresh, as pristine as they'd been made, but he knew it was artificial, a holdout, a redoubt against the tide of the twentieth century, held at bay only by bureaucratic fiat, and thus vulnerable.

It was the same all over Europe. No one drank wild water any more. You never knew what evil was being perpetrated upstream, who was dumping what, who was hiding what by throwing it into the water. He remembered his wife had told him that the last time she had strolled along the banks of the Liffey outside Dublin she had found discarded condoms, pop cans, a rotted plastic sheet, and had been afraid of needles – that she might tread or sit on one and be infected with hepatitis C or AIDS. It was where she had played as a child and swum in the river. Everything was dirty now. The grandfather supposed it was inevitable.

It was the same everywhere else, too. He had visited his daughter in Toronto the previous year and they had gone down to the beach on Lake Ontario, but there was a sign that said: "Beach

Closed. Unfit for Swimming." His daughter had shrugged: Toronto was downstream from Chicago, Detroit, and the Love Canal, though she had heard Chicago had cleaned up its lakefront by dumping its wastes into the Mississippi headwaters instead, letting St. Louis and New Orleans inherit the worry. A report late in 1998 had said the Great Lakes were "cleaner than they had been for fifty years," but that wasn't saying much. Sure, you could no longer set a match to Cleveland's Cuyahoga River and watch it burn, and the piles of dead fish on the shores have been cleaned up. The Love Canal scare had receded, but there was still toxic sediment in dangerous concentrations in all the lakes, and the beaches were often closed. Excessive bacteria levels caused by Toronto's own sewage and storm-drain runoff were blamed.

An item in the paper a few days earlier, a report by the U.S. National Resources Defense Council, said there had been 4,153 beach closings in the United States in 1997, more than double the number of the previous year. The increase had apparently come from higher rainfalls, but also because more states were monitoring their water and finding it was sub-par. None of this surprised or particularly alarmed the old man's daughter, or anyone else, for that matter. It was just what it was. You can always swim in a pool. The chlorine levels there would kill almost anything. Though sometimes, she wondered, where did the chlorine go when they drained all these pools? The same month a special report on tap water in the United States found lax standards and careless monitoring in the nation's drinking water: about 40,000 of the 170,000 water systems in some way violated purity levels. Of these, 9,500 water systems serving almost twenty-five million people had "significant" violations, defined by the Environmental Protection Agency as posing "serious threats to human health."[8]

Pamphlets handed out to Ontario visitors hoping to try their hand at sports fishing make sobering reading. "Ontario: Yours to Discover" is the provincial slogan, and one of the things to be discovered is that the province's fish should be eaten with care and in small quantities. Not because the fish are in danger of extinction, but because eating too many of them can do injury to human health. Methyl mercury, among other poisons, has been accumulating in the biomass, and several northern communities have been found to exceed World Health Organization recommended standards.

The grandfather from England who couldn't swim in the Great Lakes of America was not alone. There are more than eight thousand kilometres of shoreline on the U.S. side of the lakes, but only 3 per cent is fit for swimming, for supplying drinking water, or for supporting aquatic life. In 1993, two-thirds of U.S. "fish advisories" were issued in the Great Lakes region, most of them having to do with excessive amounts of mercury, PCBs, chlordane, dioxins, and DDT. Each year, 50 to 100 million tons of hazardous waste are generated in the watershed for the lakes, 25 million tons of pesticides alone. There are airborne pollutants, too: trace pollution has been detected from farms in Mexico and cement factories in Texas.

The 1998 International Joint Commission, in the same report that declared the Great Lakes cleaner than ever before, tracked the accumulations of a dozen-or-so noxious chemicals and found, to its dismay, a "buildup of radioactive contaminants in the lakes from nuclear power plant discharges. The evidence is overwhelming," the report said. "Certain persistent toxic substances impair human intellectual capacity, change behavior, damage the immune system, and compromise reproductive capacity." Anyone relying on fish or wildlife for food was at risk.

Elsewhere, industrial pollution by lead, mercury, and cadmium are becoming ever more serious problems. Cadmium causes renal disease, and cadmium deposits in the Rhine increased twentyfold between 1920 and 1970, from less than two milligrams per kilogram to more than forty. Mining is causing extensive mercury poisoning in hitherto pristine Amazon waterways. Lead is prevalent in many rivers in industrial countries, causing kidney trouble and damage to the nervous system. Aluminum, which earlier had been thought harmless, is now thought to contribute to Alzheimer's, and is increasingly common as industrial waste.

In developing countries matters are generally worse, and as development accelerates, the amount of water contaminated with industrial pollutants such as petroleum and toxic metals rises sharply. Perhaps the overall national water supply is no dirtier – but there's very little clean water, and the meticulous scrubbing that water gets in the cities of the developed world simply doesn't exist. Mostafa Tolba, a former director-general of UNEP, the United Nations Environment Program, admits that despite all efforts, the quality of fresh water depends not on pristine supply but on decontamination. "The number of polluted rivers is on the increase," he warns. Peter Gleick is equally blunt: "According to the latest data, most of Africa, most of Asia, and most of western South America fail the simple test of providing water for basic drinking and sanitation needs. Over a billion people have no access to clean drinking water, and more than 2.9 billion have no access to sanitation services. And the sanitation trend is getting sharply worse, to some degree only because of better and more accurate reporting, but also because of higher populations, especially in the cities."

If you drive north from the centre of the Cameroon city of Bafoussam, heading for the highlands and Lake Chad beyond, you'll pass an apparently endless stretch of ghastly slums, overcrowded shantytowns, tumble-down and wretched, hundreds of children playing in the fetid, stagnant pools and among piles of rotting garbage. On the main roads, markets are set up alongside

refuse dumps. When cars cease running, they are simply abandoned on the roadside and are immediately colonized by families
– a mother, father, and children making do in a space not much
bigger than a sofa. The soil is black with discarded crankcase oil,
half-burned debris, and general filth. Peasants come in from the
countryside and lay their piles of goods to sell in the sludge of
the sidewalks – here a small pyramid of mangos, a startling flash
of green and gold among the dun; there a row of fish, still shining
but already beginning to turn in the thirty-five-degree heat; here
an old woman wearing bunches of bananas like shackles. Diesel
fumes from badly tuned motors fill the air, and most of the
buildings are crumbling and black with soot. Off the main roads,
if you plunge into the maze of shanties, cobbled together with
strips of iron, old wire, cardboard, and flattened tin cans, the
stench can be overpowering. There are no services here: no roads,
no addresses, no sewage services, and no water. Water has to be
fetched from communal standpipes, dripping taps in a circle of
mud, ankle-deep and stinking.

Not to pick on Bafoussam: It is remarkable only by contrast with
its beautiful hinterland, rolling hills covered with emerald forests
and bright scarlet blossoms of improbable size, and because the
peasant architecture in the area is striking – steeply pyramidal roofs,
sturdy little houses of beige mud brick, their yards swept clean, an
almost Disneyfied cuteness to them. Otherwise, Bafoussam is very
like other cities in Cameroon – Douala, or the capital, Yaoundé. It
is also like the shantytowns of Cape Town, the slums of Luanda,
and the megaslums of Lagos, Kinshasa, and Dakar. Megaslums also
exist all over Asia – in China, Thailand, Laos, India, Bangladesh,
and Pakistan – and in almost every country of Latin America. In
other words, they are found throughout the developing world.

Within a year or two – no one knows exactly – the world will
pass a significant human milestone. For the first time, more people
will live in cities than in the countryside. And most of the growth
is coming from the developing world, where cities are exploding

at a rate that has made it impossible for urban services, including supplies of sanitary water, to keep up. In the early 1990s an apocalyptic literature began to emerge in the West about these burgeoning slums of the Third World – see Robert Kaplan's *The Coming Anarchy*, for example. Some of this reporting seemed obsessed with social decay and degradation, and came with odious predictions of global collapse. I don't buy into this gloomy millenarian view of the inevitability of disaster, partly because if you talk to the residents of these slums instead of just driving by, it's easy to see that they're as eager for betterment as any family in the leafy middle-class suburbs of, say, Connecticut; and partly because the writing shares so many hand-wringing attitudes with the early Victorians in nineteenth-century England who were convinced that the bringing together of millions of the brutish poor in the slums of industrial cities was bound to mean the end of civilization as they knew it.

Which is not to say that the human misery in these slums should be easily dismissed.

For most of human history, cities have been the repositories of power and wealth, but also of plagues and pestilences. It was little more than a century ago, when the major cities of the industrializing world introduced sanitary sewers and decent fresh water, that the incidence of disease began to decline. In 1852 the average age of death in the boom town of Dudley in England was seventeen, mostly because "human excrement [was] in all back streets, courts and other eligible places."[9] In the two decades after sanitary sewers were introduced, life expectancy increased by a dozen years. In French cities, life expectancy rose from thirty-two to forty-five between 1850 and 1900. Of course, advances in medical sciences and the understanding of diseases helped – a cholera epidemic in

London was famously traced to a single public faucet. But the mere fact of sanitation was the greater cause.

In the rest of the world, most of the population still lived in villages and on small land holdings; high disease rates were generally associated with the poverty-stricken life of the countryside. But the same shift that happened in industrialized Europe and America in the nineteenth century is now happening in the rest of the world, only much faster and on a hugely increased scale. It took London from 1800 to 1910 to multiply its population from 1.1 million to 7.2 million, a growth rate that was achieved in some African cities in a single generation. Many Asian cities have increased fourfold in the same period. Rapid population growth and rapid urbanization are twinned phenomena, both taking place in countries that are poor. Already, twenty-two cities in the world have populations exceeding 10 million, and seventeen of them are in developing countries. The largest city in the world, Mexico City, has more than 20 million inhabitants. The same thing is happening in Delhi, Amman, and Santiago, Chile, where the urban poor are already somewhere between 30 and 60 per cent of the population. In Addis Ababa the figure is 79 per cent; in Luanda, 70 per cent; in Calcutta, 67 per cent. The population growth rate in the slums is higher than elsewhere – in Bangladesh, for example, it is four times the rest of the country's birth rate. If the present trends continue, which seems probable, more than half the inhabitants of these megacities, well over a billion people worldwide, will live in slums and shantytowns without any amenities whatever. UN figures show that the world's rural population will peak at 3.1 billion by 2010, and decline thereafter. By 2030, by contrast, the urban population will be twice the rural one, and cities will have grown a staggering 160 per cent. This rapid increase has made it almost impossible to keep up with need, despite heroic efforts. The 1980s, the "International Drinking Water Supply and Sanitation Decade" of the United Nations, saw an impressive

80 per cent more urban dwellers gaining access to water, and 50 per cent more to a system of waste disposal. But between 1990 and 2000 an extra 900 million people were added to the planet. The actual number of those without water remained the same, and the number of those without sanitation rose by 70 million.[10]

In many cities the authorities have almost given up. Their water infrastructure is in such bad condition that it cannot possibly recover its costs from users, and the income they get is far too low to improve the service. The quality therefore declines further. There are disruptions and occasional sabotage by disgruntled consumers, or the system is damaged further by the desperate trying amateur diversions. Often, as in London, nearly half the water simply disappears through old, cracked, and leaking pipes.

Until recently, development planners seldom paid much attention to drainage and water-treatment issues. In Latin America, only about 2 per cent of human waste is treated. Untreated waste flows into the river systems, forcing cities to go further afield for their intakes; in Lima, pollution has upped treatment costs 30 per cent.[11] The same thing is happening everywhere in the developing world.

Pollution leads to disease, a straight-line computation that is self-evident. Best guesses are that some 250 million new cases of water-borne diseases occur every year, killing somewhere around 10 million people – a Canada every three years.

Cholera is the worst news story. The first cholera pandemic spread to the Americas from Asia early in the nineteenth century and, until the American Civil War, there were regular cholera epidemics in the United States. In 1970 the disease invaded Africa, and in 1991 returned to the Americas. It spiked in 1992, and is still going up. In 1990 there were about one hundred thousand cases worldwide, but in 1992 the number exploded to six hundred thousand, the overwhelming majority in the burgeoning slums of Latin America.

Bilharzia, or schistosomiasis, is the other major water-borne disease problem, currently infecting more than two hundred million people in some seventy countries. The disease is spreading and intensifying, mostly because of otherwise beneficial projects that give the host snails a comfortable new environment. In the Nile Delta, for example, the snails spread rapidly after the construction of the Aswan High Dam and its irrigation channels, leading in some areas to infection rates of nearly 100 per cent. The same thing happened in Sudan after the construction of the Sennâr Dam, and after Lake Volta was constructed in Ghana, where infection rates in children went up from less than 10 per cent to more than 90 per cent within a year.[12] Lake Malawi is now said to be the only considerable body of water in Africa that is bilharzia-free, and therefore safe for swimming. Tourists gambolling on its beaches would be unwise to bet on it.

In the developed world, the so-called legionnaires' disease is also caused through contaminated water. Its incidence is still low, but worrying – it is resistant to chlorine, and so can infect water treated to U.S. and European health standards.

Sleeping sickness and malaria are water-vector diseases, one carried by the tsetse fly, the other by the mosquito. More than two billion people worldwide are exposed to malaria, and there are about 130 million new cases every year, mostly in Africa and Asia. Malaria accounts for almost one-third of all childhood deaths.

Not all the biological news is bad, however. The other major parasitic scourge, dracunculiasis, caused by the guinea worm, is down from nearly ten million cases a year to near zero.

~

Natural water, as we have seen, is not "pure" but a dilute solution of elements dissolved from the earth's surface, or precipitated from the air, something that "mineral water" companies such as Perrier have taken advantage of for years, to the satisfaction of their share-holders. All water is in a sense mineral water, because water is one of nature's great solvents. It's just that some minerals taste better, and shrewd marketing has successfully persuaded millions of cus-tomers that they're beneficial to health. The ionic composition of typical "wild water" varies from less than one hundred milligrams per litre (rain, snow, and mountain streams) to as high as four hundred thousand milligrams per litre in saline lakes.

The rain that falls to earth seeps into the soil and percolates through organic material, roots, decaying leaves, and humus. As it passes, it dissolves minerals from the rocks and soils, and reacts with living things, ranging from microbes and bacteria to humans. The science of groundwater can become technical, and involves processes like hydrolysis, precipitation reactions, absorption and ion exchange, oxidation and reduction, gas exchanges, and biolog-ical processes such as microbial metabolism, organic production, and respiration. Water is both a solvent and a sponge: its composi-tion depends very largely on the environment in which it is found.

It is this absorptive nature, the fact that water derives its composition from the leaching of surface material, that makes groundwater so vulnerable to pollution. Groundwater is usually more concentrated than river water, and much harder to fix when it goes wrong, because it renews itself so slowly.

~

"In the second half of the 20th century," a Worldwatch report by Payal Sampat said in January 2000, "the soaring demand for water turned the water dowsers' modern-day counterparts into a major industry":

Today, major aquifers are tapped on every continent, and groundwater is the primary source of drinking water for more than 1.5 billion people worldwide. The aquifer that lies beneath the Huai-Hai plain in eastern China alone supplies drinking water to nearly 160 million people. Asia as a whole relies on groundwater for one third of its drinking supply. Some of the largest cities in the developing world – Jakarta, Dhaka, Lima and Mexico City among them – depend on aquifers for almost all their water. And in rural areas, where centralized water systems are undeveloped, groundwater is typically the sole source of water. More than 95 per cent of the rural U.S. population depends on groundwater for drinking.

The changes that humans have wrought are not trivial. The chemistry of groundwater is being modified on a global scale. Human actions have already caused the mean average dissolved "additives" to increase by 10 per cent, and in the case of salt, sodium, chlorine, and sulphate, by as much as one-third. Leachates from fertilizers, herbicides and pesticides, toxic organic and inorganic chemicals, and radioactive traces are found throughout the world. The situation is worst in the United States, mostly because the chemical industry is at its most inventive there and farming has come to depend heavily on pesticide use, and partly because there is hardly a river left undammed or an aquifer untapped. In the American southwest, as we shall see, aquifers are being recklessly mined with no hope of recovery. In the northeast, salt spread on roads in the winter has contaminated drinking-water supplies, as studies have found in Long Island. The rapid runoff, or excess water, resulting from deforestation is a problem throughout North America. Irrigation return-flows that leach salts from soils in semi-arid areas are major sources of pollution in the western United States, just as they are in Western Australia.[13]

All natural watercourses contain some nitrates, albeit in dilute form, that are generally released into the water by storm runoff, since nitrate ions travel freely through the soils. But natural concentrations of around .1 milligrams per litre are seldom harmful.

That is just as well, because nitrates in the bloodstream can inhibit the ability to carry oxygen ("blue babies" are usually caused by overly high nitrate concentrations); nitro-samites, nitrogenated compounds, are among the most potent carcinogenic substances known. The nitrate content of fresh water is rising steadily, because more and more farmers are using nitrogen-based fertilizers, and there are higher levels of industrial and urban wastes.

Rivers in the richer countries have, by and large, become steadily cleaner over the past decade. But when measured for nitrates, fewer than one in ten European rivers is any longer "natural" – most have nitrate levels four times the norms found in nature. The U.S. Geological Survey found that almost one-quarter of the wells it analyzed had nitrate concentrations higher than natural background levels, some dangerously so. In Denmark, nitrate levels have trebled since the 1940s.

As cities expand to support larger populations, roofs, highways, and parking lots increasingly replace permeable soils and vegetation. Rainwater in urban areas is channelled into sewers and drain systems instead of filtering into the ground to raise the water table. But perhaps this is just as well, considering the quality of the runoff.

In the developing countries, the picture is somewhat different. Rivers in the poorest countries have shown a substantial drop in the level of dissolved oxygen – a key indicator of increased pollution by sewage. Nine-tenths of all sewage in developing countries discharges directly into rivers, lakes, and seas without any treatment. In terms of the number of people it kills, "dirty" water is probably the world's most serious pollution problem.

On a global scale, rivers are dumping twice the amount of organic matter into the oceans that they did in prehuman times; the flux of nitrogen and phosphorus has more than doubled. The eutrophication of marine systems has become a problem for fisheries worldwide, and estuaries in the developed and the developing world are becoming polluted to a degree that is troublesome to human health. You don't have to agree with Dr. Paul (*The Population Bomb*) Ehrlich that economic growth "is the creed of the cancer cell" to know that humans are seriously skewing the very resource on which they are so utterly dependent.

~

But not all groundwater contamination is human-caused. The most tragic case of "natural" poisoning is that of Bangladesh.

For decades, donor countries and UNICEF had urged the Bangladeshis to clean up their water supply, which was badly polluted. For most of its short existence, the country was river- and stream-dependent, but throughout the 1980s and 1990s aid-financed drilling began, and by the year 2000 nearly a million tube wells had been drilled. It seemed like a triumph of good sense and good water management.

Unfortunately, there is a stratum of the subsoil, below the reach of food-crop roots, that is highly concentrated with natural arsenic, and it was into this stratum that the new wells were drilled.

Looking back ruefully, a UNICEF official said in October 2000: "Twenty or thirty years ago, people were not looking for arsenic in groundwater. It was something no one knew anything about. By sinking tube wells to enable people to drink groundwater, we were merely doing what everybody was doing, but we were better at it."

The consequences have been stark: Bangladesh is now confronting one of the largest-ever environmental disasters: the poisoning of as many as 85 million of the country's 125 million people with arsenic-contaminated drinking water. This figure may be too large, but, at minimum, 21 million people are so exposed. A UN report in 2000 described it as "the largest mass poisoning of a population in history; the scale of the disaster is greater than any seen before . . ." It is thought that one in ten people who drink the water for extended periods will suffer from lung, bladder, or skin cancers, and a large percentage of those will die. Only the capital and the fringes of the country, the hilly area of the northwest and Chitagong in the far southeast, are unaffected.

The UN gave Bangladesh an interest-free loan of some $30 million in 1998 to deal with the problem by supplying arsenic-free water. Two years later, most of the money remained unspent. "There are eighty thousand villages in the country," a furious World Bank official said. "So far, the government has covered only one thousand. If they go at this pace, it will take hundreds of years before they reach them all."

Some progress has been made. Rainwater harvesting devices and cheap filters have being supplied to thousands of villagers. But for the general population, which by and large remains innocent of the problem, none of these have the convenience of the tube wells that were so easy to sink in the country's soft alluvial soil. Many village households have several wells, just as Western countries have taps in different rooms. More are being sunk all the time – despite the disaster, the government never declared a moratorium on wells. So popular are they that a tube well is now a common item in a bride's dowry.[14]

6 | THE ARAL SEA

An object lesson in the principle of unforeseen consequences.

Even a light wind picks up the salts and the dust and turns the air hazy, like heat shimmering in a mirage. When the winds are stronger, the white clouds drift higher and deeper, and the sky is opaque, like milk made of vinyl, impenetrable and somehow ominous. Driving north along the Amu Darya River northwest of Bukhara to the Aral Sea in Central Asia – or what used to be the Aral Sea – the farms are also white, dusted with what looks like fine snow. But it's not snow. It's salt, leached to the surface after decades of careless irrigation. Keep on going, past the dreary little Uzbek town of Nukus, to Muynak, which was once the shore of the Aral, then the world's fourth-largest lake. The view from the ridge just outside town is perhaps the most famous in the depressing annals of environmentalism: as far as you can see there is nothing but sere sand, white bones of dead cattle poisoned on toxic plants, the rotting hulks of fishing vessels and barges, the pathetic detritus of a once-thriving fishing culture. In the foreground are the broken timbers of what used to be a wharf; beyond that are six or seven hulls, grounded at random, pointing no way in particular, some of them still upright but broken, the paint reduced to small scraps of rust. Further on are humps in the sand, where other hulls have been buried, and beyond that only dunes, sand, salt, and nothingness.

The Aral Basin is the catchment and drainage area of two river basins, the Amu and the Syr, which together contribute most of the water in the Aral Sea. The two rivers rise in the Tien Shan and Pamir mountain ranges of the Himalayas and flow northward through their alluvial valleys and the Kara Kum and Kyzl Kum deserts before emptying into the Aral, one into the southern tip and the other into the northern. These two rivers are the only real source of fresh water in a system that comprises 1.5 million square kilometres and somewhere around thirty-five million people. It used to be bounded almost entirely within the old Soviet Union; the Russians split the region into several ethnically based "autonomous" republics, on the tried-and-true principle of divide and rule. But when the "Evil Empire" itself split, it left a scattering of tiny, poor "independent" entities – Turkmenistan, Uzbekistan, Kazakhstan, Kyrgyzstan, and Tajikistan, all of which jealously guarded and maintained their Soviet-created "national" identities. It is these statelets that are left to wrestle with the problems Sovietism left behind. For the Aral is really an ex-sea now, a shrunken thing, and poisonous. The water-management disaster that brought it to this state has been described by horrified observers searching for an appropriate metaphor as "the quiet Chernobyl." This may be a little over-the-top, but not by much. Without any exaggeration, the Aral Sea has become the greatest man-caused ecological catastrophe our benighted planet has yet seen, an awful warning of the consequences of hubris, greed, and the politics of ignorance. It is also proof positive of how suddenly ecological catastrophes can happen, and how hard they are to reverse once they are underway.

As though proof were still needed.

⌇

I flew into Bukhara from Samarkand late one October afternoon on a rattling little Aeroflot flight out of Dushanbe. There was a dust storm in the region, and a couple of old bandits had taken refuge in the hut that doubled as a terminal building, their donkeys tethered outside and their ancient rifles, gleaming and well-oiled, propped against a fading stack of posters left over from Soviet times. The air was stifled by roiling clouds of sooty dust called a "black blizzard" by the locals. Dust choked everything in the town. There was grit on the floors, grit on the hotel beds, grit in the cars, grit on the restaurant tables, grit in the food. The sun was still up, but it was already dark, like a badly lit film noir. I went to have supper in a café someone had told me about, deep in the kasbah, a sunken well of a place that seemed to have been converted from a cistern. The only other clients were old men drinking tea from glasses and a few soldiers, one of whom got up while I was eating and stomped a mouse to death on the mosaic floor, leaving the carcass there, perhaps to be cleaned up the next day – if there were cleaners around, which wasn't obvious. The waiter had swept the table more or less free of dust with his sleeve as he plunked down a plate of lamb shashlik, grilled to stupefaction, and a Russian beer, which was warm and flat. There was sand in the glass, too.

I was actually in town on other matters – I was following the trail of Tamerlane, the greatest of the Golden Emperors of the Steppe – but the environmental degradation was so obvious it was impossible to ignore. It also seemed impossible to believe that this place, which had been desert for longer than human memory, had been chosen by the central planners of Soviet Russia to be one of the showpieces of collectivist agriculture. Burying the capitalists would be tough enough in the rich black loam of the Ukraine, I thought, but to make this desert bloom required the skills of a necromancer, not an agronomist.

There had always been some agriculture in the region, and traces of irrigation have been found dating back five thousand years. Most of the inhabitants had been nomads, but when the hordes set up their capital in Samarkand, and Tamerlane began building his magnificent palaces and public buildings, water was taken from both rivers for food crops. It was sustainable irrigation, the scale appropriate to the resource. The Aral was hardly touched.

When the Bolsheviks overthrew the czars and began to impose their command economy on a bewildered and resentful peasantry in Russia, Central Asia wasn't spared. The planners couldn't believe their luck. All that land, not being used! All those primitive peasants, not knowing which end is up! All that water, flowing uselessly northward, going nowhere! The Gosplan (state planning office) people in Centre, as Moscow was called in the hinterlands, got out their maps and their rulers and their abacuses and worked it all out. Products were needed for export. Inexplicably, despite the manifold benefits of collectivization, the Russians couldn't seem to feed even themselves, never mind growing things for foreigners. But cotton – there was a crop! White gold!

In 1937 the Soviet Union became a net exporter of cotton, and a decade later mechanized agriculture was a fact of life in Turkmenistan and Uzbekistan. Not just cotton, but wheat and other food grains too. There were cattle ranches in this place where nothing grows, and corn fields. It was wasteful, sure – as much as 60 per cent of the irrigated water never got anywhere near the crops, some of it disappearing in the unlined canals and in evaporation, and some diverted by bandits, just for the hell of it. Nevertheless, the whole scheme seemed to be working. There was lots of water, wasn't there? By the 1950s Khrushchev could say in all seriousness that he would overtake the capitalists within a decade and mean it.

In 1956, with great fanfare, the Soviet Politburo dispatched its agricultural commissar to open the Kara Kum Canal, which taps into the Amu Darya near the Afghan border and meanders through the desert towards the Turkmen town of Ashgabat, which is not far from the border with Iran. Flying over it a dozen years later was an extraordinary experience. The land was grim, black stone and dunes the colour of dirty khaki, but the canal itself was an emerald necklace on a skin of dun, and radiating out from it were the farms, laid out in Soviet fashion in immense and unwieldy blocks, the ploughing haphazard and the furrows crooked, as though the farmers had stared too much at the sun.

It was a time of optimism in world agriculture. The Indians were using new seed grains to make their deserts bloom; the Americans were diverting ever-increasing amounts of water through ever-increasing hectares of inhospitable deserts; the Israelis were exporting oranges, of all things, and had the romantic vision of turning the Negev into a garden. Anything seemed possible, and though Khrushchev had been forcibly retired to a small holding outside Moscow, his boastfulness was still Soviet policy. There were rice paddies, after all, in Turkmenistan. The Soviet system could do anything.

Soon afterwards, three other canals were opened, diverting more water into the Uzbek and Kazakh deserts, irrigating more millions of hectares. But the water that spread across the desert never reached the Aral, and after 1960 the level of the sea began to drop. Before the diversions, some fifty cubic kilometres of water a year had reached the sea; by the early 1980s the total was zero.

The village of Muynak was an island in the Amu Darya Delta in 1956. By 1962 it was already a peninsula. By 1970 the sea was ten kilometres away, and the fishing wharves and landing areas were long abandoned. A decade later the nearest water was at a distance of forty kilometres, and by 1998 it was seventy-five kilometres. The seabed had become a desert. As the Aral shrank, its salinity increased. It changed from 1,075 cubic kilometres with a salinity of ten grams per litre to 54 cubic kilometres with more than ten times that salinity.

By 1977 the commercial fish catch had declined by over 75 per cent, and a few years later, in 1982, commercial fishing ceased altogether, shutting down an industry that had caught a sustainable fifty thousand tons a year and had employed nearly sixty thousand people. A few species survive in the three small saline lakes the Aral Sea has become, but most of the fish died. The fish cannery in Murnak tried to stay in business by buying fish at great cost from Murmansk and Arkhangelsk, the Soviet Arctic ports, but when the Soviet Union collapsed, that lifeline dried up too.

The collapse of the fishery wasn't the only problem, or even the worst. There were dozens of negative secondary effects. The declining sea level lowered the water table in the region, destroying many oases near its shores. Unirrigated farms dried up. Wildlife disappeared, and the once-flourishing muskrat-fur industry collapsed too. Soviet agriculture had been pesticide-dependent – and, indeed, farms in the area still use DDT, long banned elsewhere. Heavy concentrations of pesticides had built up in the sea. Over-irrigation caused salt concentrations in many agricultural areas.

The Aral had once regulated the climate in the area. It was large enough to act as an air conditioner in summer and as a heat sink in winter, buffering the winter storms from Siberia. With the loss of the Aral, the local climate became more continental, with extremes of temperature, shortening the growing season to 170 frost-free days a year, fewer than the 200 needed for cotton. As a consequence, more farms switched from cotton to rice – which demanded even more diverted water. The exposed seabed, now over 28,000 square kilometres, became a stew of salt, pesticide residues, and toxic chemicals; the strong winds in the region pick up more than forty million tons of these poisonous sediments every year, and the contaminated dust storms that follow have caused respiratory illnesses and cancers to explode. Some of the dust has been reported as far away as Belarus, two thousand kilometres to the west, and in Pakistan to the east. As the deltas dried up and the wetlands disappeared, the local rainfall decreased by half.

If you stand on what had been the shore of this once-great lake, you can see the salts in the air. Throat cancer, the local doctors say, has reached epidemic proportions. So have kidney and liver diseases, arthritis, chronic bronchitis, typhoid, and hepatitis A. It is worst in the Amu Darya Delta, a region called Karakalpakstan, where the bio-accumulation of toxins is so great that mothers have been advised not to breast-feed their babies, and infant mortality has climbed past one hundred deaths per thousand births.

Now that the fish buyers have gone, only the foreign ecologists come, to stare appalled at the ruined landscape, impressed and depressed at the scale of the catastrophe. I have in my notebooks this thought, scribbled after talking to an Israeli hydrologist who had seen what there was to see and was returning with a message of gloom for his colleagues in Israel and America: "Maybe that'll be the Aral's only legacy. It will help make things better elsewhere by frightening people half to death."

Even the thick-headed Soviet planners came to see that something was seriously wrong.

I interviewed the Gosplan folks in Moscow in 1970, and I recently looked up my notes. In the hours I spent with them, nothing had been said that was at all negative. Officially, the line was still the Khrushchev one, and the planners dutifully trotted out the statistics: so many tons of cotton, a harvest so many tons bigger than the year before, so many hectares newly planted – you could virtually see the smiling peasant girls of Soviet myth, baskets of white gold on their burly forearms, while in the background tractors from the Volgograd (then Stalingrad) factories roared and belched, symbols of industrial might. It was all a fraud. They knew it and I knew it, but in the final days of Brezhnev, that was how things were done. Truth was in the interstices between words, and had to be inferred from silences, shrugs, and things unsaid.

The week before my interview, I had read some hints of difficulty in a Soviet newspaper, *Literaturnaya Gazeta*, which had recently strayed from its literary mandate to tackle the tricky subject of pollution. A team of reporters from the *Gazeta* had found pollution staining some of Russia's finest rivers and had traced the origins of one mysterious slick on the Volga to a toothpaste factory, which had dumped its entire year's production into the river when it ran out of tubes.

I shoved the clipping over at the planners. "It says the fish are dying in the Aral," I said. "What's happening?"

They wouldn't say, and fobbed me off with some lame story about natural cycles. But they knew, all right, for down the hall others were already working on a remedy, a "solution" typical of the thinking that could set farm quotas without ever looking at the farmland itself. On the map, it seemed obvious. There was lots of water. Siberia had plenty of rivers it wasn't using. Take the five-thousand-kilometre-long Ob, for example. Flows into the Ob Gulf in the Arctic, a chilly and sand-choked inlet unfit for any purpose. The Ob and its tributaries rise near the Kulunda Steppe, to the northeast of the Aral Basin. Why not simply divert it south-ward? Brezhnev, who was then sort-of in charge of the Kremlin, had another thought: the Yenisei and Lena rivers are even bigger, even further away, with even more water. They would restore the Aral, giving it as much water as it had once received from both the Amu Darya and the Syr Darya combined. Despite dire warnings about the ecological consequences, this lunatic scheme, not unlike some of the more preposterous diversion schemes proposed by American water planners, was kicked around by Soviet bureaucrats for a decade or more, while the Aral continued its descent into obsolescence.

It wasn't until the Gorbachev era of glasnost that the scheme was put to rest. By then, Russian newspapers were vying with each other to find atrocious stories to tell, and the Green movement was springing up everywhere, easily finding fodder for its entirely justified indignation.

In 1988 the Central Committee decreed that cotton production was to be reduced, so that the Aral could receive water in ever-increasing amounts, through to 2005. Two years later, having seen that this remedy was not nearly enough, the Supreme Soviet declared the Aral a disaster zone and decreed an emergency relief effort. But by the time the Soviet Union itself collapsed a few years later, nothing much had been done. The water diversions had been marginally reduced, but that was it. By 1994 Moscow was officially out of the game, and the Aral crisis was now in the hands of the five brand-new Central Asian nations. The Amu Darya and the Syr Darya had become international waterways, which in turn created an urgent need for international co-operation to deal with the environmental degradation.

The economic numbers are not trivial. Water diverted from the Aral still supports agriculture worth billions of dollars, employing millions of people. Although Uzbekistan is as far north as Maine in the United States, and probably would never have grown cotton had it not been for those Soviet planners with their maps and rulers and impenetrable ignorance, cotton is Uzbekistan's now chief crop, making up one-third of all exports, and the country's leading hard-currency earner. Agriculture accounts for 43 per cent of gross national product, and employs one-fifth of the population. Turkmenistan, for its part, depends on agriculture for 46 per cent of GNP. In Kazakhstan, the figure is 33 per cent, employing one out of every five Kazakhs. The World Bank estimated that the total population of the five states will more than double to 86 million by 2025, before stabilizing at some later date at 186 million. There is not enough water to handle more than a fraction of the agriculture required to feed all these people.

In 1992 the five Aral Basin states signed an agreement pledging co-operation, using as a measure water quotas allocated under the Soviet regime. Little, however, was done. The following year, with World Bank support, they signed a General Agreement on the Aral Sea Crisis in the town of Kyzyl-Orda. Russia took part in this meeting as an observer, but never promised any money – no one believed it had any, and the routine demands that were made were half-hearted. Another meeting, in 1994, resulted in a variety of offers to reduce water consumption. Again, nothing much happened, except that Turkmenistan surreptitiously began to extend the Kara Kum Canal towards the Caspian Sea to the west, sneakily extending its own farmland.

In June 1994, the five states met in Paris to see if they could hammer out a solution. An unstated but important objective of this meeting was to "alert the international donor community [the rich countries] to the Aral disaster." Viktor Dukhovny, head of the Scientific Information Committee of the five-state Interstate Water Commission, remains optimistic that, despite all the obstacles, progress will be made. In an interview in Paris in 1998, he laid out an "action program" that would begin around 2002 and terminate somewhere around 2040. This program sounded thorough and sensible, on paper. It involved an emphasis on permanent water conservation by all parties; the joint management of transboundary water resources "based on national parity of rights and responsibilities"; the systematic development of a regional water database; the involvement of the public and NGO community; the gradual shifting to a more user-pay system of water distribution; and the popularization of water conservation among the general public.[1]

It all depends, he admitted a little glumly, on "a constant consensus among participants." Whether that was likely, he declined to say.

In June 2002, a member of the International Water Management Institute, in Tashkent, Uzbekistan, did a new assessment of water quality in the Aral region. The results were not encouraging.

Cotton, the major crop, required the admixture of large quantities of fertilizers and pesticides; as a result, some two cubic kilometres of toxic runoff entered the Amu Darya River every year, "dramatically worsening" drinking-water quality.[2]

The Aral continues to shrink and may soon be entirely lifeless. Even if the five states succeed in stabilizing water withdrawals, this restraint would do nothing to deal with the pollution problem. Full restoration would require wholesale changes. Agriculture would have to be de-emphasized, for one thing – a possibility, since the region has oil and gas to spare and is urbanizing rapidly. With World Bank help, the remaining irrigation could switch to supplying less water-demanding crops and itself become more efficient. Israeli engineers experimenting on an Uzbek cotton farm recently claimed they had increased yield by 40 per cent while reducing water consumption by two-thirds. The Israelis had a self-interest in being there: to observe first-hand how to cope, or not cope, with an overstressed resource. They have an eye to their own future, as the Dead Sea shrinks and the Sea of Galilee comes increasingly under stress.

But no one believes this full restoration will happen. The primary obstacle is money. The Central Asian republics are the poorest of the former Soviet Union and have little cash to invest in rebuilding irrigation infrastructure. The GNP per capita ranges from $2,030 in Kazakhstan to only $980 in Uzbekistan. Instead of being scaled back, agriculture is in fact expanding. Everyone is squeezing as much water from the system as possible. Competition replaces co-operation, and angry noises are emanating from all five capitals.

Tajikistan, where both rivers rise, is the key jurisdiction. It has plenty of water, but its own water is filthy, and corruption in the capital, Dushanbe, is endemic. In 1997 a typhoid epidemic killed more than one hundred people and put nine thousand in hospital. Water-treatment plants are in disarray, with no money even for chlorine, while the fleet of new limousines for government cronies continues to grow. The state government has so far refused to vote the 1 per cent of GNP it is supposed to put into solving the Aral crisis; on the contrary, as the upstream owner of the water, it is demanding payment from downstream users for "its" water, particularly from the Uzbeks, traditional rivals. As a result, the Aral Fund is completely empty. The Tajiks point out that the Uzbeks don't scruple to charge them for the natural gas they supply, and will turn off the taps if payment isn't prompt. Why one thing for gas and another for water?

Uzbekistan has troubles of its own: a catastrophic thaw in a glacier in mid-July 1998 produced flooding that destroyed a dam on the Kuban-Kel Lake, causing it to collapse; forty-three people were killed. The Uzbeks, who are entirely dependent on Tajik water, are in turn accusing the Tajiks of exporting pollution and disease in the guise of fresh water. There have been threats from radical Tajik nationalists to deliberately poison the water leaving the country, and on three occasions Uzbek water bandits crossed the frontier to steal tanker loads of water and to dynamite Tajik storage facilities. Posses have been seen patrolling the borderland irrigation canals on both sides to combat water poaching. Outright fighting has already occurred between Kyrgyz and Uzbeks, Kyrgyz and Tajiks, and Turkmen and Uzbeks. The Uzbek president, addressing the central committee of the local Communist Party, admitted that "we have already had dozens of conflicts over land and water." A running battle in the Osh region of Kyrgyzstan between resident Uzbeks and Kyrgyz killed almost three hundred people. The Uzbeks have also admitted drawing up a contingency plan to invade northern Turkmenistan to obtain water.[3]

Conflict specialist David R. Smith, writing in the journal *Post-Soviet Geography*, identified two possible immediate war zones. The first zone is Kyrgyzstan and Uzbekistan, "given Kirgistan's riparian position as an upstream country and Uzbekistan's great interest in access to its water." The second zone is the two downstream countries, Uzbekistan and Turkmenistan, in which there have already been sporadic outbreaks of local violence against perceived water diversions.

Complicating the politics of the whole region is the Islamist threat. The Uzbeks have already condemned Pakistani schools for allegedly training young Uzbek dissidents in the use of guns and explosives, with the excuse that they are needed in Afghanistan; they have charged that Pakistan is fuelling unrest, especially in the volatile region of Fergana, to the southeast. Uzbekistan is not freedom-loving to begin with; its intolerance for dissent has been spreading to crackdowns on ecological groups trying to change Aral-related policies, and to the Karakalpaks, who live around the south end of the sea and are disproportionately affected by the Aral disaster. The Birlik political movement, which called for diversification of agriculture "on both ecological and economic grounds," has been suppressed.

There are other players in this already difficult multistate dispute. Kazakhstan, for example, has already demanded that China stop "over-using" the Irtysh River, which flows through China to Kazakhstan; and Moscow has on occasion cut off electricity supplies to northern Kazakhstan in an attempt to get its overdue bills paid. Afghanistan, which itself occupies nearly one-quarter of the Amu Darya Basin, currently uses little water from the region. Whatever happens in the post-Taliban dispensation, it's very likely that the Afghan regime will eventually demand the right to upstream withdrawals, further stressing an already overstressed system.[4]

The 1992 conference that allocated quotas to the five states also set up an international body, the Interstate Coordinating Committee for Water Resources, which still meets five times a year to re-examine water quotas. This commission has in turn established what are known as BVOs, from the Russian letters for River Basin Commission, which in theory have supranational powers, and therefore constitute the basis for international agreement. The BVOs have supposed control over water resources across state borders, including canal headwaters. A recent World Bank report was positive about the quality of the people running the BVOs and their serious intentions.

But the BVOs, too, are being starved of the money they need to do their work, and meanwhile the sea – and the people who live around it – continue to suffer. In practice, planners have reduced the amount of money to be spent on water management sevenfold since Soviet times. The Karakalpaks, hardest hit by the crisis, declared their sovereignty from Uzbekistan in 1990 and began to set up their own water-distribution regime. No one took either action seriously, since Karakalpakstan has even less money than everyone else in the region.

"We probably only have ten years left before the sea is beyond hope and the whole region is turned into desert," a Karakalpak bio-ecologist, Dr. Akmed Hametyllaevich, told a researcher from the Green organization, People and the Planet. A local physician was even more stark: "We may very well be witnessing the death of a nation as a result of human folly," she said.[5]

7 | TO GIVE A DAM

Dams are clean, safe, and store water for use in bad years,
so why have they suddenly become anathema?

Early in 1997 the people who build dams, the people who fund them, and the people who hate them got together in Switzerland to see what could be done about them. The co-sponsors were the World Bank, which had put up the money for many of the huge dams in developing countries but which, under the presidency of Jim Wolfensohn and the hidden guiding hand of his then-adviser Maurice Strong, had been turning skeptical; and the International Union for the Conservation of Nature (IUCN), usually known as the World Conservation Union, a network of conservation agencies and organizations, whose views on dams had always been a model of transparency. To the IUCN, dams were almost always wholly bad. Perhaps subliminally, an Internet press release issued by the IUCN after the meeting had misspelled *dam* as *damn*.

The subject matter of the two-day Swiss symposium was a review of fifty World Bank–funded dams carried out by the bank's Operations Evaluations Department. Clearly, the hapless officials in Operations hadn't yet got the new signals being given off at the top. Their review had stated that "the benefits of large dams far outweighed their costs." This wouldn't do. As an announcement issued later by one of the participants, the International Rivers Network, put it in terse bureaucratese, "participants at the workshop largely agreed that [the study] was based on inadequate data and flawed methodology."

There were a few more measured voices than that, among them Aly Shady, a vice-president of the International Water Resources Association and president of the International Commission on Irrigation and Drainage, whose paper asserted that "the survival of the human race in the next millennium will be tied to the success in managing fresh water. One cannot isolate large dams from the overall water management of the planet. Existing and new dams will continue to play a major role in the management system." In the end, everyone at the meeting agreed to set up an International Commission on Dams, made up of "eminent persons, acceptable to all sides," whose charge would be to develop guidelines for funding, constructing, and managing any dams that were still contemplated.

A series of self-congratulatory press releases followed. Patrick McCully, campaigns director of International Rivers Network, said, "We are greatly encouraged that the World Bank and other dam builders have accepted that dam building has caused many problems, that there is a need for an independent review of dam impacts, and that existing practices through which dams are planned and built are in serious need of improvement." Shripad Dharmadhikary, a leading activist with India's Narmada Bachao Andolan (Save the Narmada Movement), said, "While we warmly welcome the willingness of the dam industry representatives to work with dam opponents in establishing an independent review, this does in no way mean that we will lessen the intensity of our campaigns against dams, for justice for dam-affected people, and for the implementation of equitable and sustainable alternatives." Peter Bosshard, secretary of the Switzerland advocacy group Bern Declaration, said, "We are delighted at the call for the establishment of an independent review; however, we are also aware that we will need to be constantly vigilant to ensure that the review is truly independent and that its terms of reference are as comprehensive as agreed." Bosshard and Dharmadhikary clearly interpreted the word *independent* as meaning "that which agrees with us."

The dam builders, for their part, maintained a glum silence.

Shortly afterwards, McCully and his allies were back on the Internet, roundly condemning the World Bank for appointing all the wrong people to the commission, and expressing, in the language of communiqués, their "grave disappointment." More silence followed, and then a third series of press releases, once again warmly welcoming the (new) commission.

What was going on? Why have dams — for so long accepted as a benign way to save water for a not-so-rainy day and to produce clean, non-polluting power — suddenly become anathema?

Just off the roadway on the Nevada side of the highway that crosses the Hoover Dam into Arizona is a commemorative frieze and a monument to the men who died while building it. The monument is really to the heroic spirit of the age that produced the dam; the men who "fell," in the measly cant of politicians, are presented mostly to demonstrate the enterprise, courage, and true grit of Americans and to eulogize the skills of engineers. The structure is actually quite moving. It is made of brass and granite, concrete and steel; if you turn your head you can see the waters of Lake Mead lapping against the curving concrete of the dam and, on the other side, on the concave side of that sweeping mass of man-made stone, the dizzying drop to the Colorado riverbed, more than two hundred metres below. It's hard not to feel puny next to that immense artifact, and even harder not to feel awe. The dam is so massive . . . but also graceful, a sweeping curve across the canyon, even the 120-metre intake towers rising from stone niches carved out of the canyon and designed to look elegant and slender. Hidden from sight are tunnels blasted through the rock, first the diversion tunnels that created a new and temporary Colorado River while the dam was being built, and then the spillways, designed to handle twice the flood load of the river, each with huge brass doors incised with Art Deco devices.

Everything about the dam is improbable. It was built by a consortium of small-time contractors, most of whom were road-pavers and railway-track layers who had never built anything like this before. Not that anyone had built anything like this before: the temporary coffer dam would alone have been one of the largest dams in the world had they bothered to leave it standing; the dam itself was filled with 66 million tons of concrete; the cooling pipes laid in the dam would have made a refrigeration plant that stretched all the way to the coast, more than two hundred kilometres away (the dam would have taken a hundred years to cool without those pipes, so great was the pressure), and so on and so on. The contractors were mostly venal men with maniacal obsessions, with not an ounce of aesthetic sensitivity among the lot of them, yet they managed to build a massive structure that was also beautiful, with a sense of design that matched the glories of great cathedrals. And it was built by men earning four dollars a day, in the blinding heat of the hottest place in North America, in the depths of the Great Depression, in just three years. It cost less than fifty million dollars (U.S.).

The Hoover Dam is important in another way too. Humans had been building dams for thousands of years, mostly from mud and stone, and often from wood. In the century before Hoover, many dams had been built across many significant rivers, but none so wild and apparently irresistible as the Colorado. The completion of the Hoover unleashed a flood, if that's the right term, of other massive projects. In Russia, Stalin, not to be outdone, began construction of the system of canals and waterways that were to reach from the White Sea to the Caspian and the Sea of Azov, and began planning to re-engineer Mother Volga, which ran through Russia's heartland. When he had finished, the river was not really a river any more but a series of gigantic stepped reservoirs, some of them hundreds of kilometres long, which came to be called the Volga Cascade. The same thing happened on the Yellow and the Yangt'ze in China, the Zambezi and the Niger and the Nile in Africa, the Paraná in South America, and a binge of dams elsewhere. One of the things that Hoover set in motion was a change in the character of the world's waterways, permanently altering the ecosystems of entire drainage basins, in some cases changing the local microclimate, and in at least one case, that of the Nile, permanently changing a flow-pattern that had sustained civilization for five thousand years. The numbers are startling. In 1900 there were no dams in the world higher than fifteen metres. By 1950 there were 5,270, two of them in China. Thirty years later there were 36,562, of which no fewer than 18,820 were in China. The binge is slowing now, mostly because there are few rivers left worth damming – and because the ecologists are now, at last, beginning to count the costs.

~

I've been collecting dams – mostly, I think, because of my grandfather's obsession with them. The rationales for building these mammoth projects include the production of hydroelectricity (a "clean" energy source, after all), flood control, navigation, and water storage – in wet periods for use in dry ones, for irrigation, and for direct human consumption. Drought insurance was my grandfather's overwhelming reason, and I like to visit dams whenever I can because of it.

Some years ago, in the last years of Sovietism, I spent some time at the furthest-downstream of the Volga dams, just above Volgograd and only a few hundred kilometres from the river's mouth in the Caspian Sea. I had become interested in a small stream called the Akhtuba, which was the real boundary between steppe and desert; the hot winds from the east break on the white hills that fringe the river, and shield the Volga embankment from all but the worst of the desert storms. All the way to the delta at Astrakhan the two rivers are linked by hundreds of shifting channels; the land is marvellously fertile, consisting of silt as rich as that of the Nile. The Volga Akhtuba Water Meadow, as the area was called, has been diked and tamed, just like the Volga itself; eighty-five tons of vegetables per hectare were produced at that time in the irrigated fields of the meadow, phenomenal by Soviet standards, where a quarter of that was the norm.

But the Akhtuba, I found, was no longer there. It has been engineered out of existence to make way for a massive hydro-electric scheme across the Volga with the old-style moniker of "Twenty-Second Congress of the CPSU Dam." I was quite taken with this dam, though it was nothing like the Hoover. A brochure in its visitors' office described it as a "miracle of Communist construction," which after the fall and disgrace of Communism was given a considerably more sardonic meaning than the one intended, communist construction being well known for its shoddiness and ugliness. The brochure goes on in great detail to list the tons of "ferro-concrete" that built the dam and how very big it is: "Each of the twenty-two turbines puts out twice the power needed by Volgograd itself." The reservoir the dam created is almost a thousand kilometres long and up to twenty kilometres wide. Some hundred thousand hectares were flooded and fourteen thousand buildings submerged. This expropriation was done in the usual peremptory Soviet way: the residents were told the water was coming and they'd better be off, and off they went.

I spent a morning, my last in Volgograd, leaning on the railing at the fish elevator at the Twenty-Second Congress of the CPSU Dam, watching for sturgeon. The Volga and its tributaries have always been spawning grounds for the Caspian sturgeon, as well as herring, bream, roach, and other Caspian fish: Mother Volga was mother to caviar too. That I didn't see any sturgeon wasn't so surprising — it wasn't the spawning season. But even if it had been, the dams of the Volga Cascade have disrupted these ancient rhythms, breaking the breeding cycle. What were the fish to make of these barricades, especially this one, the first in from the sea? Fish moving upriver to spawn navigate by sensing the current flow, and locks confuse them. So the dam's engineers provided a large movable chamber containing a turbine, which was surrounded by a fine mesh to prevent damage to the fish. The turbine is supposed to propel the water at exactly the speed of the Volga current, and when the fish get to the dam they dart about until they find where this "current" is coming from. They swim into the chamber, are caught, elevated the twelve metres to the upper dam, and released.

That's the theory, anyway, but the Volga is now so polluted that the effort to preserve it as a hatchery in these parts has been abandoned. The elevator is still there, still going up and down, still manned, still working, still carrying no-doubt-bewildered fish into the Volgograd Sea, but no one pays much attention.

"So why is it still in use?" I asked a grizzled worker I found near the elevator's gates.

"Why not?" he replied.

"How often does the hoist operate?"

"As often as needs be," he said. "We don't hoist for one fish, but in the spawning season we're going up and down all the time." He smirked. "I wouldn't eat the caviar from one of these fish in any case. None too clean around here: you should see what else floats into the chamber." He laughed humourlessly. "And God knows what's in here you can't see." He peered down into the water. The railing on which he was leaning was polished smooth.

"So your job is to watch for fish, and raise the elevator when you see some? Even if no one believes in the Volga sturgeon fishery any more?"

"Yes," he said, "that's what I do."

"Not a bad way of making a living," I said, "watching for fish all year."

The old man spat into the tank. "It's a job," he said.

~

A few years later I stood – very carefully – on the lip of the sheer two-hundred-metre gorge on the Zambezi River, below Victoria Falls. I peered gingerly over the crumbling edge. Far below, an eagle was cruising, hunting. And below that I could make out a yellow raft as whitewater thrill-seekers went hurtling down the river.

There was a furious debate about that gorge in the freewheeling newspapers of Zimbabwe and Zambia, which share it as a national frontier. A consortium of engineers had proposed the construction of a dam about twenty kilometres downstream. The dam would flood the gorge, but not affect the falls themselves, one of Africa's greatest sights and a major tourist draw for both countries. It would, however, do away with the famous rapids downstream from the falls.

Green opposition to the dam centred on the nesting sites of eagles, which would be destroyed by the reservoir. This focus was a tactical mistake, I thought then, for it trivialized the debate, allowing opponents to dismiss protesters as people who would oppose something grand on the grounds it would disrupt a few birds' nests. The region was hardly short of eagles, in any case. And, indeed, the press in Zimbabwe was having a field day, laying the sarcasm on thick. Zimbabwe wanted the dam. The other Zambezi River dam, the one four hundred kilometres downstream at Kariba, had been built at great cost to both countries, but most of its hydro power had gone to Zambia. Zimbabwe wanted this one, although it had zero chance of being built. Neither Zimbabwe nor Zambia had the money, and the World Bank was less enthusiastic about funding large dams in developing countries than it had been, having become skeptical of their real value, and having better uses for the vast sums of money they inevitably consume.

I had seen the dam at Kariba a few years after it was built. Not quite Hooverian in scope, it was big and impressive enough, 128 metres high and more than 600 metres along its curving crest. Lake Kariba was nearly 300 kilometres long and some 30 wide, 100 metres at its deepest point, with lush shoreline and dozens of islands. These islands had become famous, or notorious, as the reservoir filled in the late 1950s and "Operation Noah" became a worldwide alert for the thousands of stranded animals that would otherwise have drowned. Independence movements at the time – for these were still colonies, white-ruled under the short-lived and widely hated Federation of Rhodesia and Nyasaland – pointed out snidely that more concern had been expressed in Europe and the Americas for elephants and leopards than for the fifty thousand Tonga-speaking people who'd been forced to evacuate without an airlift.

By the 1990s, ecologists were divided. The natural habitat of millions of creatures had been fatally disrupted, but Lake Kariba had become home to a hugely successful population of hippos and other animals, and it provided reliable water for the wildlife, as well as the farmers and residents of two countries. Tilapia fish, a species alien to the Zambezi, had been introduced to Kariba and had flourished, driving out some native species and pushing others to extinction; on the other hand, they provided a brand-new fishing industry that employed thousands of small-scale fishermen and yielded more than twenty thousand tons of fresh protein, annually and sustainably. On balance, I thought then, Kariba was probably beneficial, just as the dam at Akhtuba was. It drowned some farmland, forced people to move, changed habitats, and permanently altered ecosystems. But it also provided clean power for industry and cities, and made possible the cultivation of land that had been barren.

Of course, that was before ecologists began to understand the true economics of dams.

～

There's no doubt where ecologists stand on the Aswan High Dam, on the Nile near the Egyptian border with Sudan. In their view, it is an unmitigated disaster. A product of overweening ambition and Cold War competition, it has damaged far more than it has benefited, and has placed the Egyptians in a policy bind: they cannot destroy the dam, but nor can they tolerate it. (To be fair, this view is not shared by the people who manage the dam or by many of the people who use its water.)

If the Aswan High Dam is indeed an ecological disaster, its worst perils are hidden. On Lake Nasser, the six-hundred-kilometre-long reservoir formed by the dam, everything looks peaceful. The feluccas ply the waters as they have done the Nile for forty centuries, identical in style to the boats depicted on the frescoes at Luxor and Karnak. You can sit by the Nile at Aswan of an evening and, if you're lucky, you may see the feluccas drifting by in the light of a blood-red tropical moon. But below the surface, and in the lands north of the dam, problems are accumulating.

There had been an earlier dam at Aswan, completed in 1902. At the time, it was one of the largest dams in the world, more than two kilometres long and pierced by 180 sluices. Those sluices, and the fact that it wasn't longer and higher, meant that the dam's net effect was relatively benign. It held back about four million acre-feet from the tail of the annual Nile flooding for later distribution, which was a benefit in dry years, but it passed on the bulk of the flood, with its critical load of silt.

The High Dam was different, a lot bigger and a lot longer: one hundred metres high and nearly four kilometres in length. Gamal Abdel Nasser had skilfully played off the Soviets and Americans to get the funding he needed for this grandiose scheme (more than $1 billion, twenty times the cost of the Hoover Dam), and then persuaded the United Nations to pay for the shifting of the Great Temple at Abu Simbel, Ramses II's monument to his own deification. The moving of this monument out of the way of the waters of Lake Nasser rivalled in its engineering skills the original conception. The dam, at least in the view of its builders, was one of the world's great engineering works. The problem, however, is that the famous Nile silt doesn't pass the dam. It stays there, slowly accumulating.

At first, this problem wasn't recognized. A Western commentary on the dam in the mid-1970s had this to say: "It yields enormous benefits to the economy of Egypt. For the first time in history, the famous Nile flood can be controlled by man. The dam contains flood waters, releasing them when needed to maximize their utility on irrigated land, to water thousands of new acres, to improve navigation both above and below Aswan, and to generate electricity."[1]

In hindsight, it's true that the dam has virtually eliminated the potential for flooding, thereby preventing damage to buildings, farms, villages, and roads. It's also true that it preserves water for those years in which the Nile barely stirs at all. But Egypt was able to sustain agriculture for so many centuries, unique among irrigating cultures, entirely because of the annual deposit of fresh silt from the Ethiopian highlands and the marshes of Sudan. Now that same silt is piling up behind the massive walls of the Aswan High Dam – slowly but surely making it obsolete. The nutrient-rich ooze no longer makes its way to the Mediterranean, leading to the collapse of the sardine fishery in the Nile Delta – a collapse that more than offsets the new fishery introduced with such fanfare into the reservoir behind the dam itself.

The changes to the system's water deliveries has meant that the shoreline of the delta is eroding at about three metres a year. Egyptian farmers, now able to irrigate all year round instead of waiting for the flood, have been protected from the periodic droughts and famines to which the Nile system was so prone. But another consequence of intensive irrigation has been a steady rise in the salt levels of the soil, a rise so critical that Egypt hired an international team of engineers, including the former commissioner of the U.S. Bureau of Reclamation, Floyd Dominy, to advise it what to do. The panel's only suggestion, after it had finished shaking its collective head, was a radical and improbably expensive drainage system under the irrigated lands.[2] No one thinks this is financially practical, though they all accept it as necessary.

As the Egyptologist Peter Theroux puts it, "for more than 30 years the Aswan High Dam . . . has kept the river from flooding and depositing renewing sediment at its mouth. The delta has instead been inundated with catastrophic superlatives: it is among the world's most intensely cultivated lands, with one of the highest uses of fertilizers and highest levels of soil salinity."[3] The delta is slowly sinking, a condition that will affect the groundwater. Egypt's precious soil is being buried beneath the concrete of Cairo's relentless sprawl. The coast is eroding and chemical pesticides are killing marine life.

At the same time, the irrigation channels and the new permanence of the water have led to an epidemic of bilharzia, approaching 100 per cent in some areas. Egypt is not alone in finding that dams increase disease. Bilharzia and malaria have increased after the building of every large reservoir in tropical regions. But that's just one of the problems associated with dams.

In earlier years, when dams were still regarded as being ecologically benign, the common-sense view was that their most serious problem had been the displacement of people. Yet dams cause worse problems than that. Ironically, it has been reliably shown that malnutrition often follows dam-building instead of the expected bounty resulting from newly irrigated lands. The reasons are complex, but stem to some degree from the failure of engineers to understand the ecological benefits of flooding and the positive effects annual floods have on ecosystems. In many regions the local people developed over the centuries an intensive and varied flood-dependent agriculture. Often, after dams are built and irrigation follows, the broad and reliable array of local foods is replaced by irrigated monoculture, often aimed at export markets. Diversified agriculture has been found to produce more per hectare than irrigated monoculture.[4]

Dams also alter the flow, and sometimes the temperature, of rivers. They almost always lead to elevated salt levels in the surrounding soil, sometimes infecting groundwater. The change in sediment loads affects major resources, as the Aswan High Dam did the Nile Delta's sardine fishery. The same thing has happened elsewhere. Dams on the Niger dropped catches by one-third. Even after decades of retrofitting and tinkering, the number of wild salmon returning to the Columbia River is less than 6 per cent of what it had been before the dams were built. In those reservoirs where new species are introduced, catches almost inevitably drop, sometimes after an initial surge.

Dams also radically affect the nutrients and the saline mix in downstream deltas. Michael Rozengurt, a Russian aquatic ecologist, has estimated that estuary areas can withstand alterations in their nutrient fluxes to about one-quarter, after which catastrophic changes take place. In south Florida the fish catch has been reduced to 42 per cent of its former level. In San Francisco Bay about four-fifths of the migration and spawning areas of salmon, shad, and striped bass have disappeared. The Snake River's coho salmon run is extinct. Most of the fisheries in the Black Sea, the Sea of Azov, and even the Caspian are near death.

In Canada, the W.A.C. Bennett Dam, a monster on the Peace River, is fatally disturbing fisheries in the Peace–Slave–Athabasca Delta, a World Heritage site and one of the largest freshwater deltas in the world. The Athabasca Chipewyan First Nation complained to the Indian Claims Commission that their way of life had been damaged and, in a trenchant judgment, the commission had this to say: "As is glaringly apparent from the evidence in this case, it is more than the First Nation's treaty rights to hunt, fish, and trap for food that have been affected; the First Nation's very way of life and its economic lifeblood were substantially damaged as the government of Canada, armed with full knowledge of the ecological destruction that would follow, did nothing."[5] Reacting to an earlier complaint, the Canadian government had weakly maintained that it had no power to affect what the provincial government of British Columbia got up to, a notion that was contemptuously dismissed by the commission.

All over the Canadian Arctic, resources are in trouble. A report by Dianne Murray of the Hudson Bay Research Group at Ottawa's Carleton University mentioned the Bennett Dam problem, and also had this to say:

The whitefish fishery of Southern Indian and Notigi Lakes, once Manitoba's most commercially important, collapsed because of the Nelson–Churchill megaproject. Much of the fish in the Churchill River Basin is now too contaminated with methyl mercury to be eaten or sold. Fish from the estuaries of the Nelson and Churchill rivers contain as much as four times the safe limit. In the lower Churchill River in Labrador, fish stocks have been rendered inedible. The James Bay 1 project rendered much of the fish in the LaGrande River unusable. And some scientists now suspect that mercury found in the belugas of Hudson Bay is due to contamination caused by downstream migration of mercury from JB1, Ontario's Moose 1, and the Nelson–Churchill.

Murray observed that "Canada has the most freshwater withdrawals from coastal habitats of any country in the world."[6]

And then there is silt. Dams – all dams – silt up, for dams have a natural lifespan, depending on the silt level of the rivers they contain. In certain rocky areas with poor soil, dams will last for a long time, perhaps for a thousand years, though no one knows for sure. In other cases, like the Nile at Aswan, rivers that once ran brown with silt are now running clear beneath the dams. The retained silt, which is really just mud, is accumulating at the dam wall, millions and millions of tons of mud. For a generation or two, and possibly longer, dredging and silt removal can work, but in the end they get too expensive and the dams fill up – often much faster than the amortization rate for which they were planned. A true cost-accounting for dams almost always shows the bottom line in red.

In parts of the world, careless agriculture – deep ploughing of mediocre soil, clear-cutting woodlands to make way for fields, heavy grazing by stock animals, cheap fertilizers that allowed fields to be worked year after year and thereby increased the rate of erosion – have dramatically increased the rate of siltation. An extreme case, cited by Marc Reisner, was the Sanmexia Reservoir in China, which was completed in 1964 and taken out of commission four years later. Reisner also cites the Tehri Dam in India, the sixth-highest in the world, which had its lifespan reduced to thirty years because of deforestation in the Himalayas; and the Dominican Republic's Tavera project, completed in 1973, which by 1984 had accumulated silt to a depth of eighteen metres. "In thirty-five years," Reisner wrote, "Lake Mead [behind Hoover Dam] was filled with more acre-feet of silt than 98 per cent of the reservoirs in the United States are filling with acre-feet of water." This mud will never be removed. There is too much of it, and there is nowhere to put it even if anyone could afford to remove it.[7]

There are two more unforeseen consequences of heavy silting. The first is the destruction of deltas. The Nile Delta is the most obvious example, but the Mississippi Delta is under threat too. The silt that built up the Mississippi–Atchafalaya Delta no longer reaches the river mouth, because it is being accumulated behind a series of dam walls. Eventually, unreplenished, the delta lands will wash away. The second problem is that silt-free water flows faster. River basins are scoured more thoroughly, becoming deeper and more dangerous. The floods of the Missouri and Mississippi rivers in 1993 and the Red River in 1997 were entirely predictable. As Janet Abramovitz put it in her Worldwatch paper "Imperiled Waters, Impoverished Future," "they provided a dramatic and costly lesson on the effects of treating the natural flow of rivers as a pathological condition." Humans had tried to wall the rivers in and had colonized their natural flood plains for farms and towns. The floods were an attempt by the rivers to reclaim what had always been theirs. They were not flood control but flood-threat transfer mechanisms.[8] "What earlier would be a seven-day flood is now a seven-month flood because the water doesn't know where to go," was the way New Delhi's Anil Agarwal put it in 1998. Agarwal is the director of the capital's Centre for Science and Development, and was reacting to a disastrous year in which more than seven thousand people were killed in Asian floods, including China's Yangt'ze and India's Rapti, floods that were often thought to be wayward, capricious, and inexplicable, but which more and more were seen as a reaction to human-caused environmental changes, especially deforestation.

The Rhine is the most severe example of this process. Although eleven million people still get their drinking water from the Rhine, it is no longer really a river but an engineered waterway of levees, concrete embankments, locks, flow-control devices, hydro plants, weirs, and channels. Once the Rhine meandered over an extensive flood plain; now, cut off from its natural controls, it flows twice as fast as before in a channel up to eight metres deep, dramatically increasing the power of its floods. The Netherlands is forced to absorb the full effects of the Rhine's more powerful flow, largely caused by the Germans and the French, neither of which has offered to pay for the damage the faster Rhine has caused. So far the Dutch have been politically polite, though there are signs their impatience is growing.

Sometimes, though, the imperatives of politics are on conservation's side. In Nebraska, the National Wildlife Federation and the

State of Nebraska won a $7.5-million lawsuit against dam builders on the Platte River. The money was used to purchase and restore river channels and their surrounding wetlands. So far, the trust has picked up about four thousand hectares of land, including more than twenty kilometres of river frontage.

~

Nor are dam collapses exactly unknown. Chinese government reports indicate that, in 1975 alone, one of the wettest years on record, dam collapses killed a quarter of a million people and brought famine and disease to more than eleven million others. Even in the United States, a country that prides itself on its engineering skills, dam collapses occur. The Teton Dam on the river of the same name collapsed in 1976, even before it was completely finished, wiping out three towns and hundreds of thousands of acres of farmland, which the flood scrubbed down to bare rock; fortunately there was enough warning that casualties were avoided.

At the Glen Canyon Dam on the Colorado River in Arizona, a section of spillway was destroyed by floodwaters in 1982. The spillway runs directly through the sandstone underlying the dam, and the writer Philip Fradkin described the event this way: "Two thousand tons of water per second soared from the spillways and the river outlet works on both sides of the dam in the most spectacular display of cascading water ever seen on the artificially controlled river. Rumbling noises were heard on June 6. The jets of water became erratic. Water the colour of diluted blood, chunks of rocks, and pieces of concrete issued from the mouths of the spillway tunnels, as if the dam was mortally wounded." The level of the reservoir peaked, Fradkin points out, at 3,708.34 feet above sea level on July 15 – six-hundredths of a foot below the point where engineers forecast control might have been lost. Afterwards, the Bureau of Reclamation declared to its satisfaction that the dam itself was never in danger, but written documents at the time showed that bureau engineers "feared for the safety of the dam and its foundation."

Should we be reassured by the embedding in the structure of "1,142 strain meters, 264 joint meters, 74 resistance thermometers, and 60 stress meters"? Or that to "measure the dam's movement," five plumb lines were installed inside the dam, "each consisting of a 637-foot stainless-steel wire holding a 26-pound weight immersed in a barrel of oil so as to damp its movement"?[9]

British Columbia's W.A.C. Bennett Dam, one of the world's largest earth-filled structures (it is 2 kilometres across the top and 183 metres tall; its Williston Reservoir covers 166,000 hectares and holds 70,308,930,000 tons of water), had some uneasy moments in 1997 when two large sinkholes were discovered just downstream. Emergency repairs were undertaken and in 1998 BC Hydro, which operates the dam, spent $4 million reinforcing its drainage system by gouging out a 5-metre-deep trench and backfilling it with rock. Ron Fernandes, Hydro's area manager, assured everyone who would listen that the dam was perfectly safe. "Geophysical tests confirmed the success of the sinkhole repairs last year," he said, "and we plan on conducting these tests on an annual basis as part of our ongoing surveillance program." It was unclear whether this assurance – and the closing of the road across the crest of the dam in the summer of 1998 "to install additional monitoring instrumentation" – made the people living downstream feel better or worse. It's undoubtedly unfair, but when a bureaucrat talks about "ongoing surveillance," the assurances of politicians come to mind and the tendency to skepticism can be irresistible.

If huge dams are such a menace, then why are we still building them? Sometimes, as with the Okavango pipeline in Namibia, need trumps caution. Sometimes inertia is the culprit. And sometimes, despite their manifest disadvantages, dams still make sense.

When I was writing the first edition of this book, in 1998, up in the high hills of Lesotho – the Worlds Without Horizon, as the Basotho call them – the bulldozers and earth movers were scraping away the brush, soil, silt, and shale to get down to bedrock, hard granite into which they could anchor the first of a series of dams that comprise the Highlands Water Project. The World Bank was funding this scheme, one of the last of the massive dam projects it agreed to support, and every few months an anxious delegation flew into Maseru, the mostly sleepy (if occasionally riot-torn) little Lesotho capital, to see how things were going.

The country, at least in the government's view, had little choice. Lesotho does not have a lot of farmland. The country is largely mountain, austere and beautiful, wonderful for hikers and possibly shepherds, though no paradise for farmers. But the country had nothing to export but its labour, and half the country's men were away at any one time, most of them working the gold mines in Johannesburg. The new South Africa, however, is a more militantly nationalist South Africa, operating under the notion that jobs should go to South Africans first. As a consequence, the Lesotho unemployment rate headed up past the 60-per-cent range, and Maseru's one casino wasn't nearly enough to offset the disruption. It was one reason why unrest spread across the tiny kingdom late in 1998, leading to a bungled South African military incursion.

One other thing Lesotho has in abundance is water. It rains a lot there. The hot air sweeping in across KwaZulu–Natal from the humid Indian Ocean coast condenses as it crosses the Drakensberg Mountains, and rain falls in sheets in the hills. Why not build a series of dams to catch the water, install generators, and export both electricity and water to ever-thirsty South Africa? I thought this a good scheme when I heard it; my grandfather's parched farm had been no more than a hundred kilometres from the Lesotho border, and he would have paid most of what he had for reliable water. No doubt, the dams would fill quickly. No doubt, little would be lost: Lesotho would never be able to feed itself anyway, and since it was going to have to import food, it might as well have the hard cash to pay for it. Engineers had said the water in the hills was almost silt-free; the dams would last for hundreds of years.

Traditionalists opposed the scheme; not only would villages be flooded but burial grounds too. There were a few violent demonstrations and at least one gruesome ritual murder that the news quickly suppressed. The World Bank grew ever more anxious about its investment. Traditionalists making common cause with environmentalists had opposed other dams, successfully, in Nepal, on the Narmada in India, and elsewhere. In 1997 the Malaysian government had finally announced it was cancelling the $5.5 billion Bakun Dam, which had been strongly opposed by environmentalists who insisted the dam would be a financial as well as an ecological disaster. Patrick McCully of the International Rivers Network had been exultant: "The Bakun cancellation highlights the economic unviability of large hydro projects. Bakun was the dam industry's flagship for privatized hydro projects – and the flagship has sunk." The dam's contractors, a German–Swiss–Swedish–Brazilian consortium, maintained another in the series of glum silences to which engineers were becoming accustomed. Often, time had proved the traditionalist-environmentalist alliance right, even when they lost. And did a worsening economy in South Africa, where the rand collapsed in 1998, mean that no one would be able to buy Lesotho's newly produced electricity? The World Bank was right to be cautious.

Nevertheless, at the end of 1998 the project was proceeding. And in 2000 the first water flowed through a three-hundred-kilometre pipeline into the reservoirs of industrial Johannesburg, and the hard currency – albeit devalued South African rands – flowed comfortably into Lesotho's not-very-capacious coffers.

Still, the project is having its effect on dams worldwide, for curious and somewhat ironic reasons. One of the companies involved in its construction, Acres International of Toronto, was found guilty in September 2002 of passing $260,000 as a bribe to the chief executive of the project – who in turn was convicted on thirteen counts of bribery and sentenced to fifty-seven years in jail. Late in the year, legal actions had also been launched against a number of other major construction companies that were influential in the world of dams, among them Lahmeyer International of Germany and Balfour Beatty of Britain. There were two immediate consequences. The World Bank pronounced that the Lesotho project had been "moderately unsatisfactory" (Bankspeak for atrocious) and its Operations Evaluation Department, which by now had got the message about megaprojects, trenchantly urged that the Bank avoid big building projects and concentrate instead on poverty alleviation. The second consequence was more piquant. The Lesotho government, which launched the anti-corruption drive that resulted in the prosecutions, was feeling especially virtuous. As a government spokesman put it, "This turns on its head the perception in First World countries that consultants are obliged to pay bribes in Africa. It was the company that offered money. . . . People here are very aggrieved when the rich world writes this off as just an African problem."

Across the world, literally and in sensibility, another huge project was taking shape, this one with neither the advice nor the money of the World Bank. The Chinese were financing the Three Gorges Project entirely on their own.

This project has set off an uproar among international conservationists, ecologists, political dissidents within China, and artists. When the Three Gorges Dam on the middle Yangt'ze is finished, it will submerge one of the most picturesque places on the planet, an inspiration to Chinese painters and poets for three millennia, and the site of many shrines and temples. It will end forever the extraordinary and extraordinarily difficult boat trip through the awe-inspiring gorges. The six-hundred-kilometre reservoir will, not incidentally, mean the removal of 1.5 million people, and the submerging not only of villages but of a major industrial city. And this on a river prone to flooding and in a country where dam collapses, as we have seen, are hardly unknown.

Flooding is one of the primary reasons the Chinese give for wanting to build the Three Gorges Project in the first place. According to Chinese statistics, the Yangt'ze has seen 214 "major floodings" in the past two thousand years. This century alone there have been four "disaster level" floods; the recurrence frequency is now about one flood every ten years, as deforestation and the draining of wetlands have made the river ever more powerful. The plains along its middle and lower reaches are highly developed, with industrial cities and intensive farming. The river's own dis-charge capacity is only seventy thousand cubic metres per second, but it routinely reaches one hundred thousand in high-water floods, and the dikes built between Jingjiang and Hunan are not always strong enough to protect against flood crests that can reach seventeen metres above normal.

In 1998 floodwaters crested at fifteen metres above normal and threatened to overwhelm the city of Wuhan and its 7 million inhabitants and, by the time the crisis was over, more than 3,000 people were dead and millions more homeless. As many as 240 million people, fully one-fifth of China's population, had in some way been affected by the flooding, and Chinese economists figured it would shave almost a percentage point off the GDP for the year. Even the normally docile provincial legislatures were vig-orous in their criticisms of the policies – or lack thereof – that had brought the disaster about. Lu Youmei, the chairman of the Three Gorges Development Corporation, said rather tactlessly at the height of the crisis that "if the Three Gorges Dam had been built by now, the problems of flood control would already have been solved." The project's opponents, in contrast, maintained that the dam was partly at fault. So much money and resources were being diverted to it, they said, that the building of levees and dikes and the clearing of the riverbed had been neglected.

The other reasons for going ahead with the project, or the "benefits," as Chinese hydrologists prudently call them, are power generation, navigation improvement, aquaculture, tourism, ecological protection, environmental purification, development-oriented resettlement, water supply and irrigation, and the transfer of water from south to north – all of which, in the enthusiastic encomium of C. Yangbo, of the University of Hydraulic and Electrical Engineering in Yichang, make it a "unique super project worldwide." This is the same shopworn list trotted out by the builders of the Aswan High Dam and the propagandists of the U.S. Bureau of Reclamation. It is perhaps doubly ironic in this case that tourism be included, since one of China's major tourist attractions will simply cease to be, and another ordinary-looking lake will take its place. As for the "ecological protection," hardly anyone has figured out what it means in this case.

When I was in China in May and June of 2002, I interviewed a score of hydrologists and engineers concerning China's ecological record. Almost all of them were candid about almost everything, but when it came to the Three Gorges, the most politically charged project of them all, they were mum. No comment was forthcoming, either for my notebook or on camera. But when the camera lights were off and the notebooks tucked away, most of them would shake their heads and look depressed. A big mistake, was the consensus. A disaster in the making. And politically inevitable.

It is true that, downstream, there are one-and-a-half million hectares of fertile farmland, several major railway lines, and fifteen million productive citizens, all of whom could use a little protection from flooding. And, of course, all are vulnerable should the dam ever fail.

The final reason for building the dam is the notion of transferring water from the middle Yangt'ze to Beijing and the parched northern plains of China. But it will take another seventeen years to get the water there, far too late to avert the looming water crisis. And even then, will it have been worth its $29 billion cost?

Medha Patkar, an unassuming social worker from Bombay, is now an international commissioner on the question of dams, appointed after the World Bank–IUCN meeting in Switzerland in 1997. This appointment was neither arbitrary nor gratuitous political correctness on the bank's part, but a neat act of co-option. The bank had been financing a huge dam called Sardar Sarovar on the Narmada River in India. According to plans, it would be the second-largest dam in the world, providing irrigation water for three hundred thousand people and electricity for half a dozen major cities, at a cost of some $5 billion. One of the villages to be submerged was a small farming community called Manni Belli, but the villagers had not been consulted. Angry, they called the environmental activist Medha Patkar in to help, but by then the dam was half-built and there wasn't much left of Manni Belli any more: only the tip of its temple roof was showing above the rising water. That didn't matter. Medha Patkar and her organizers put an immediate stop-work order to the whole thing.

In March 1998 a court order halted construction of the dam for a review period of three years. The builders were furious: the dam was already half-finished, and the millions that had been spent would be wasted if the dam were never completed. The protesting villagers of Manni Belli, who had previously been ignored, were now roughed up. Some were tortured by hired security forces, and several of the women were raped. Patkar was herself beaten and jailed. The dam builders made their greatest mistake in allowing this to happen: it made her, if possible, more implacable than ever, and even more saintly to her supporters. The BBC, which caught up with her after the Supreme Court decision, found her resolute. "Gigantic powers were against us," she said, "but we were never pessimistic. If one has no hope one must not fight the battle."[10]

A year after construction was halted, villagers in the basin of the proposed reservoir were still getting eviction notices. Patkar knew the fight was not over. But she also knew the tide of world opinion was turning in her favour. Massive dams are suspect, and the funding is drying up. Other solutions will have to be found. Money, she felt, will now have to be spent where it is more useful, in saving water instead of wasting it. Dam builders would be sensible to listen.

In the year 2000, the World Commission on Dams issued a long-awaited major new report, *Dams and Development: A New Framework for Decision-Making*. As a policy document, the report was something of a bust, because no one quite understood what it was recommending, if anything. It spent most of its time evaluating the performance and achievements of major new dams around the world, which was all very well, but it came under furious assault from the Greens, because it resolutely refused to take sides. Even the most sympathetic readers were left with the feeling, Okay, but now what? That one of the report's major outcomes was the setting up of yet another bureaucratic unit, the Dams and Development Unit of the World Commission on Dams, to operate under the UN Development Program, was not reassuring. Dams were silting up and operating at fractions of their designed efficiencies all over the world, but the response, it seemed, was . . . much more study. As a rousing call to action, it left a little something to be desired.

8 | THE PROBLEM WITH IRRIGATION

Irrigated lands are shrinking, and irrigation is joining dams on an ecologist's hit list.

Pessimistic projections made in the middle of the twentieth century often indicated that the world would run out of food within decades, that agriculture could not possibly keep up with explosive population growth. Instead of global famines, however, we had the Green Revolution, a biped with one foot in chemistry (pesticides and fertilizers) and the other in managed water distribution, or irrigation, which caused a quantum leap in agricultural efficiency and productivity. As a result, India and China, the focus of much of this Malthusian angst, not only fed themselves but contributed to world surpluses, selling grain on the open markets. India, particularly, seemed bent on irrigating every possible hectare – 113 million, at last count.

Worldwide, irrigated lands are only about one-sixth of total area farmed, but account for more than one-third of the global harvest. But now the per capita extent of irrigated land is shrinking. Why is this?

The first irrigation scheme I ever saw was very small, a single, narrow furrow on my grandfather's farm, which led from the corrugated-iron storage tank out back of the house to a patch of ground in front. It meandered a bit – the workers had skirted rocks rather than moving them or breaking them up – and it wasn't more than one-third of a metre wide and about twice as deep. Narrow, yes, but wide enough for my sister, little more than a baby, to plunge in one day. I still remember the shrieks and the running farmhands, and how one of them plucked her out just in time, seconds before the swimming snake got to her, a poisonous adder, its sinuous body driving it through the water, straight as an arrow from a Bushman hunter. She got out in time, but for a long time afterwards I dreamed of that evil black head striking through the water, its tongue flicking, a creature with a sinister purposefulness. Or so I thought then. Later, when I was older, another thought interested me: Where did the damn thing come from? In this dry and arid land, how did a water adder find its way to water that had never been there before?

The corrugated tank was fed by a clanking windmill. Alongside a third of it was a concrete trough for the cattle to drink from. A stopcock along the side filled the furrow, and the furrow led to a small strawberry patch in deep and well-drained soil close to the house. My grandmother never persuaded her husband to take the water to the house itself: all her life she was obliged to fetch her own water for cooking, cleaning, and washing. My grandfather – my Oupa, as we called him in Afrikaans – believed water was too precious to waste on taps in the kitchen. Water was for the crops; humans could fend for themselves.

The strawberry plants were watered directly, by bucket. There were no sluices in the furrow: there was not enough water for that. Each day the farm workers gave each plant exactly what it needed, and twice a year my grandfather could hitch up his wagon and take the crop to the farmers' market in town, five hours away by horse-drawn cart, whence it would go by train to the provincial capital, Bloemfontein, one hundred kilometres away. What happened to it there I never knew. We also ate the berries at home, fresh and sun-ripened and watered by moisture from the centre of the earth, and we drowned them in fresh Jersey cream from Oupa's herd, perhaps after a meal of home-grown roast mutton and asparagus spears from the bed next to the strawberries. We never thought much about it. It was just how we lived.

Oupa never learned to read and write. He was a Boer soldier fighting the British when he should have been in school, and he remained unlettered all his life. But he was not a fool. He was a careful farmer, with a scrupulous love for natural things, and through an intuitive understanding of how plants grew, he had devised the very model of a modern irrigation scheme: his water was stored in a tank, not an open dam; the furrow was lined with stone and filled only when needed; the plants were watered directly; and the soil was properly drained.

The manuals at the agricultural colleges now recommend schemes very like it. More massive, of course – sometimes hugely more massive – but in essence the same, closed systems with little wastage. It was, in all the best senses of the word, a sustainable practice. And just as important, the good drainage and careful watering meant there was little chance he was poisoning the soil with increasing salinity.

Much bigger farmers, and much more sophisticated farming operations, were not paying the same attention.

Irrigation is an ancient technique, going back five thousand years in Central Asia and a great deal further in Mesopotamia. It was used in Imperial China, in old Laos, in ancient Africa, in Tanzania and Zimbabwe, and in America before the Spanish ever laid eyes on the place. For millennia, farmers have used aqueducts and ditches to take water to where it was needed, and furrows with crude sluices to divert the water to their crops, giving them increased yields and lessening their dependence on the caprices of the weather. But on any real scale, irrigation is not much more than a century old; it needed the massive dams and water diversions of the modern era to make it possible.

The numbers are revealing: two hundred years ago, total irrigated land probably amounted to as little as 6 or 7 million hectares, not much bigger than Long Island, New York, or metropolitan Los Angeles and its exurbs. By 1900, the total had jumped to some 50 million hectares, and that number nearly doubled in the half-century that followed. Of the estimated 230 million hectares currently under irrigation, half was added in the three decades from 1950, and dozens of countries, ranging from little Israel to massive China, rely on irrigation for much of their domestic food production – Pakistan up to 80 per cent. Most of North America relies on rain-fed farming, but without irrigation, agriculture in California's Central Valley, one of the world's most productive vegetable and fruit resources, would hardly be possible. Only 15 per cent of the world's cultivated land is irrigated, but irrigated land accounts for almost 40 per cent of the global harvest. Without irrigation, yields in the world's major breadbaskets – on which the feeding of the planet is dependent – would drop by almost half.

But this rate of expansion couldn't go on: there was only so much water available, and by the 1980s the rate of construction of new dams was slowing. All the easy rivers had already been dammed, for one thing, and the expense had escalated dramatically, to the point where many of the proposed new irrigation schemes had priced themselves out of possibility: they would have delivered water at a rate not much cheaper than the farmers could sell their commodities. In parts of Africa the cost of delivering water from some of the irrigation schemes on the drawing boards in the 1980s was projected to reach more than $10,000 per hectare, and, as Sandra Postel has pointed out, "not even the double-cropping of higher-valued crops can make irrigation schemes at the top end of this spectrum economical."[1] In fact, she says, for the past two decades the population has been outstripping what irrigation increases there are. Projected population growth rates for the next thirty years will require an increase in food production equal to 20 per cent in developed countries and 60 per cent in developing countries to maintain present levels of food consumption. Yet per capita irrigation peaked at forty-eight hectares per thousand people in 1978, then dropped 6 per cent in the following decade, to forty-five hectares per thousand people. The trend is accelerating.

The situation is unlikely to reverse itself, even if commodity prices rise high enough to catch the attention of water engineers. The costs of damming the remaining undammed rivers is prohibitive.

But the situation is worse than just a lack of expansion opportunities. Not only is no more land being brought under irrigation, but a good deal of irrigated land is being taken out of production. Salination of the soil, to a degree that inhibits farming, is spreading at the rate of more than a million hectares a year. In Pakistan alone, two million hectares have been decommissioned, the soil poisoned by high salinity, and farm yields are down 30 per cent. Egypt is showing similar declines. In the U.S. Imperial Valley, more land is being decommissioned than commissioned, again because it is overly salty. Improved drainage might save the fields, but there may not be enough water or money to do the job. Ironically, the Imperial Valley sits on an enormous geothermal aquifer which contains great quantities of salty water that could possibly provide a solution to the problem. Water temperatures in the huge underground pool go as high as 280°C. Researchers are developing methods to use the steaming brine to generate electric power.

They believe it's feasible that the clean waste water produced by a power plant could be used to dilute the salty water in the fields.[2]

In the United States as a whole, yield reductions due to salinity occur on an estimated 30 per cent of all irrigated land. Worldwide, crop production is limited by the effects of salinity on about half of the irrigated land area, and nearly two-thirds of the total needs expensive renovation just to keep working.[3] The consequences for world food production are still unknown, but are sure to be significant.

That pesticides and chemical fertilizers have their own set of evil consequences is well known. That irrigation – the delivery of clean water to the plants that need it – contains within it another unexpected set of difficulties has come as a nasty surprise. But it is becoming more and more apparent that salinity and the increasing concentrations of noxious substances in water is the vulnerable underbelly of the irrigation revolution, threatening huge areas of otherwise-productive lands. The problem is simple to state, though devastatingly difficult to fix: without great care and skilful management, irrigation almost inevitably causes waterlogging, depletion and pollution of the water supply, and rising salinity in the soil. Left unchecked, these problems can eventually kill the soil altogether.

~

Anyone who has driven for any time through Nevada, Central Asia, or the southern savannahs of Africa will recognize a salt pan without any difficulty.

Not all "pans" are saline. Some are shallow depressions that fill with water and greenery when the rains come, attract a wide variety of wildlife, and, over the centuries have maintained a sustainable balance. The Etosha Pan in the northern part of Namibia, for example, is now a major tourist attraction. Dry for most of the year, it is a green paradise for a few months, a series of shallow lagoons fed by the same drainage system as causes the Okavango wetlands far off to the east. The vegetation, dormant for so long, springs into life; the pan can become carpeted in green virtually overnight, and it fills up rapidly with a bewildering array of animals and birds. The muddy water churns with catfish spawned by parents that survived the dry season by burrowing into the mud, living off body fat, and surfacing only to use their peculiar lunglike auxiliary organs to obtain oxygen directly from the air.[4]

A true salt pan, in contrast, is a place that no longer turns green, a place where there is so much salt in the soil that nothing will grow there. The processes that caused such pans are perfectly natural. The problem is, they are the same processes that, man-made, now bedevil irrigation schemes. A summary sheet put out by the U.S. Salinity Laboratory outlines the problems succinctly:

> Application of irrigation water results in the addition of soluble salts such as sodium, calcium, magnesium, potassium, sulfate, and chloride dissolved from geologic materials with which the waters have been in contact. Evaporation and transpiration (plant uptake) of irrigation water eventually cause excessive amounts of salts to accumulate in soils unless adequate leaching and drainage are provided. Excessive soil salinity reduces yields by lowering plant stand and growth rate. Also, excess sodium under conditions of low salinity and especially high pH can promote slaking of aggregates, swelling and dispersion of soil clays, degrading soil structure, and impeding water and root penetration. Some trace constituents, such as boron, are directly toxic to plants.

The first problem is one of concentration. Because water is such an efficient solvent, it picks up salts and minerals from the soil on its journey down from the hills. When it collects in the shallow lakes in the valleys – salt pans to be – or when it is applied to irrigated land, some of it seeps into the ground, and some of it evaporates.

If the soil is well-drained, the water that seeps downward might eventually trickle into an underlying aquifer, in which case it causes little harm. But if the soil is not well-drained, or is compacted clay, or there is no natural outlet for the water, it will become waterlogged and the salts will begin to accumulate. Over time, heavy concentrations of salts get closer and closer to the surface, eventually reaching the root zones of vegetation. For a while, there is a struggle – some plants are less efficient at handling salts than others, as early irrigating cultures learned to their cost. But eventually even the salt-resistant plants succumb and the last vegetation dies.

Not all the water seeps downward. Much of the water evaporates – in incipient salt-pan formations, for every hundred cubic metres of water, some seventy-five will typically be lost to evaporation. If the wind conditions are right, some of the salts will be taken up into the atmosphere too, through aerosols. But nine-tenths of the salts will remain behind on the surface, and the pan becomes ever more saline. Eventually, the salts in the root zones meet the salts on the surface, forming a crust, and the lake will become permanently dead.

In many cases in the past, these dead zones have also become arid. Less water runs into them than before, partly through changes in the local climate, and partly by increased radiation of heat from land stripped of its vegetation.

The second problem is accumulation – concentration in another form. This, too, occurs in nature, but seldom to a troublesome degree. Rivers are naturally efficient flushing mechanisms, and the salts that accumulate in them tend to make their way to the sea. It's true that where groundwater runs through naturally occurring saline formations, the water that enters rivers might contain a higher concentration of salts, and it is also generally true that rivers running through naturally alkaline or saline soils tend to became saltier the further they travel. Nevertheless, the concentrations are usually not critical, and over the eons the local wildlife has adapted to deal with it, to such a degree that many delta fish habitats have become dependent on specific salt concentrations.

But where river water is taken up by humans for use in irrigation, the method of concentration – and the concentrations themselves – changes. For irrigation, river water is either diverted to canals or directly taken up by pumping. Much of the water so diverted is lost in the process. As a working average, almost 60 per cent of the water intended for irrigation never gets to the croplands. Leaking pipes, unlined canals, evaporation from open reservoirs and canals, and poorly directed spraying cause much of the water to be wasted. Some of this wastage returns to the groundwater, so it is not "lost," except in the sense of adding to farmers' costs. Much of this missing water seeps through the soil, picking up some of the already-accumulating salts as it goes, and returns to the river further downstream. There, it is recaptured in another irrigation scheme, and the same thing happens. If the river is intensively managed, it will pass through several reservoirs before eventually making its way to the sea. At each stage, in each reservoir, water is lost to evaporation, as much as two metres a year, especially in dry climates, which further concentrates the salts. In certain rivers, water may be "used" more than a dozen times, in each cycle becoming more and more salty. By way of comparison, in parts of Spain, for example, where the soil is not particularly alkaline, river salinity can increase from around 400 or 600 parts per million – or less at the river's source – to somewhere around 1,200 to 2,000 parts per million before it discharges into the sea. In the American West, where the soils are poor to begin with, and there is a good deal of natural salt, the intensifying power of the process is much higher, and salt concentrations can be almost twenty times higher at the mouth than the source.

Even the most pristine water has some salts, typically about 200 parts per million, compared with the 500 considered the highest level safe for drinking. (The ocean, by comparison, contains about 35,000 parts per million.) As Sandra Postel points out, applying 10,000 cubic metres of irrigation water per hectare, a fairly typical rate, would add around five tons of salt to the soil annually.[5] The salt is therefore being "stored" in the land and in river basins, instead of being flushed out to sea. As a consequence, productive land is becoming salt-encrusted and barren.

None of this has to happen, and none of it means that irrigation inevitably poisons soils. Where drainage is good and the soils not naturally alkaline, carefully managed irrigation can persist for centuries without harm. In areas where irrigation is used as a water supplement rather than the whole diet, little harm is likely to befall the land.

It is where irrigation is intensive, the soils naturally poor, and the drainage either inadequate or nonexistent that the most serious problems will occur. The American West – the turning of the Great American Desert into one of the planet's breadbaskets – is the most notorious example. In some areas in the San Joaquin Valley, in central California, the salts are so obvious that the earth looks as though it has dandruff; not even weeds grow there, and – shades of the Aral Sea – the salts drift on the winds, sometime for hundreds of kilometres, exporting the problem to distant communities. Unless something drastic is done, and done soon, almost a million hectares of the most productive land in the world will be irretrievably doomed.

But what can be done? More efficient irrigation would help – the large-scale application of the methods my grandfather had worked out intuitively. The Israelis, always searching for better techniques, have improved efficiencies sixfold by using laser technology to get fields absolutely level, and by reusing surplus water from one crop on other, more salt-resistant crops. Israeli water engineers have taken "water-stress-management" of crops to a high art, and water only when absolutely necessary. Drip irrigation, once expensive and thus only useful for high-value crops, is becoming cheaper and more accessible, as we will see in Chapter 17: Solutions.

The other solution is expensive – to install master drains under agricultural valleys that have poor natural drainage of their own. These drains would take the unused water to a reverse-osmosis desalination plant somewhere downstream, where it would be scrubbed before being dumped back into the distribution system. By most estimates, though, drainage and desalination systems would cost several thousands of dollars more per hectare than the land is actually worth. So would most of the other solutions suggested by agricultural engineers, such as chemical filtration or detoxification by genetically engineered soil microbes.

The most notorious example of a conduit that was built with nowhere to go was the San Luis Drain, built to take waste water from the West-lands Water Project in the San Joaquin Valley to . . . well, to somewhere. Unfortunately, the somewhere turned out to be nowhere, and the water was instead dumped into a man-made lake called the Kesterton Reservoir. For a while, this seemed like a good solution. It even attracted millions of wildfowl, which began to use it as a stopover on their migration routes, and, to help the water authorities write off some of the cost of the drain, it came to be called the Kesterton National Wildlife Refuge. Alas, a decade after it first became a wild-bird reserve, the birds started dying or developing grotesque birth deformities. A furious fight erupted between conservationists and farmers over whether pesticides had anything to do with the kill-off. Biologists finally pinned the blame on selenium, a mineral not toxic in small quantities, which was naturally present in valley soils. Irrigation and poor drainage had, unknown and unseen, been concentrating the selenium into lethal doses.

~~~~~~~

My grandfather died and the farm was sold to some cattle barons from the Transvaal. They consolidated it into an agribusiness, running great herds of beef cows on land that was marginal for the purpose. After twenty years, they went broke and the land was abandoned. I went back to see it a decade ago, but I will not return. The old farmhouse where I had sat on the stoop with my Oupa, staring down the lane of blue gums into the imagined future, was now just a ruin. The stock sheds had crumbled. The corrugated iron tank was still there, but the windmill was broken. And the furrow that had taken the water to the strawberry beds had filled in. You could still see where it had been, a slight depression in the dusty soil, but the furrow itself had gone and, with it, the precious water. I squinted down where the berry patch had been and, for just a second, I thought I could smell the newly wet earth and see the ripe strawberries glistening from the moisture. But my imagination wasn't really up to the task and, after I watched a dust devil dancing in the particles beyond the broken barn, I turned and left.

9 | SHRINKING AQUIFERS

If the water mines ever run out, what then?

I grew up in an arid zone, the austere but beautiful landscape north of the Great Karoo in South Africa, on the approaches to what my ancestors called the Thirstland. The rivers there seldom flowed. For most of the year they were reduced to muddy puddles, and sometimes these, too, vanished. My grandfather was a farmer in this uncertain land, and water – the idea of water, the longing for water – consumed his life. With sweat, fanatic labour, and the help of teams of oxen, he and his workers built a number of mud dams, but sometimes years would go by with no rain whatsoever, and the dams would shrink and his livestock perish. When it did rain, it rained in cloudbursts, and he would patrol his dams gloomily, watching for faults that would tear the retaining walls apart.

Towards the end of his life, my grandfather brought in a drilling rig and sank a borehole 300 metres into the rock and shale of the substrata. At around 250 metres the drillers found an aquifer, and although I was just a child I still remember the cold, clear drops of water from the centre of the world mingling in the dusty earth of the farm with the old man's tears. To my grandfather, it was a miracle, a purely good thing, life extracted from rock.

In this modern, more highly populated, and less optimistic age, what he was doing is less well-regarded. It is called "water mining," which is the unsustainable withdrawal of water from an aquifer that is no longer being "recharged" or refilled. My grandfather operated in a sparsely populated part of the world, but if there had been enough people doing what he did, the aquifer would have begun to shrink, and eventually it would have run dry and all farming would have had to cease. This may seem like a small thing in a faraway place, but aquifer depletion is happening in many more difficult and important places in the world, including in many parts of the United States and in Libya, but also in Mexico, in the Middle East, in India, in a large swathe of the former Soviet Union, in northeastern China, and in Southern Europe.

For the first century of the modern irrigation age – from 1850 to 1950 – new water supplies came mainly from rivers. For the next fifty years, cheap diesel fuel and pumps and better well-drilling technology allowed farmers to sink millions of wells into subterranean water sources. In China alone, the number of irrigation wells went from 110,000 in 1961 to almost 2.5 million in the 1980s; in India, the land irrigated from wells shot up from 100,000 hectares in 1961 to 11.3 million hectares in 1985, a 113-fold increase. Since some 40 per cent of the global harvest comes from the 17 per cent of cropland that is irrigated, the fact that irrigation is going into hydrologic debt should worry planners everywhere.[1] Groundwater depletion is one of the great unseen but looming crises facing our planet, with all its implications for falling food supplies, human misery, famine, conflict, and war.

Late in 1998 the editor of *Worldwatch*, the monthly journal of the Worldwatch Institute, asked a critical question. Ed Ayres had been noticing – as had anyone who cared about water – the reports coming in from all over the globe about falling water tables, most disturbingly in major food-production areas of the world. Northeast Chinese water tables had been dropping by more than a metre a year. In the Punjab, the drop had been greater than that. The water table throughout the Mexico City region, notoriously, had dropped twenty metres in fifty years, causing large parts of the city to subside. In California's Central Valley there had been a similar though smaller subsidence. A quarter of U.S. irrigated cropland was experiencing drops – from twenty centimetres to more than one metre a year. Some irrigation wells ceased to draw water at three hundred metres, and are now down to almost a thousand metres. Everyone, Ayres noticed, was tracking the falling water levels, but no one seemed to be asking the critical question: If the top of the aquifers is dropping, where's the bottom? The answer, he found very often, was, We don't know. "Mostly, it seems, we find out where the bottom is when a large number of farmers' fields go dry," he said. "If the people taking out the water aren't anticipating when they're going to hit bottom, they're in for a staggering shock."

Ayres is professionally concerned with the environment, and he had a larger point to make:

> In our general distraction with the complications of managing even the here and now, we forget to focus on determining at what point that shock will come, and what we will do then. We know for example that you can't go on killing off the species in an ecosystem indefinitely; at some point [it] will fail like a building in which a key beam gives way and the whole system collapses. Yet we keep pushing the limits . . . we keep forgetting about that inevitable point of

shock where the water supply is abruptly gone – where the lake goes anoxic, the forest turns to desert, or the river turns to dust.[2]

The process of taking more water out from an aquifer than naturally returns to it is called "groundwater overdrafting." It is a modern phenomenon, depending on reliable pumps and cheap energy. Before mechanical pumps, societies might build dams and irrigation canals and aqueducts, but they were dependent on gravity for directing the water to its intended destination. Cheap oil changed all that. Among other things it made deep wells possible, and these deep wells exploit underground aquifers that have, in some cases, slowly accumulated for thousands of years. The limits are clear, both environmental and economic: either the water runs out or the aquifer is so deep even that cheap energy at some point becomes too expensive to pump the water to the surface. At that point, the pumps are turned off, and the users are forced to move or to find alternative sources of water. Sometimes this is not possible, and the farming just stops.[3]

There are other problems with falling water tables. They may cause rivers to dry up. And in many critical cases the withdrawal of too much water in coastal aquifers causes sea water to intrude, ruining the aquifer permanently for human use.

~

Like surface water (lakes and rivers), subsurface water, usually called groundwater, is constantly on the move. Over the years or decades or centuries it eventually finds its way into the channels of rivers and streams and then back to the sea. It, too, is replenished through rain. Somewhere around 10 to 20 per cent of rain finds its way into underground water systems; in return, groundwater contributes, on a global basis, about 30 per cent of total river runoff.

It has a stabilizing effect on rivers, minimizing the differences between wet and dry seasons. The movement is much slower than runoff on the surface, but it does move.

The process is well understood, though its measurement is still haphazard. Rainfall – at least that part of it that doesn't directly join free-flowing surface water or become absorbed by plants – seeps into the soil zone, the upper metre-or-so of earth. At first, there is a "zone of aeration," in which the small interstices between soil particles are filled with a mixture of water and air; if the precipitation continues, complete saturation of the soil zone occurs and the water continues to descend until, at some point, it merges into a zone of dense rock.[4]

Some rocks are too dense to allow penetration by water; these impermeable rocks are called aquicludes. Others are more porous and store considerable amounts of water. These rocks are called aquifers. The geology can get quite complicated. Aquifers can sometimes lie underneath layers of aquicludes, and are then called confined aquifers. In other cases, called unconfined aquifers, there is nothing around the aquifer but unsaturated and permeable material, such as gravel, shale, or sand. The boundary between the unsaturated stuff and the water-bearing rock is called the water table. In water-rich regions like Canada, Northern Europe, and parts of the United States, the water table might be relatively close to the surface. Depending on the subsurface geology, the aquifer underlying the water table might be several hundred metres deep, in others only a few metres. In certain arid zones, where the aquifers are either not being replenished or being replenished very slowly, the water table can be hundreds of metres below the surface. Where no permeable rocks exist, or where they exist, but climate changes have allowed no recharge, there is no water table at all. In other cases, water rises without pumping; these "springs" happen in confined aquifers where the hydraulic potentials are greater than the energy required to bring the water to the surface, and where there are small natural fissures. A well drilled into one of these pressurized aquifers will cause the water to gush to the surface, and it will continue to gush until the pressure has been equalized. Some of these "artesian wells" have been lifting water to the surface for decades. Eventually, though, the pressure will equalize, the potentials will decline, and the well owner will have to resort to pumping.[5]

Groundwater is tracked by remote sensing and tracer techniques, but its movement is exceedingly difficult to follow. It is known that groundwater migrates slowly, sometimes only a few millimetres a day, though occasionally a few metres. Near the water table, the average cycling time of water may be a year or less, while in deep aquifers it may be as long as thousands of years. It's easier to measure water tables. Through test wells and controlled pumping, it's not difficult to measure the recharge rate and flow behaviour around a particular site. The difficulty comes in sensing movement in the aquifer as a whole. Water can be stored in the pores of rocks, but it can also be stored in cracks and fractures, and sometimes these fissures can cause water to travel quickly and great distances. Sometimes these "preferred pathways" become subterranean streams, or even rivers, which can help swell surface rivers dramatically during heavy rains. That is why aquifers are almost impossible to clean up after they have become polluted, and why pollutants, leaching through the soils from waste dumps, landfills, and agriculture, can occasionally be found alarmingly long distances from the polluting source.

No one knows how many unexplored and unexploited aquifers exist, but the amount of water stored in them is thought to be considerable. Water exploration companies have not yet made a major impact on the stock markets of developed countries, but they wouldn't be a bad bet for adventurous investors.

The slow replenishment of groundwater is called "groundwater recharge." It happens, obviously enough, mostly during the rainy season, or during the winter in temperate climates.

When the water table is deep underground, the water of an aquifer may be exceedingly old, sometimes left over from a previous climatic epoch. The best-known example is the "Nubian sand-stone aquifer," which underlies what is now the Sahara Desert, but which was even in historical times (maybe eight millennia ago) verdant grasslands and occasional swamps – fish and hippopotamus fossils have been found in many parts of the Sahara. Radioisotope dating techniques have shown that this water is many thousands of years old. This ancient aquifer is now being exploited by Libya's Colonel Muammer Gaddafi in the planet's most grandiose episode of water mining.

A small canvas bag of water, sweating in the heat, hangs from a hook in the cab of a huge truck, swaying as the vehicle lurches across the desert, its great tracks as wide as a house, scraping across the gritty stone and drifting sand. From the cabin high over the sand, the driver and his companion see nothing but more sand and scrub, and great gouges in the desert floor where the earth movers have left their traces. If the driver were to look up from his treach-erous road, he would still see nothing – nothing but the same sand and stone, stretching to the horizon. And if he drove to that horizon, there would still be nothing – nothing but the Great Emptiness for a thousand kilometres of the greatest desert on earth, stretching all the way down to Chad in the south and, far to the southwest, Agadez in Niger and the fading Saharan termini of Gao and Timbuktu.

The small canvas bag is cool to the touch, damp. The driver, Ahmed, wipes his hand across his forehead, leaving beads of mois-ture. He finds the bag reassuring. It holds five litres of fresh water. In fact, it was trucked in from Tripoli, hundreds of kilometres to the north, but it could just as easily come from the desert oases of Tazirbu, Sarir, or Al Khufrah. Just off the road near Tazirbu is a small well that has been known to travellers for centuries, the

Birbu Atla, not much more than a metre deep, the clear water mesmerizing to a thirsty passerby. It bubbles as you look into it, little bubbles of air coming up from . . . where? Traveller's tales often speculated, though no one knew for sure. In the past, camel caravans took eighty days to plod their way from Timbuktu to Marrakech on the western route, or from Cairo to Gao on the eastern, threading through a scanty network of oases, following the veins of water.

Ahmed is only sixty kilometres from the Birbu Atla now. He manoeuvres his huge vehicle – a crane, really, a mobile crane capable of lifting up to two hundred metric tons – until it is parallel to the gouge in the ochre earth that runs arrow-straight to the north, vanishing up the road to the coast, so many hundreds of kilometres away. On his back – Ahmed thinks of the cradle of his immense carrier as his own flesh and blood – is a section of pipe. He lowers it gently into the waiting trench, and another machine, made just for this purpose, nudges it into place with a solid thud.

The section of pipe Ahmed is locating is seven metres long and four in diameter. He knows something about this pipe. They all do, for it looms large in their lives, as it has done for the four years since they first began laying it. There is no other pipe in the world like it. It is bigger, and better engineered, using the most modern of manufacturing techniques imported from the West and adapted on-site by the swaggering South Korean contractors. A whole factory had been built just to make this pipe, with its alternating layers of steel and concrete. When the sections were laid together, there would be no leakage and hardly any sweating; it wouldn't even feel damp to the touch. That was as well, for what it would transport was more important than anything, more important than oil, although oil was paying for it in the end. This pipeline would be one tributary of what Colonel Gaddafi was calling, with no hyperbole at all, the Great Man-Made River. What it would transport was, in this dry country, the most precious thing of all: water.

The pipeline was to take water from the deep desert and carry it to the populated coast. The Sahara was giving up its last secret, something it had kept hidden for ten thousand years. Down there, deep down, was water – lots of it.

~

The Sahara is the greatest desert on earth, the apotheosis of dryness, but even on or close to the surface it is not entirely without water. Several major rivers originate outside its boundaries, flowing through or disappearing into the desert itself, affecting in various ways its surface water and groundwater. Some of these rivers rise in the tropical highlands well to the south. The Nile, of course, is one of them, meandering out of Uganda and Ethiopia and flowing along the desert's eastern border to the Mediterranean. Several rivers flow into Lake Chad, including the Chari. The lake is shrinking now, but gives up its water in two ways – through evaporation and by seeping northeastward to recharge underlying aquifers. The Niger River rises in the sodden hills of Guinea but traverses the southern end of the Endless Desert before turning south to the sea. To the north, the Saoura and the Drâa flow from the Atlas Mountains, as do many other streams and occasional streams, or wadis. Within the Great Desert are shallow, seasonally inundated basins called *chotts*, or *dayas*, and many other wadis, most of them remnants of systems that existed in the distant past, when the Sahara was still verdant, but some still occasionally flow in the rare desert rainstorms. Tamanrassat, Algeria, was destroyed by a flood in 1922, to the terror of its inhabitants and the bewilderment of the geographers. There is a network of wadis and remnant pools in the Tibesti and Ahaggar mountains; even the great sand dunes of the desert store water, sometimes in considerable amounts, and seeps have been known in desert escarpments. Saharan aquifers have been recharged in historical times; in the time of Europe's "little ice age" (from the sixteenth to the eighteenth centuries), there was heavy rainfall along the southern and northern fringes of the Sahara, and very likely in the desert itself.[6]

But the greatest "pools" of water in the Sahara lie far underground, fossil water from when the desert was last verdant, some ten thousand years ago. This water was first discovered in the 1970s, in the Al Khufrah region of Libya, far away in the southeast quadrant, where it converges with Egypt, Sudan, and Chad. It was an accident – the prospectors were looking for oil, not water. The massive oil fields in the Sirte (Surt) Desert to the northwest were good for generations to come, it was believed, though it never did any harm to find some more. And they found a gusher all right, but it was clear, clean water instead of black gold. A few more wells were test-drilled in the region, and they all told the same story: underlying the desert, in the sandstone caverns and underground canyons, was an immense pool, which came to be called the Nubian Aquifer.

A few years later more water was found, this time to the west of the country in the Marzüg Basin. It was, the geologists reported, another massive aquifer, probably unrelated to the first.

These discoveries came at an opportune time. Libya was parched. It was, after all, 93 per cent desert, and its renewable water resources were less than two hundred cubic metres per person per year, about one-fifth of what the United Nations considers water-stressed, and below what all hydrologists think of as a water-critical state. The population was concentrated in two coastal regions, the Gefara Plains around Tripoli and the northeastern Benghazi Plains. In the middle, sparsely populated, are the oil fields; water shortages there have always imposed boundaries around possible development.

The growing population along the coast, and the paucity of rivers, meant that groundwater aquifers were used for drinking water and sanitation. Even the Libyans, notoriously reluctant to admit anything other than Triumphs of Socialist Construction, have confessed to severe overdrafting of their coastal aquifers. A report by Saad Al-Ghariani, of the Department of Water Sciences at Al-Fateh University in Tripoli, found that these aquifers had been exposed to "unacceptable levels of piezometric declines and sea water intrusions with disastrous environmental and socioeconomic impacts" – pretty straightforward talk for bureaucratese, a frank admission that many of the coastal wells had turned saline and become unusable. New water, then, was critically necessary.

The Libyans knew there were only three options: radically expand the desalination of sea water along the coast; set up economic and agricultural zones in the newly discovered groundwater basins in the deep desert; or take the desert water to where it was most needed. The latter was not, they knew, unfeasible. Hadn't the Americans already built massive interbasin water transfers, what with their California State Water Project and the Colorado River diversions? It went against the grain to admit that the Americans had done anything worthwhile, but these were, after all, useful precedents. And hadn't Turkey just proposed its Peace Pipeline, which would take Euphrates water all the way to the Gulf States, refilling the Jordan River and watering the deserts of Iraq, Syria, Jordan, and Saudi Arabia as it went? Of course the American transfers and the proposed Turkish pipeline were different in one significant way: they involved re-engineering rivers, and not the one-time mining of an unrenewable aquifer. But needs must as needs do.

To outsiders, the Great Man-Made River seemed mad, a product of senseless ambition combined with endless flows of money. To build the largest civil-engineering project on the planet, to spend $32 billion to take water from the desert to the coast, to create new agricultural zones, and to allow the population to increase and industry to establish itself as a result of ready access to water – to build all this, knowing that in thirty years, or forty, or perhaps fifty, the water would inevitably run out . . . it seemed insane.

But what were the options, where the alternatives? A British firm hired to do the analysis found, to its own evident surprise, that the Great Man-Made River was the most cost-efficient of the three possibilities. The average unit cost of transferred water, their report said, would be about twenty-five cents per cubic metre. Not cheap – but best estimates for sea water desalination still came out somewhere between $3 and $5 per cubic metre, even using considerable amounts of cheap Libyan fossil fuels to do it. The third option, moving the people and the industries, made no sense. There are no seaports in the desert. And when the water ran out . . . they'd be stranded.

Of course, that applied to the Great Man-Made River too. When the water ran out, as it must – what then?

How much water is there? Geologists estimate that in the Khufrah Basin alone there are something like half a million cubic kilometres of usable water. Based on a transfer of some forty billion cubic metres a year, the local water table would drop about a metre a year, which would mean it would last for a minimum of fifty years. Then, of course, it would be gone.

Nevertheless, by a series of interesting logical leaps, the Libyans consider this a sustainable project. How? As Al-Ghariani put it,

the question of sustainability . . . depends on water production costs and management skills rather than available water supplies, which are apparently sustainable for hundreds of years even in the absence of natural recharge of the aquifers. Sustainability can be assured if the transferred water is utilized in such a way as to provide the national economy with the means and strength that enable it to develop alternative water supplies when the GMMR sources become uneconomical to pump, or are exhausted altogether.

WATER

There were five "stages" to this immense project. After a decade
of construction, the South Korean contractors finished Stage
One of the project at a cost of less than $4 billion – under budget
and ahead of schedule. It was a 1,900-kilometre waterway, 4
metres in diameter, connecting a network of pipes and ditches to
two wellfields consisting of 234 new wells drilled into the aquifer.
A river indeed: it could carry 2 million tons of water a day, each
wellfield contributing some 350 million cubic metres of water to
the flow each year. Stage Two, finished by late 1999, was to construct an even bigger wellfield to tap into the western aquifer at
Marzüg. This one consists of 484 wells, and two 4-metre pipelines
to take the water north. One of these lines carries up to 700
million cubic metres a year to the coast and the farms of Gefara,
mostly to stimulate growth there, but also to redress the environmental balance, restoring water tables and improving water quality.
The other line is to transfer 175 million cubic metres a year to the
communities along the northwestern mountain range.

Three more stages are planned before the project can be considered finished. By 2001 Stage Three had already developed a
further wellfield in the Khufrah oasis region, connected to Stage
One's pipeline, adding another 560 million cubic metres a year to
its capacity. Stage Four will connect the eastern branches to the
western branches and take a further 350 million cubic metres to
the Gefara plain. The final stage, Stage Five, is to extend the
eastern branch of Stage One to the city of Tobruk. More than
four-fifths of the new water will irrigate new agricultural zones
deep in what used to be desert.

There have been some grumblings from all three of Libya's desert
neighbours about what the colonel is up to. The presumption of
most hydrologists is that the aquifers being mined spill across
national boundaries, if these boundaries were extended vertically

downward. But there is no proof: only if the water tables in Egypt, Chad, and Algeria begin to sink, and their aquifers become depleted, will evidence be to hand, and then it will be too late. So far, pumping tests and simulation models show no lateral flow from other countries, but all hydrologists acknowledge that it is too early to tell. Official Libyan policy is that groundwater aquifers, unlike naturally running surface water, should be considered "as any other natural resources of vertical utility," such as oil or minerals. And, as Al-Ghariani pointed out in his paper, nations cannot really consider, at this stage of international law, whether "migrating resources" like aquifers harm other nations using the same resource. Otherwise, he says, "the world would be overwhelmed by geopolitical conflicts that may arise among cohabitant nations as a result of induced pollution, whether due to circulation, geomorphological changes and migratory herds, birds, or humans." In any case, Algeria has its own vast sub-Saharan aquifer, under the plateau that slopes upwards to the Ahaggar Mountains. Is that "theirs"? And if not, why not?

There was also, a few years ago, a brief flurry of paranoia among Western security buffs that the $34 billion wasn't really being spent on water tunnels at all, but that the "pipelines" would be used, instead, to hide and transport troops. This curious conclusion left unstated, for obvious reasons, why Colonel Gaddafi would want to send an army haring off into the middle of the Sahara. Later, a more sophisticated version of this notion surfaced: some of the subsurface "reservoirs" along the coast were not really water-distribution plants at all, but deep cover for chemical warfare factories, built with Iraqi assistance. Adduced as evidence was an Iraqi passport that had been "clearly seen" in a government building in Tripoli. Despite the paranoia, of course, it might even be true, but meanwhile the water is flowing, and when the Libyans are asked what they will do when the water runs out, they speak of the American experience. The Americans, they point out, are mining their own aquifers in an unsustainable way. They've known for years that the Ogallala Aquifer under the High Plains states is being seriously overdrafted. What will they do when their water runs out?

It's a good question, and it needs a little backing up to answer properly.

The Ogallala Aquifer lies deep in the shale and gravel beneath some 580,000 square kilometres of the Great Plains region of the United States, and particularly the portion known as the High Plains, ranging from West Texas, Oklahoma, and New Mexico in the south, through Kansas, part of Colorado, and Nebraska, and tailing off in South Dakota. The plains have always been a place of extremes – bitterly cold in winter, with great driving blizzards, and torridly hot in summer. They are difficult and treacherous to farm, for, while the High Plains were grasslands and not part of the Great American Desert proper, they had poor soils and only meagre and erratic rainfall.

Some two thousand years ago the Woodland Indians began moving west, planting settlements along the eastern edge of the plains. But it wasn't until the bow and arrow arrived from the north that serious hunting began. When the Spanish romantic and explorer Francisco Vásquez de Coronado passed through in the mid-sixteenth century, the plains were grasslands, home to millions of bison, wolves, and grizzly bears. The Great Plains cultures of the Cheyenne, Sioux, Comanche, and Apache came into being only after the Spanish brought European horses from the southwest in the 1700s. The first permanent white settlers brought cattle, which meant the end of this nomadic native culture. For a few short decades, during the flowering of the mythical Wild West and the Lonesome Cowboy, there were great cattle drives from West Texas to the railheads in Kansas, but the ecological consequences were not nearly so romantic: overgrazing, soil depletion, the beginning of desertification marked by invasions of mesquite and arid-land weeds, and prolonged droughts. When they went broke, the cattle barons made their way to California or back east. Their ranches were broken up, and only a few hardscrabble farmers remained, stubborn, tenacious, and poor.

That period ended during the First World War, when a series of wet summers and a booming economy attracted more prosperous farmers. Within a few years, the plains were a sea of wheat. But a few years after that . . .

> The crop has failed again, the wind and sun
> Dried out the stubble first, then one by one
> The strips of summer fallow, seared with heat,
> Crunched, like old fallen leaves, our lovely wheat,
> The garden is a dreary blighted waste,
> The very air is gritty to my taste . . .[7]
> ("A Farmer's Wife," Edna Jaques)

The farmers had come to the High Plains from the high-rainfall areas of the east. They brought with them eastern techniques, the worst possible for farming in the west, including the shallow ploughing of marginal land, the destruction of ground cover and windbreaks, and an ignorance of what wind would do to sandy soil when it had nothing to anchor it.

In the 1930s, the rains failed. Or, rather, the rains became normal, for over the centuries they had often failed and the landscape had adapted. But the farmers could not adapt. When farm prices collapsed early in the decade and the drought persisted, the unanchored topsoil began to move. Blowing topsoil drifted across roads and railroad tracks, keeping the towns and cities bathed in dust and grit inside and out, causing yellow twilights at midday. The roads were impassable, with deep drifts of sand that built up until they covered the fences, choked out the few remaining shelterbelts and gardens, and reached the roofs of chicken houses. On

May 12, 1934, the Associated Press reported huge clouds from the Great Plains Dust Bowl at ten thousand feet over the Atlantic, and amateur statisticians began calculating the amount of arable soil that had been removed in this storm alone, which reached more than three hundred million tons before the exercise became pointless. The Black Blizzard, the name ironically foreshadowing the laments of Uzbek peasants a decade or two later (who called it Black Snow), swept from the Rocky Mountain states to Washington, D.C., and New York, and deep into the thoughts of Congress.[8]

When the Dirty Thirties were over, a few stubborn farmers remained. But the weather continued poorly, the droughts stubbornly extending themselves. The farmers survived by sinking boreholes deep into the earth. A windmill could keep a household going, a few head of cattle, and a few hectares of land. A major study of the area, the Great Plains Report of 1936, blamed poor land practices for the disaster. It recommended that the most vulnerable areas be taken out of cultivation and put into rangeland for livestock. The Civilian Conservation Corps, at the report's urging, planted millions of shelter-belt trees and shrubs to break up the scouring wind and help anchor the soil.

Then came the invention of the centrifugal pump, along with cheap power from Texas oil wells and the hydroelectric plants at Hoover Dam. Suddenly, pumps that had brought up a few litres a minute were replaced by devices that brought up thousands, easily enough to irrigate fifty hectares of cropland. Hydrologists soon confirmed what the farmers already suspected: that underneath the plains was a massive aquifer, the size of one of the smaller Great Lakes, that had been lying there, steadily expanding and accumulating its precious water, unmolested, since the last ice age. Its area is around 453,000 square kilometres, and (at least prior to its heavy exploitation) held as much water as the annual flow of two hundred Colorado Rivers.

Farmers who once feared bankruptcy faced it no longer. The crops would no longer fail – irrigation water would see to that.

There was no thought then for how long the water would last. A Traveler's Insurance study of the High Plains in 1958–59 concluded that the future for irrigators was bright; there was no mention that water mining must have a finite end. But the numbers speak for themselves: at the end of the Second World War there were only a few thousand irrigated hectares in the whole basin; by the 1980s, there were more than 7 million. A federal study of the same area, commissioned with great fanfare and a budget of many millions of dollars, was undertaken in mid-1982. It introduced the first note of foreboding in the hitherto sunny vision of irrigation's future. By 1914, the report said, there were only 100 irrigation wells in West Texas. In 1937 there were more than 1,000. When the report was written, there were 74,000. While understanding the need for drawing on the aquifer ("the dependence on ground water introduces an element of control by society over water resources and their use"), the report noted that the "drawdown" was excessive and that the aquifer as a whole was "in serious overdraft." Almost 14 million acre-feet a year were being taken out and not replaced – as much flow as the Colorado River in a good year. Falling water tables and higher fuel costs have forced many farmers to abandon irrigation already, and while 5.2 million hectares were irrigated by the aquifer in 1978, a decade later the number had dropped to 4.2 million. In 2002, Ogallala water still accounted for one-fifth of the total irrigated land in the United States, but best estimates were that another 40 per cent, or 1.2 million hectares, would have to be withdrawn from irrigation by 2020.

So, when? When will the Ogallala give in? The 1982 report, grudgingly admitting the probability of catastrophe, predicted it wouldn't happen until that magic number of 2020, which seemed a decent interval away. More recent studies think this projection was overly optimistic. The real answer is complicated by geology and politics. The aquifer is not uniformly thick, and the rates of drawdown vary; also, the aquifer as a whole is too dispersed and fractured to act as a single pool of water would.

Not all states, or even regions within states, are facing the same degree of crisis: Texas has already decommissioned nearly a million hectares, one-third of its total irrigable land, and west Texas is almost in panic mode, but other parts of the aquifer are falling much more slowly, and Nebraska, to the northern end of the aquifer, has a positive recharge rate and is not much concerned. The varying levels of concern affect local politics too. Members of Congress from the area want to raise the national consciousness of the issue but retain control in local hands; and local people prefer local responses to local changes in the water table, and generally prefer conservation to bans on irrigation.

The conflicts have reached down to individual householders, and the puzzling facts of ecological life have been upsetting even the well-meaning.

For example, the magazine *Natural Home*, in its May/June issue of 2000, carried an enthusiastic piece on rooftop catchment systems for people living in the arid regions of the American southwest. "Right as Rain," the piece was entitled. "As more people discover the benefits of collecting their own rainwater, catchment systems are catching on."

Alas, it isn't so simple. Only a year later, more than a dozen water authorities in Texas and Arizona issued cease-and-desist edicts against householders who had taken *Natural Home*'s conservationist advice. These hitherto benign devices – designed to employ runoff from roofs for household use – were suddenly anathema. The reason for the new prohibition? Streams and aquifers were being depleted at a faster rate that previously thought, and demand was growing: any action, therefore, that prevented water returning to the aquifers was now to be regarded as anti-social.

A pattern of unusually dry weather since 1992 accelerated the depletion rate in the aquifer, only temporarily derailed by the El Niño of 1998. The average annual drop in the water table was about fifteen centimetres through the 1980s and into 1991, but thereafter it began to accelerate. In 1994 it was sixty centimetres, in 1995 almost a metre. The manager of the Ogallala Underground Water Conservation District, Wayne Wyatt, was not optimistic in 1998. Drought increases irrigation and decreases aquifer recharge, he pointed out. "It's a pretty serious change," he said. "It's impossible to gauge exactly when the aquifer will run out, but the current lack of rain will hasten that day."

More likely, the pumps will never suck air. The cost of raising the water the extra metres required will have driven many marginal farms out of existence and imperiled others before that happens. But there's another problem. Most of the farmers have already cut down the shelter belts planted after the Depression – they got in the way of the mobile irrigation sprinklers – and have been shallow-cultivating marginal soils again. It's a perfect recipe for another Dust Bowl when the water does run out.

If there is not to be a large-scale collapse – if three hundred thousand people are not going to be destitute, and a substantial American industry consigned to the trash can of history – something will have to be done. But what?

A report issued in 1989 called *Forecasting by Analogy: Societal Responses to Regional Climatic Change*, edited by Michael Glantz, put it tactfully: "Policies developed in response to depletion of the aquifer that may be technically and economically feasible must also be politically and socially acceptable. For example, the large-scale inter basin water transfers to this region from the Great Lakes or some other river basin may be sound technical projects, but they face considerable social and political opposition." To put it mildly.

If there is no water in the aquifer, and six states and a dozen cities depend on it, where is the water to come from? The Colorado and its tributaries are already spoken for. There was some notion that "surplus" water from the Mississippi Gulf area could somehow be lifted 1,000 metres and transported 1,500 kilometres to the High Plains, and it was seriously enough mooted that Texans were asked to consider it in a referendum. This mad notion would have taken the Mississippi water across 4 major rivers and 150 minor ones, and the energy required to lift it the required distance would have consumed the entire output of 12 medium-sized nuclear-power plants. No one any longer believes that the untold billions this plan would cost to rescue half a million farmers and some minor-league cities was politically or even economically feasible, but that didn't stop the water planners floating, if that's the right word, an even more hare-brained scheme – to redirect Canada's Arctic-bound rivers southward into the Great Lakes, and, essentially, re-engineer the entire North American continent. No one thinks that feasible either. Perhaps the Canadians could be bullied into allowing it – though it's doubtful – but who would pay the billions it would cost, and who would benefit?[9]

That the water of the Ogallala will, at some point, simply run out is a given. Like all mines, it will exhaust its resource. The only real answer is rigorous conservation and the virtual abandonment of agriculture by the six states.

Meanwhile, farmers in Texas were bringing some of that famous American ingenuity to bear on the problem of water survival. If there was no more water to be had, or if the flow had to be reduced to sustainable levels, well then, they would have to use less, wouldn't they?

The High Plains Underground Water Conservation District in Lubbock, Texas, has put together "a package of technologies and management options that has boosted the region's water productivity . . . the effort has involved a major upgrade of the region's irrigation systems."[10] Initiatives include a shift away from furrow irrigation, which has typical efficiencies of around 60 per cent, to surge-pulse irrigation, which raise efficiency to around 80 per cent. Low-energy precision-application spraying, which delivers water in small nozzles just above the surface, nearly eliminates evaporation and wind drift, and can raise efficiency to 95 per cent – often cutting water use from 15 to 40 per cent over other methods. More recently the district has begun experimenting with drip irrigation for thirsty crops. Overall, the High Plains program has allowed farmers to boost productivity by nearly 75 per cent, especially for crops like cotton.

But even here, not everything is going well. The U.S. government, which has historically loved shovelling pork at farmers, still subsidizes the wasteful use of water; farmers in the Ogallala area get a break on their federal income taxes for the cost of diesel, and thus depleting the aquifer, an allowance similar to those received by oil companies depleting their reserves. In practice, the money saved goes not into saving water or into better technologies, but into planting thirstier crops.

No state in the Union has a policy governing or controlling private access to groundwater, except Arizona. Arizona's comprehensive groundwater law has mandated sustainability – a balancing of recharge and withdrawals – by 2025, which sounds good in theory, except that the preferred method of getting there is not conservation but the importation of more Colorado River water through the Central Arizona Project, a federally subsidized canal system. The state is thereby, as Sandra Postel points out, merely replacing one type of excessive water use with another.

All over the West – all over America – water overdrafting continues, perhaps not on the scale and at the rate of the Ogallala, but fast enough. Long Island, for example, gets its water from a closed-basin aquifer rapidly being depleted and poisoned by industrial runoff. And the water laws of the United States, which seemed so sensible in pioneer times, now only make things worse.

In the 1980s, Canadian businessman Maurice Strong became embroiled in a water dispute in Colorado. He emerged from it with his honour intact and his business almost so, but for a time he was portrayed in the local press in the San Luis Valley as a blood-(or rather water-) sucking Canadian, intent on siphoning the valley's precious resource and exporting it to Canada – a bizarre reversal of the usual Canadian paranoia about Americans.

Strong, through a complicated series of stock swaps and deals, became a principal owner of the Baca Grande Ranch, one of the historic ranches of Colorado, dating from a grant by the king of Spain to Luís María Cabeza de Baca, who had accompanied Coronado on his expeditions to what was then Spanish America. The ranch dominates the northwest end of the San Luis Valley, including the peak of the highest mountain, Mount Kit Carson, and a corner of the sand dunes, an extraordinarily picturesque piece of desert nestled at the base of the peaks that has been made into a national park. During the gold-rush days of the 1800s, the area had seen one of Colorado's most active and controversial mining booms. The village of Crestone, which now has fewer than a hundred inhabitants, was a thriving and boisterous mining town, and the Baca Grande Ranch adjoining it attracted a small army of prospectors panning for gold in the streams and digging along the front of the mountains. One particularly rich lode became the Independence Mine, for a time one of Colorado's most prolific producers. Water, says Strong ruefully, is an issue which, if any-thing, arouses even stronger passions than gold.

Underneath the valley is a massive aquifer – massive enough that some have suggested it might rival the Ogallala itself in volume. It is smaller in area, but deeper, possibly up to nine thou-sand metres deep, though the volume of water it contains, and its exact structure, is still unknown. There is thought to be a smaller aquifer overlying a much larger one, and early assessments of the amount of water underlying the valley have been revised radically upwards. The U.S. government, in the form of the Bureau of Reclamation, wanted to export the valley's groundwater, to ease the demand on the Colorado River. Strong's company, too, wanted to sell the water to the Californians.

The laws governing water use in the United States are a unique product of the early struggles over water rights. One basic principle that actively encourages waste is the "use it or lose it" rule, under which the owner of water rights will lose them if the water is not put to constructive use. Strong disagreed with his own company's proposals. He wanted to keep the water in the valley and suggested a variety of "use it" proposals, including irrigation and a local brewery, before he had a falling-out with his partners. He withdrew his cash, and shortly after that the Water Court, and subsequently the Appeals Court, rejected the company's application and it had to be liquidated at a loss of well over $20 million to the remaining partners.

~

Water mining is occurring in many more places than the United States and Libya.

Saudi Arabia is not just marginal desert, like the American West, the Aral Basin, or South Africa's Karoo. The Rubʿ al-Khali, the Empty Quarter, is the real thing – 640,000 square kilometres (bigger than France and the Low Countries) of hard-core desert, containing steep dunes more than 240 metres tall and reaching temperatures in the high forties in the summer sun. Yet the Saudis – and the Omanis to the east, on the Persian Gulf – were briefly, in the early 1990s, actually exporting wheat and fruit.

There is some surface water and a few streams around the periphery of the Arabian Peninsula, particularly close to the old capitals of the Yemeni Empire in the south. But even in Oman, where the government has recruited Western engineers to reinvigorate an ancient system of *qanats* (locally called *falaj*) that siphons water from a subterranean aquifer, most of the water comes from groundwater.

The Bedouin nomads of the deep desert needed little water, nothing more than the occasional oasis was enough. They lived off animal husbandry, some agriculture, and warfare. It's not surprising that the Arabs of the Empty Quarter were the models for the Fremen, the fierce desert-dwellers of Frank Herbert's novels about the planet Dune. Once settled in the cities, they became masters of conservation. Some of their sophisticated catchment and distribution systems can still be seen in Yemeni cities, though they probably reached their apotheosis further north, in the Jordanian rock city of Petra, where catchment basins hewed from bedrock distributed water in an intricate system of stone channels.

But the modern Desert Kingdom, awash in oil money, decided in the mid-1970s to become self-sufficient in food. The result has been what you would expect if a state throws serious money at a problem: the production of food doubled in the decade before 1982, quadrupled in the decade after that, and, by the mid-1990s the Saudis were exporting dates, dairy products, eggs, fish, poultry, vegetables, fruit, and flowers worth half a billion dollars. The most dramatic change, and the most surreal, has been to transform the country into a major exporter of wheat. It wasn't until 1978 that the first grain silos were built; six years later the country was self-sufficient in wheat, and in 1992 it exported 4.2 million tons – all of it bought by King Fahd, at a price four times the open-market price, and then resold at a loss.

At Jizan in the southwest (admittedly the best-watered place in the country), an experimental station is producing watermelons, pineapples, pawpaws, bananas, and mangos. Almost all of it was being grown with fossil water, mined on an unprecedented scale and at a dizzying rate. Water demand is increasing steadily: it is now 1.5 billion cubic metres a year, and will rise another half-billion or so in the next few years.

What surface water exists is being meticulously harvested, most of it from seasonal floods. There are now more than two hundred dams with a cumulative storage capacity of more than half a billion cubic metres. Saudi Arabia is the world's pre-eminent desalter of water; Saudi government figures show that desalination capacity will reach 2.5 billion litres a day by 2010, more than enough to supply the country's drinking needs.

The fossil aquifers that supply almost all the country's water lie deep in the rock, three hundred to five hundred metres underground. The thousands of wells drilled into the sedimentary rock have shown no signs of exhaustion. But groundwater depletion has been averaging more than five billion cubic metres a year, and at that rate the water will run out altogether in less than fifty years. Muhammad H. al-Qunaibet, a hydrologist and government adviser, estimates that only a third of what the country uses for agriculture is replaced through rainfall. The rest simply disappears. How much is left? No one knows, but probably half the total reserve has disappeared by now.[11] Meanwhile, research on desalination goes on apace. And if they have to, they can switch agricultural water back to the cities, and the country can go back to importing food – if they can find any to import.

And more examples: Indonesia has so depleted its underground aquifers that sea water has seeped 15 kilometres inland; although the country has no money, pipelines to bring the needed water to the cities are expected to cost more than $1 billion. In Mexico City, depletion of the aquifers has caused a dramatic 4-metre collapse in ground levels; the Mexicans have been forced to find new water supplies 250 kilometres away and several hundred metres lower, building an expensive new pipeline to pump it uphill to the city.

～

I have asked myself many times what my grandfather would have said if he had known about Colonel Gaddafi, about the desert cities tapping into the Ogallala, about the whole notion of unsustainability. He was a large man, nearly 140 kilos when he died, and I remember him best sitting on the stoop of his little farmhouse, drinking large mugs of strong black coffee, staring down the row of blue gum eucalyptus trees into the dusty, arid distance. I loved him fiercely, as only a grandchild can, and was never happier than when I was trotting at his side, visiting his small and usually empty dams, his sheep-shearing sheds, the pens where the milk cows were kept. Afterwards I would sit there on the stoop with him and he would let me sip from his mug, and together we would watch the distant dust clouds march across the horizon, as a neighbour's cart passed down the sandy rural roads. My grandfather was a man who understood the soil and knew how natural systems worked, and he probably had an intuitive understanding of how that water had been secreted away in the interstices of the subterranean rocks, so many centuries before. He would have considered it a gift of God, I think, to be used wisely and not squandered, but in any case used as you use the rain that is also given by God, to produce things as farmers must. He never knew of its frightening finiteness. And thus are ecosystems ever changed, more by the cumulative deeds of good men doing what they believe than by the rapacious actions of Green demonology.

My grandfather had enough troubles in his life. I'm glad he never had to wrestle with this one.

But . . . is "water mining" always a bad thing? There are those who think otherwise, among them the Spanish hydrologist Ramón Llamas.

In a paper presented to a UNESCO conference in Tripoli in 1999, he pointed out that groundwater use has substantially increased over the past half-century, mainly because it has certain inherent advantages over dammed water: it does not suffer from evaporation losses, it does not flood usable land, it is more reliable in its flows than rivers, and it is, generally, cleaner than surface water, being inherently more difficult to pollute. Irrigated use of groundwater is often more efficient than using surface water, he pointed out. An assessment in south Spain (an area with an irrigated surface of almost one million hectares) demonstrated that a cubic metre of water in groundwater irrigation produced five times more revenue and three-and-a-half times more jobs than the average cubic metre applied in surface-water irrigation. "If a similar situation exists in a good number of other arid or semi-arid countries, it will be necessary to change most of the [commonly accepted data] on future water crises, needs, and solutions," he claimed.

He criticized his colleagues for so often concentrating on the negatives of groundwater use, instead of examining its possibilities. A commonly accepted number that would make global withdrawals unsustainable is two hundred cubic kilometres of stored water. Llamas believes the actual withdrawals total less than a third of this. If this is true, hydrologists would have to radically revise their assessments of the imminence of a water crisis, something Llamas is not shy about pointing out.

In a roundtable discussion in Montreal in 2001, Llamas went even further, maintaining that, "in many regions, groundwater development can probably solve most of the so-called water crisis issues."

His views, expressed on many occasions at water conferences, acknowledge that water tables are dropping in many places. "But we know nothing of the cycles. Maybe replenishment cycles are much longer than we assume, and therefore these overdrafts we're seeing now are just temporary. It is what I call a hydromyth. A hydromyth leads to the notion that groundwaters are fragile, and shouldn't be developed. It's not always true that a falling groundwater table is a sign of something wrong. A steady state can take a hundred years to arrive at.

"And take land subsidences. Extracting water does lead to ground collapses, it's true. But is this the problem it has been assumed to be? Mexico City has dropped fifteen metres in places, which has damaged some buildings, but is it really the doom we have been told? In the Central Valley in California the land has subsided ten metres in fifty years, without catastrophe – people have just adjusted. The real hydromyth is that all anthropogenic actions are bad."

I first heard him making his case to an audience at a UNESCO conference in Paris, and was received in stony silence. This didn't faze him at all.

"Look, the Saudis are using twenty cubic kilometres a year more than the replenishment rate. But why is it right to tell them not to use it now, just because they might run out in two hundred years? Maybe we will have other solutions by then. And why should the present generation suffer for an unproved and assumed suffering ahead? Is Sudan's use of their own water a *causus belli*? Why shouldn't they use their water? Spain is the most arid country in Europe, so we will not plant unsustainably. But why should that give us the right to tell China what to do?"

When the stony silence continued, he said, with some exasperation: "I know of no place where overexploitation has caused catastrophe. Catastrophe is always in the future."

Which is where everyone devoutly hopes it will stay.

10 | THE MIDDLE EAST

―――――〜〜〜〜〜〜〜〜〜〜〜〜――――――

If the water burden really is a zero-sum game, how do we get past the arithmetic?

Not long ago I had a conversation about water with a member of the Kabbalah, the mystical Jewish sect whose notion of a universal life force has been tried on for size by an assortment of lightweight Hollywood stars and heavyweight intellectuals. We had come down from the hilltop town of Safad, as it's rendered in English, the Kabbalist stronghold, where the body of the seer Rabbi Isaac Luria is buried. On the steep western slope is a wonderful little fourteenth-century synagogue, a holy place of the Kabbalists, where the son of a friend had gone for his bar mitzvah. The great Rav's body lies in a simple grave lower down the slope, and at the bottom is a perennial spring, where first the men and then the women go to be cleansed.

This spring is very old. It has been running, as far as anyone knows, forever. It was here when the town was built, which was before Solomon cut the first rock for his temple. It was already old when Alexander the Great clattered through on his way to glory; the inhabitants had rolled rocks down the hillside, but the passing army paid them no attention and the spring hadn't been tampered with – it was hardly worth the effort. The town was already old when Jesus walked these hills, and ancient by the time the Prophet Mohammed was bringing about the Great Awakening down in the deserts of the Arabian Peninsula. The locals have hewn a bath from the rock, and the spring water is clear, and cold as a Greenland glacier.

I liked the Kabbalists; their worship is joyous, full of song and rhythm, but their devoutness runs to prayer sessions of eight hours or more, and I was relieved to be able to get away. My friend Isaac from Safad volunteered to drive me down to Lake Kinneret, a.k.a. the Sea of Galilee.

What an incongruous setting! Just down the coast, at Tiberias, the fishermen who became Jesus' disciples had been caught in a storm and had seen the Master striding across the lake, walking on water. Now, as I stared across the sea, two water-skiers zipped by; nearby was the Sea of Galilee Parasailing School, and you could learn to hang-glide over the place where Christian pilgrims come to venerate the Miracle of the Water. A little later we went to have lunch at the St. Peter Restaurant in the Ron Beach Hotel, and I started to laugh.

"Better order fish," I said, but the Fisher of Men was a non-starter with the Kabbalists, and Isaac just looked puzzled.

After a while we went back down to the shore.

"This is where we get our water," Isaac said. "More than half the country's water comes from here. Tel Aviv and Jerusalem could not exist without Lake Kinneret. This is where the Jordan waters are stored. It makes Israel possible. Without water, we are dead."

"Without water, we are all dead," I said.

"No, yes, I mean . . . that's true, we're all dependent. But it's starker here. I mean politically, not personally. This little river, and this little sea, give us nearly two-thirds of all our water. Also about three-quarters of Jordan's water comes from this valley."

He showed me a clipping he'd torn from the local paper. Since I couldn't read Hebrew, he paraphrased. "They're worried," he said, "about the water in the lake. There are some signs of saline intrusion. I don't understand the engineering, or where the salt is coming from, because we're a long way from the sea, but the water people are worried that, if the level of the lake drops too much — and the drought combined with increasing demand is threatening to do just that — the hydraulic pressure will ease and salt water will intrude."

He thought for a moment. "They're diverting the Jordan River above the lake, to take water to the Negev," he said. "The problem with water in Israel is that we're already using all of it. There is no more. And the population is growing."

"Zero-sum game," I said.

"Yes, if we get more water, which we'll have to, it will have to come from somewhere else. But where?"

～

The conversation turned to other things, and after a while we'd returned to Safad for the evening feast, a groaning-board banquet with music and song and dancing, and the notion of water slipped away. Later that night I heard a rumble, which I took to be thunder, but it could have been military jets. It could also, Isaac told me the next morning, have been artillery. Safad is only fifteen kilometres from the Lebanese border, and if you stand at the lookout on the summit you can see the blue hills of Lebanon and the Golan Heights, once Syrian territory. Israel patrols Golan now, but the Syrians are still there. Everyone in Safad can feel their restlessness. There were Syrian troops in Safad itself not that long ago, when the war was on. One never forgets about security in Israel — or water.

The day before I had driven down to the north end of the lake to see the Jordan River itself. It was not much, perhaps a hundred metres across, and sluggish. I had wanted to see the intake pipes for the National Water Carrier, but water is a security matter in Israel and I had been forbidden. The Water Carrier takes Jordan

water to the communities of the coast, including Tel Aviv, and
through an intricate network of canals to the irrigation schemes
of the arid south. Israel, famously, is making the desert bloom, and
the Water Carrier is how they do it. No one knows how much
water the carrier is capable of transporting; popular assumptions
are that it could carry the full capacity of the Jordan River, but
officialdom is not saying.

A few weeks after I left Safad, early in May 1998, I read a dis-
patch in a European newspaper. It was datelined Tel Aviv: "Early
yesterday morning, the main water pipe serving the south of the
country was severed, cutting off water to more than a million
people as temperatures soared above 30 degrees in the hottest April
in 30 years. Israeli police were not sure whether it was a politically
motivated attack."

They never did say, then or later. No one in Israel forgets that
two-thirds of the water Israel uses originates in territory it now
controls through military conquest, in the Golan Heights and the
West Bank. The Fatah, among other groups, has been targeting
Israeli water institutions for thirty years, and so has Hezbollah and
a long shopping list of organized and disorganized terrorists. I
thought back to Isaac's notion: "We'll have to get more water.
But where?"

That was still the critical question.

By the mid-1990s, Israel was over-exploiting its water, drawing
down its aquifers at beyond replenishment rates by about 15 per
cent a year (2,100 million cubic metres, against a supply that ranges
in good years from 1,950 million cubic metres to 1,600 in
drought years). By mid-2000, the cumulative "overdraft" was
approaching 2 billion cubic metres – that is, the Israelis had
extracted 2 billion metres more than nature was putting in, and
demand was increasing steadily, at about 40 million cubic metres

a year. The absence of good rains in the area meant that the streams that fed Lake Kinneret were lower than they had been for decades, and the water authorities responded in the only way they could – by lowering the "red line," the minimum level below which extraction would be forbidden, down another 20 metres, to 215.5 metres below sea level. Early in the winter of 2002–2003, the level was still dropping, and engineers were worrying that saline brine would seep into the lake, threatening to turn it into another Dead Sea and making it useless for human consumption. Jordan was doing even worse: it was using 20 per cent more water than it was receiving. The coastal aquifers in the region, especially in the critical Gaza area, were seriously overpumped, and seawater intrusions were becoming a major problem – and a major political problem, given that wells drilled by Jews were allowed to be deeper than wells drilled by Arabs or Palestinians. The already potent Palestinian grievances were being ratcheted up by the brutal politics of water: if your tap runs only a day or two a week, and the Jews' taps run all week, well, that's an easy grievance to nurture, isn't it?

By 2010, according to Israel's own figures, it would have an annual water deficit of 360 million cubic metres. Jordan's deficit would be closing in on 200 million, and the West Bank's on 140 million. Considering that the Jordan River in good years yields only 1,400 million cubic metres a year and is already overstretched, where, indeed, is the water to come from?

Zero-sum game: the peculiar and fragile ecology of the region has always collided ferociously with the complicated, fractious, perilous politics of the region, and also with the population dynamics and urbanizing economies. The Middle East has always been the place where water wars are most probable. Indeed, Israel did have a shooting war with Syria over water, and it is now widely accepted that the 1967 Arab–Israeli war had its roots in water politics as much as it did in national territorialism. Israel controls the Golan Heights for its water as well as for reasons of military security. Of course, there are potent psychological and political reasons for keeping the Golan Heights, but Israel doesn't need them to keep snipers away; modern artillery doesn't need line-of-sight, and modern "snipers" could easily be twenty kilometres into Syria. Similarly, Israel until recently maintained its military presence in South Lebanon at least partly because that's where the water was, not just because it was a home to terrorist bases. In fact, the boundaries of the state of Israel are to some degree the result of water considerations.

The water resources of the region are simple to outline, if difficult to control. The Jordan River, the key to the system, begins in three headwater streams. The Hasbani River, which contributes about a quarter of the Jordan's water, originates in Syria and has at least a part of its outflow in Lebanon. The Dan and the Baniyas rivers originate in the Golan Heights, occupied in 1967 by Israel and annexed in 1981; both flow into the Jordan above Lake Kinneret. The Syrian announcement in 2002 that, should they regain the Golan, they would settle an addition half-million people there, made it almost certain that Lake Kinneret would suffer irreversible pollution. The lower Jordan River is fed from springs and runoff from the West Bank and Syrian and Jordanian waters, and by the Yarmük River, which rises in Syria, borders Jordan, Syria, and the Golan Heights (and so Israel), closely parallels the Jordan for several hundred kilometres, and empties into the Jordan at Adam (Damya) Bridge. The Jordan Valley is thus an international drainage basin, a naturally defined area that cannot be artificially subsectioned.

Still, only about 30 per cent of the water in the region is surface water, from rivers. Groundwater, from the Mountain, Eastern, and Coastal aquifers, accounts for the rest. The Mountain Aquifer is the most substantial of the three, and gives Israel almost one-quarter of its total water supply. It consists of several major drainage basins: the Western, which falls almost entirely within the boundaries of Israel; the Northeastern, which is located inside the West Bank; and the Cenomanian–Turonian, under the northern West Bank. By 2002, the entire Mountain Aquifer was being red-lined – that is, engineers said it had been depleted to an extent that permanently imperilled the resource. The Eastern Aquifer basin contains a number of smaller aquifers, all located within the West Bank area; 90 per cent of the water in the area comes from wells drilled into these sources. The Coastal Aquifer, of which the Gaza Strip Aquifer is a part, has been continuously overpumped for many years, not only by the refugee population in the area but by Israeli settlers tapping into it from outside Gaza itself. It is not hyperbole to say that the Gaza water system is in crisis: the pumping is far above the recharge rate, and there is already enough sea water in the wells that most of it is undrinkable.

Pollution is adding to the country's woes. Municipally added chlorine and ill-treated sewage are both finding their way into the groundwater; traces of nitrates, soluble organic material, and heavy metals have been found in groundwater throughout the country. Sewage water has long been used in Israel for agricultural purposes. Once, this was regarded as evidence of the country's genius for maximizing its resources, but now its deleterious effects are becoming only too apparent: destruction of the soil through salination and alkalization, reduced crop yields, and pollution of the groundwater previously used for drinking.

Water was critical to the founders of Israel and to Zionist planners. A viable state needed substantial numbers of immigrants, and immigrants needed water for living, for farming, and for industry. As early as 1875, Charles Warren, a researcher with the British Royal Society, declared that, with proper control of water, Palestine and the Negev could easily absorb fifteen million people. Chaim Weizmann had demanded as early as 1919, at the Paris Peace Conference that followed the First World War, that the boundaries of any future Palestine should include the headwaters of the Jordan (Mount Hermon) and the lower reaches of the Litani River, where it turns sharply west to head for the Mediterranean, just above the city of Tyre (Sür). In a letter to the British prime minister, David Lloyd George, Weizmann had asserted his desire to have control of the "valley of the Litani, for a distance of about twenty-five miles above the bend, and the western and southern slopes of Mount Hermon." Control of the Litani, the Jordan, and the Yarmük had been seen as critical to the future state's security. David Ben-Gurion in 1948 demanded the same thing: "The most important rivers of the land of Israel are the Jordan and the Yarmük," he said. But he also demanded that the boundaries of Israel include the southern banks of the Litani.

In the end, Israel didn't get the Litani. The League of Nations rejected the Zionists' claims, and the Litani became part of Lebanon instead. It is a constant frustration: only a narrow ridge separates the Litani from the headwaters of the Jordan, and it has always tempted Israeli planners to correct nature's mistake in this regard. Until Israel unilaterally departed from South Lebanon in May 2000, the westward leg of the river was in the Occupied Zone, but Israel did not exploit the waters to any degree.

From the beginning, Arab plans to protect their water necessarily contradicted Zionist plans to divert it to immigrants. Rising political tension in the region and the lack of a solution acceptable to all parties exacerbated the situation.

The first time water politics impinged on the modern Middle East was in 1926, when the British High Commissioner granted the Jewish-owned Palestine Electricity Corporation, founded by

Pinchas Rutenberg, a seventy-year concession to scoop up the Jordan and Yarmük rivers to generate electricity. It was an early harbinger of trouble: the concession denied Arab farmers the right to use either river for any reason without permission from the Palestine Electricity Corporation, permission that was never granted.

A decade later, the British assigned a hydrologist called M.G. Ionides to study the water resources and potential of the Jordan Valley basin. Ionides recommended that the Yarmük floodwaters be stored in Lake Kinneret; that the lake, plus another 1.76 cubic metres per second of the Yarmük, be diverted through the East Ghor Canal, Jordan's main water-management facility, which parallels the Jordan on the east side, to irrigate some thirty thousand hectares east of the river on the Jordanian side; and that the secured irrigation water of the Jordan River system, estimated at a potential of 742 million cubic metres, be used primarily within the Jordan Valley. Nothing much came of this advice: the Palestine Electricity Corporation still controlled both rivers and refused to let go.

In any case, the Zionists didn't like the Ionides plan. They much preferred another study, this one commissioned by the U.S. Department of Agriculture. The author, Walter Clay Loudermilk, held there was much more water than Ionides had said – almost double. His scheme called for the irrigation of the Jordan Valley, the diversion of the Jordan and Yarmük rivers for hydroelectric power, the diversion of water from the north to the Negev Desert in the south, and the use of the Litani River in Lebanon. He also recommended a canal to join the Mediterranean Sea with the Dead Sea – the first sighting of that bird of rare plumage, the Dead–Med Canal (later to be distinguished from its even more popular counterpart, the Red–Dead Canal). All these waterworks were to be controlled by Jewish immigrants, a notion that went down well with the Zionists, but doomed his ideas elsewhere in the region. Any Arab who didn't like the plan, Loudermilk wrote, was to be deported to the Euphrates and Tigris rivers.

The subsequent Hays Plan of 1948 was based on Loudermilk's notions, from which it differed only in its softer rhetoric.

～

If you take the twisty mountain road from Safad down toward Rosh Pina and then north to the Benot-Ya'akov Bridge, and if you go north from there on Secondary Road 918, you'll find yourself winding into the Huleh Valley. Here, ecology and ideology have more than once collided, a swirling nexus in the anxious water politics of the region.

One hundred years ago, and presumably for millennia before that, this area was a swamp. By the time of the Palestinian Mandate in the late 1940s, it was still called the Huleh Marshes, and it seemed a fit place for the sturdy agricultural settlers of Zionist legend. The swamp was good for nothing but breeding disease and mosquitoes; malaria was rife, and so was yellow fever. It was once Syrian territory, but it became part of Israel under the Partition Plan of 1948. A year later, Syrian troops were evacuated, and the valley became part of the demilitarized zone. In the Syrian interpretation, this meant nothing was supposed to happen there. To the new state of Israel, however, it was an opportunity. Israeli settlers would irrigate the Negev with water from the Jordan River, drain the marshes, and the kibbutzim, the self-sufficient agricultural collectives, would turn the valley into productive farmland. The Seven Year Plan, published in 1953, adapted the Hays variant of Loudermilk's plan and confirmed the intention to divert the Jordan River south towards the Negev. The plan included a comprehensive water network that would cover the whole country, the first mooting of the National Water Carrier. The Huleh Marshes were an integral part of the system, and the first settlers arrived there in 1953, the same year construction began on the Jordan diversion at the Benot-Ya'akov Bridge.

A month later, after furious Syrian objections – as well as U.S. and Western economic sanctions against Israel – construction was frozen.

The Arabs, for their part, adopted a notion put forward by an American engineer, M.E. Bunger, to construct a storage dam along the Yarmük River at Makarin, where three valleys join together. Bunger was a political innocent, ignorant of the buzzsaw of Israeli politics. Indeed, he largely ignored Israel, and set about trying to solve Jordan's and Syria's water needs, at the same time trying to resolve, at least to some extent, the Palestinian refugee problem by increasing the productivity of available agricultural lands in the East Jordan Valley and parts of Syria. Work on this project actually started in 1953, but Israeli opposition exerted pressure on both the United Nations and the United States to withdraw support. A short while later, Jordan and Syria discovered, to their fury, that work had simply stopped.

The political tensions over Huleh and the demilitarized zone escalated rapidly. Neither side felt it could give in. To Israel, the right to farm the land and use its water were bedrock values, as basic as any in the Zionist philosophy. To the Syrians, it was theft, pure and simple. The Americans intervened with one more water plan. This one was produced by a special envoy, Eric Johnston, who modelled his scheme after the Tennessee Valley Authority. Johnston's expressed desire was to be more inclusive. He wanted to find what he called "an equitable distribution for existing resources" and a method for the countries to co-operate. Naturally, his plan was promptly shredded by all sides. No one liked it – not Israel, not Jordan, not Syria, not even Lebanon. (In later years, this wholesale condemnation was interpreted as a sign of the plan's essential fairness, and Israel largely adopted it.)

Later in 1953, Israel scrapped the Seven Year Plan in favour of the more grandiose Soviet-style Ten Year Plan. Among other things, the revision resurrected the Jordan diversion plan of the

National Water Carrier project. Prudently, the diversion point was shifted to Eshed Kinnoret, at the northwest corner of Lake Kinneret, safely out of Syrian reach. Also prudently, the diversion was carefully designed not to overdraw Israel's water allocation under the Johnston plan.

That second reservation, if it was intended to soften Arab animosities, was a bust. In 1959 Syria took the case to the Arab League, then chaired by Colonel Nasser of Egypt (Syria and Egypt were at the time collectively known as the United Arab Republic), and the League approved a retaliatory measure: storage dams on several tributaries of the Jordan and Yarmük rivers, the Baniyas and the Dan. This was all entirely legal – the work was well within Syrian territory – but it turned on the Israeli buzzsaw just the same. The Israelis calculated that their quota from the Jordan would shrink by almost 35 per cent, and they issued carefully measured threats. For a while, nothing much happened. Construction work was approved in 1961 and actually started in 1964, but that was enough for the Israelis. Deputy minister of defence Shimon Peres had already declared that water issues would be a cause of war, and now the new prime minister, Levi Eshkol, was even more explicit: "[This] water is like the blood in our veins," he declared. If the Syrians persisted it would be they, not Israel, who had declared war. Israel wouldn't exceed its allotment under the Johnston Plan, but was determined to go to war to protect its vital national interests. In response, the Syrians deployed forces along the border, and several bloody skirmishes resulted. Construction on the dams continued, however, and after Israel threatened "massive retaliation" on several occasions without result, it finally sent in the planes to bomb the diversion installations, destroying them without effort. This was two months before the outbreak of the 1967 war.

With its occupation of the West Bank, the Gaza Strip, and the Golan Heights, and the outright annexation of Golan in 1981, Israel tightened its grip on regional water resources as far north as Mount Hermon, where the headwaters of the Jordan are located.

After the 1982 invasion of South Lebanon, Israel extended its command even further, to include part of the Litani.

Israel has promised to return the Golan Heights to Syria, on a contested timetable. Seen from Safad, it doesn't look like such a good idea. Who controls the Golan Heights controls the water, and control of the water is key.

For another fifteen or twenty years after the original settlers first tilled the soil, farmers continued to move into the former Huleh Swamps, now safely under Israel's control — "safely" if you didn't count the occasional shelling from guerrilla bases in Lebanon, but most of the farmers were used to those. As it turned out, however, there was a much more insidious enemy at work.

As the wetlands drained, the groundwater tables began to drop. Small streams and springs dried up, and wells had to be extended downward almost every year. Some of the farmers gave up, but others began to irrigate more heavily, adding salts to the soil in hazardous quantities. To deal with the new salinity, they switched crops to more salt-resistant varieties, but the same thing happened again, as the saline levels increased past the plants' ability to adapt. Then the ground started to collapse. As the underground aquifers drained away, the earth subsided into the cavities left behind. In some parts, the collapses were up to seven metres deep, and occasionally whole houses disappeared. Agronomists and hydrologists probed at the soil, but it didn't take a genius to see what had happened: the now-depleted aquifers had been essential for the health of the whole ecosystem; now that the water was gone, the region was dying. But if it was easy to see the problem, it was difficult to see what could be done about it. The farm lobby in Israel has always been a potent political force. Part of the Zionist ideal had been self-sufficiency, and the kibbutzim had become part of the Israeli identity. How could you shut down any of them?

In the end, it was beyond politics. There was nothing anyone could do. The subsidences continued. In the mid-1980s dust storms began to be seen in the area, salty soil picked up by the swirling summer winds, carried far to the south. Leachates started to appear in the Jordan, and drifted down towards Lake Kinneret.

By the 1980s, the political dynamic had changed. Water politics and rapidly increasing urban populations dictated new priorities. No longer were the farmers to be given everything they wanted. It was better to use water to manufacture goods, which could then be sold for hard cash, and to buy food. In other words, industry gave more value to the water available than farming did, and the inevitable happened: the last farmers were pushed from the Huleh, and the hydrologists moved back in.

The Jordan diversions for the National Water Carrier have also had far-reaching and unintended ecological effects. In 1953 the Jordan had an average flow of 1,250 million cubic metres at the Allenby Bridge, near the Dead Sea; it now records flows fluctuating between 160 and 200 million cubic metres, an eighth of what it was. And the water in the national water system itself is not as pristine as it should be: it contains mineral concentrations higher than is considered safe in Europe or the United States.

The reduced flow of the Jordan – its conversion from a river into what is essentially a drainage ditch – has had its effect on the Dead Sea. Already the lowest land surface on the planet, about three hundred metres below sea level, its surface has dropped more than ten metres since the beginning of the century, and the drop is accelerating. The Dead Sea is important not only for its minerals (potassium and bromine factories mar its landscape) or its tourism potential (its famous ability to prevent humans sinking has drawn thousands of tourists a year), but because its very volume has an ameliorating effect on the microclimate.

~

In the summer of 1998 the rattle of machine-gun fire could be heard at several points along the highway to Amman, Jordan's capital. There were no human or animal casualties: herdsmen were shooting holes in a state water-supply pipe to get water for their stock. On the roofs of the capital, there were frequent shouting matches and occasional fights, as arguments erupted over accusations of theft from housetop water tanks. There were reports of assaults on private water sellers, who had tripled the usual price for 5,000 litres of water to 20 Jordanian dinars, or about $30 – in a country where the per capita annual income is $1,100. The wells in the oasis of Azraq, once the home of Lawrence of Arabia, have been dropping for years, and in this long, hot summer some of them ran dry; nevertheless, corrupt officials handed out dozens of new licences for well-digging. Rationing was imposed: most citizens could get water from public sources only two days a week. Each household was to get 85 litres a day – one-third of the quota in Saudi Arabia, and one-quarter of that allocated to each Israeli household. Meanwhile, in the wealthy districts (where water was not yet rationed), there were private swimming pools. And in the capital's public parks, the lawns were, and still are, kept alive year-round with sprinklers. In all, 50 per cent of pumped water is lost, either to outright theft or to leaky pipes, some of them dating back to the time of the Palestinian Mandate.

It's not hard to see why there is rationing. The Jordan Basin (the Jordan and Yarmük rivers) is the country's sole source of running surface water. Rain-charged subsurface aquifers are the only other source of fresh water. The kingdom began pumping water from this handful of underground aquifers in 1989. Its renewable water resources account for 650 million cubic metres each year. It is currently using 990 million.

In the 1950s, Jordan tried to gain control of its water resources by building storage systems on the Yarmük. The Israelis destroyed them in the 1967 war and have, on repeated occasions since, prevented maintenance being done on the Jordanian dams and canals. Withdrawals were seen as threats to Israeli farmers in the Jordan Valley.

In the 1970s, Jordan tried again, and came up with a plan to build a dam at Makarin, on the Yarmük, near the border with Syria. But Israel, in the technical jargon of the negotiators, was a "downstream riparian" and, under World Bank financing rules, had to be consulted. Israel refused. It explained it had nothing against the notion of a dam, but the one under discussion was too large and would take too much water from the system. Harking back to Johnston's notion of equitable distribution, Israel brazenly maintained that its quota should now be larger than originally allocated, because its territory had increased. Jordan, for its part, didn't see why unilateral annexation of the Golan should oblige the other parties to divvy up differently. From their point of view, Johnston had allocated water to sovereign nations, not to an arbitrary number of square kilometres, and occupation of territory shouldn't automatically entitle a country to more water. Israel, for its part, continued its objections to the East Ghor Canal, which had been damaged by air attacks in the 1967 war and had been only partly rebuilt. In an attempt to persuade Jordan to react against the escalating violence of Palestine Liberation Organization factions based in its territory, Israel once again threatened the canal and its works. The threat worked. King Hussein expelled the PLO in 1971 after private negotiations with Israel which led to both sides "affirming the principles of the Johnston allocations." The canal was repaired, though it has never functioned efficiently.

In 1990 Jordan applied to the World Bank for money to build another dam called Wahda, on the upper Yarmük. Once again, Israel objected. There was no doubt that the dam, if built, would affect Israel's ability to cope with ever-escalating domestic water demands. King Hussein, usually a careful man accustomed to using diplomatic language, was driven to say that "water is the only reason that could bring Jordan to war with Israel."

In 1994, with the signing of the Jordan–Israel Peace Accord, there was, for a change, a dollop of good news. Water issues took up a substantial part of the text of the treaty. Among other things, the two parties approved financing for a Unity Dam on the Yarmük, and Israel agreed to divert water from the upper Jordan River to Jordan during the dry summer months. Entitlements to Jordan River water – still loosely based on the Johnston plan – were agreed upon. Water flow to Jordan would increase; jointly, the two countries would repair the Dead Sea by importing sea water from the Red Sea via the Red–Dead Canal, or from the Mediterranean, via the Med–Dead Canal. Either would cost around $5 billion, and their ecological effects were unknown. But the drop to the Dead Sea would generate huge amounts of electricity, which could be used to desalinate water. The desalinated water would flow in pipelines from Israel to Amman – peace water.

Alas for good intentions. The water crisis within Israel meant that, by 2002, the allocation of water to Jordan under the Peace Agreement was no longer being delivered, and desalinated water seemed further off than ever. Authorities on both sides of the border regard this unilateral abrogation of the accord as a "catalyst for broader confrontations," as a diplomat delicately put it.

Apart from Turkey, Lebanon is the only other country in the region that has plenty of water. Lebanese farmers are chronically short of water, and there is rationing in the capital, Beirut, but this problem has more to do with the collapse of state authority in Lebanon following the disastrous civil war of 1976 to 1990 than with natural supply. Estimates in 2002 were that the country had 2,854 million cubic metres of usable water, and in the year 2000, the latest figures available showed a consumption of only about half this amount, 1,412 million. Most of this was used for agriculture.

But a UN agency, the Economic and Social Commission for
Western Asia, projected the demand would reach 4,000 million
cubic metres by 2025.

This sounded dire, but the same agency pointed out that almost
a third of the county's water was simply wasted, and additional
millions of metres were made unusable through pesticide contamination, outdated irrigation methods, and industrial pollution.

The Litani is one of Lebanon's major rivers, with a flow of
about 580 million cubic metres a year, smaller than the Jordan but
with better quality (salinity at a purely notional 20 parts per million,
compared with Lake Kinneret's 300). It rises in the central part of
the northern Biqa'a Valley, a short distance west of Ba'albek, flows
southward between the Lebanon and Anti-Lebanon mountains,
enters a thirty-kilometre-long gorge at Qarun, and abruptly turns
west, emptying into the Mediterranean north of Tyre.

The Israelis have never really given up thinking of the Litani as
rightfully theirs. And since almost half the water used in Israel is
already captured, diverted, or pre-empted from its neighbours,
why stop there? After all, Moshe Dayan, Israel's defence minister
during the 1967 War and a notorious hawk, said that Israel had
achieved "provisionally satisfying frontiers, with the exception of
those with Lebanon." It is difficult to know how much it was water
that motivated Israel to invade Lebanon in 1978 and again in 1982,
and to maintain its "Security Zone" there. The given reason was
its need to control guerrilla and terrorist bases in Lebanon, but
water undoubtedly played a significant role.

There are conflicting reports, never commented on by Israeli
authorities, that during their occupation they were diverting water
from the Litani southward. An FAO report in the mid-1970s
declared, on thin evidence, that Israel had built underground
pumps along the Lebanese frontier to suck water across the border.
In 1984, the Lebanese government lodged a formal complaint with
the United Nations about Israel's supposed diversions, not only
of the Litani but of two smaller rivers that conveniently fell into

the Security Zone controlled by the Israeli military. Some Western researchers have declared that Israel has been diverting southward about a hundred million cubic metres a year from its own part of the Hasbani River, and also some, "in insignificant quantities," from the Litani. Israel has consistently denied all these allegations, and it is fair to point out that most of the more indignant accusations about Israeli water grabs in Lebanon have come from politicized Arab sources.[1]

International law, for what it's worth, is on the Lebanese side. In principle, water within one catchment area should not be diverted outside that area – regardless of political boundaries – until "all needs of those within the catchment area are satisfied." Lebanon puts the upstream Litani to good use: some 35 per cent of Lebanon's total production of electricity comes from the Litani, and nearly 230 million cubic metres a year of Litani water is diverted through the Markaba Tunnel to the Awali River, which is used to supply Beirut and other coastal areas. In the early 1970s Lebanon planned to irrigate the Shi'ite territories in southern Lebanon through a series of dams and weirs, but the United States and Israel blocked this plan by pressuring project lenders.

Domestic political considerations inside Lebanon are also a factor: the Lebanese political buzzsaw is just as sharp, and as busily employed, as that of Israel. There are still dozens of Lebanese villages without running water, and were the country to allow Israel to siphon off part of the national supply, there would be an uproar. Even without that, the Shi'a Muslims in the Litani watershed area have been angry for years, and in 1974 rumours that water from the Litani was being diverted to Beirut to meet shortages there set off massive anti-government demonstrations.

Vast quantities of Litani water flow unimpeded (uselessly, by some definitions) into the Mediterranean. Since 1984, the situation has improved somewhat. The domestic political rhetoric is still extravagant, as are the mutual recriminations between Israel and Lebanon, and there are endless arguments about who could better use the Litani water and who needs it more. But, as every-one agrees, rhetoric is less lethal than artillery.

Twice since the Israelis withdrew from southern Lebanon, the Lebanese government, prompted by agitation from Hezbollah, attempted to divert the Hasbani to prevent its reaching Israel. Twice they were thwarted by international condemnation and American pressure. The third attempt was made in late fall 2002, when they announced their intention of diverting the water from the Wazzani river, a main feed of the Hasbani and therefore the Jordan, to supply its own towns along the border. The scale of the intended diversion was tiny (it would have lowered the level of Lake Kinneret by about an inch) and perhaps only in the Middle East would such a trivial issue have escalated to threats of war. But in October, a U.S. State Department official was in Beirut trying to work out what he diplomatically called "an acceptable compromise." Ariel Sharon, for his part, took to the public airwaves to threaten war if the status quo changed. "We are deploying maximum efforts to keep our water resources, and Israel always has and always will do whatever it takes to protect its vital resources," he said, a comment the United States deemed "unhelpful."

At present, Israel controls all the aquifers in the West Bank, as good a reason as any for not handing the region "back" to whomever it once belonged. About one-fourth of the country's water currently comes from the West Bank, an ecological happenstance that has induced the minister of agriculture, Rafael Eitan, to declare on Israeli radio that the country would be in "mortal danger" if it lost control of the Mountain Aquifer.[2] In the mid-1990s, there was still considerable confusion about how much water there was in these underground reservoirs; majority scientific opinion seems to believe that with equitable distribution, there would be enough water in the region for both the Israelis and the Palestinians, but this advice is hotly disputed, as resource analysis gets snarled in politics.

The control of water is as complicated as any other aspect of Israeli politics. After the Six Day War, all water resources in the area were centralized under the control of the military. In 1982, however, the then-minister of defence, Ariel Sharon, handed responsibility over to the National Water Carrier, which was supposed to integrate the Mountain Aquifer with the rest of the country's water resources. At the turn of the millennium, though, water management was scattered among eight ministries, with none holding ultimate responsibility. Finance and Agriculture wield the most influence, but so does Health, and even Religious Affairs has its say, insisting that no water be pumped from Lake Kinneret in the two months before Pesach (Passover) for reasons to do with kosher dietary laws.

Despite the urgent nature of the crisis, the Water Commissioner has no professional expertise in water management; the office is hounded by allegations that it routinely suppresses negative news. Until recently, the entire department saw fit to employ only one engineer.

Israeli water law still imposes restrictions on water use. For example, on anyone wanting to drill a well: "It is permissible," the law says with bland bureaucratic precision, "to deny an applicant a permit, revoke or amend a licence, without giving any explanation. The appropriate authorities may search and confiscate any water resources for which no permit exists, even if the owner has not been convicted." In 1998, the number of licences issued for new wells for Jewish settlers far outnumbered those for Palestinians. Also, Palestinian wells may not exceed a depth of 140 metres, though Jewish wells could go down 800 metres. Similarly, more than half the Palestinian villages still had no running water, whereas all the Israeli settlements did, another point in the litany of Palestinian grievances.

The basic fact of the region is simple enough: Israel could not exist without West Bank water, and the Palestinians are entirely dependent on it. The meagre available water is critical for both sides.

~

In Gaza, the home of the discontented, birthplace of the PLO and the Intifada, water has always been a critical issue. Recent salination of the water supply has only intensified the problem.

Gaza is small but densely populated, only forty kilometres by ten, but geopolitically important far beyond its size. It lies on the Mediterranean Coast where Israel meets the Sinai, with an eleven-kilometre boundary with Egypt and a fifty-one-kilometre boundary with Israel. Fewer than 1 per cent of the people in the enclave are Jewish settlers; almost all the rest are Sunni Muslim Arabs. About one-third of its people live in UN-run refugee camps.

The climate is temperate, but dry. Two-thirds of what rain does fall is lost to evaporation; the rest recharges Gaza's only natural freshwater supply, the Gaza Aquifer. This sandstone sponge, only a few metres below the surface, is thought to store a renewable 42 million cubic metres a year. Wells and boreholes currently extract 45 million cubic metres, and overpumping lowers the water table about 15 centimetres a year. As the water table falls, salt water from the Mediterranean and other nearby saline aquifers is introduced. Sea water has been detected up to 1.5 kilometres inland, and continues to threaten the entire aquifer. In many parts, the water is so saline that it is damaging soil and sharply limiting crop yields. Citrus, Gaza's main agricultural product, is highly salt-intolerant, and is suffering from declining quality and crop yields. In addition, uncontrolled use of pesticides and fertilizers has chemically contaminated the groundwater. The refugee camps, most of which have no proper sewage control, have made things worse. One-tenth of the population simply uses the sand dunes as latrines, and poorly maintained septic systems frequently overflow.

Palestinian control since the Gaza–Jericho Accord of August 1993 has been rocky, and the Palestinian Authority has had little luck in reducing the ecological degradation. Ongoing talks since 1998 between the Palestine Authority and Israel's government mentioned "national territory," terrorists, and land-for-security in great detail, but hardly mentioned water.

But here, too, water scarcity, or what Thomas (Tad) Homer-Dixon, director of the Peace and Conflict Studies Program at the University of Toronto, calls "demand-induced scarcity," has played a large part in the continuing violent conflict. Population growth is the principal culprit. The mass exodus of refugees to Gaza after Partition in 1948 more than tripled its population, which is now grown to almost a million people. Despite high infant mortality, Gaza has one of the highest growth rates in the world, perhaps higher than 5 per cent, and, as a result, per capita water availability has decreased dramatically.

When, in 1967, Israel declared all water resources to be state-owned, strict quotas were imposed on Palestinian water consumption. As a consequence, thousands of citrus trees were uprooted, cisterns demolished, and natural springs and existing wells blocked. Israeli water consumption in Gaza, on the other hand, is subsidized, encouraging overuse and misuse. In 1995, Gaza Palestinians paid $1.20 per cubic metre for water, while Israeli settlers paid 10 cents.[3]

As in the West Bank, water problems can be solved by neither one side nor the other, but only by co-operation – however improbable that may seem.

Still, there are small but positive signs. Such was the publication of a book in 2000 called *Management of Shared Groundwater Resources: The Israeli–Palestine Case with an International Perspective.* This could have been (and alas in some chapters was) another one of those interminable jargon-filled treatises of interest only to specialists, but a clue to its importance were the editors, Eran Feitelson and Marwan Haddad, whose ethnicities are obvious from their

names. And indeed, the book is the product of a full seven-year collaboration between scores of Israeli and Palestinian scientists, who
were able, crucially, to devise an "implementation plan" for the
management of shared groundwater resources that transcends
the Middle East and is of relevance to water skirmishes everywhere,
as well as to suggest an agenda for a full-scale Israeli–Palestinian
water summit. That such a summit is unlikely in the short term
detracts nothing from the book's originality and bravery.

The Peace Process, such as it was, appointed a multilateral working
group to deal with the contentious issue of the region's water.
From the beginning, its deliberations were made moot by two
facts: first, Syria and Lebanon refused to take part, on the grounds
that doing water deals with Israel could be taken as de facto recognition of Israel's right to exist; and, second, Israel, treating the
matter as a security issue, refused to discuss water allocations,
except bilaterally. The working group was therefore reduced to
suggesting ideas for enhancing supply and water management, but
without much encouragement. The Declaration of Principles
between the PLO and Israel, signed in Washington in 1993,
was vague, but at least allowed the issues to be dealt with through
an Israeli–Palestinian Continuing Committee for Economic
Cooperation. That so far there hasn't been much co-operation,
economic or otherwise, isn't the committee's fault.

    If there are not to be further water wars, some minimum steps
are necessary. The Gaza Aquifer will have to be supplemented, and
soon, for that situation is perhaps the most critical. The allocation
of water for residents of the West Bank – all of it, including the
new 13 per cent to be reassigned to Palestine – will have to be
raised. This issue is also urgent: some areas of the West Bank have
been without water for periods of two months or more. The
Eastern Aquifer on the West Bank should be reallocated, and the

Palestinian portion upped from its current twenty-five million cubic metres per year to somewhere better than a hundred million. Above all, it should be recognized that the crisis is not so much one of supply as one of inequitable distribution, complicated by historic grievances, ideological quarrels, and questions of national security. Everyone agrees on all these matters. It's implementing them that's the problem.

In 2002, the Sharon government announced massive plans to desalinate sea water, and reached an agreement (almost immediately abrogated) to import fifty million cubic metres a year from Turkey.

Late in the year, tantalizing tidbits of information leaked out about Israeli negotiations during the Camp David meetings. The Israeli side apparently suggested that, if a peace agreement were to be reached, the Americans should, in their own and everyone else's interests, be prepared to stump up $35 billion to help rebuild the Middle East. The American reaction to this mind-boggling sum was not recorded, but it seems that $10 billion or so was to be earmarked for desalination facilities, which would have assured regional supplies for the foreseeable future. Since no confirmed details are available, and there have been bland denials on all sides, it is not known whether the Red–Dead or Med–Dead canals played any part in these projections.

Without new water, hydropolitics in the Middle East remains a zero-sum game, and in such games someone wins and someone loses – unless the rules are changed.

11 | THE TIGRIS–EUPHRATES SYSTEM

*Shoot an arrow of peace into the air, and get a quiverful of
suspicions and paranoia in return.*

The Sumerians came first, as far as anyone knows, as an
urban civilization. They were the first to use the Tigris
and the Euphrates rivers as lifelines, as a storage bank of
energy, and therefore of power, and therefore of prestige, and
therefore of history, and art. The Sumerians invented irrigation,
and the crescent from Baghdad to the gulf bloomed, and they
became rich. Then came the Assyrians, and the Babylonians, then
the conquering Greeks, and the short-lived Romans. Capital cities
rose, then fell, rose again, decayed. The Ottomans followed, and
Islam. At the start of the third Christian millennium you had Assad
the Younger in Syria; the egregious Saddam Hussein and his even-
tual American nemesis; a succession of more-or-less mediocre Turks
following the splendid Atatürk; Kurdish rebels (or "Mountain
Turks" as the Turkish authorities call them) up in the Anatolian
highlands; PKK guerrillas still finding safe harbour in Syria; and
Saddam's twin persecutions of his own Kurds in the north and, in
the south, the Marsh Arabs, through whose villages his tanks rolled
on their way to Kuwait. It's hard not to feel that somehow the
Fertile Crescent has gone steadily downhill in the several millennia

since the Persian Darius the Great exhibited his extravagant management skills by spreading his intricate irrigation net into the Syrian desert. But even if it hasn't skidded much – even if history misleads, remembering only the achievements and none of the sins – it's still true that the Old Mesopotamia has been in turmoil as long as there were historians to recount it. Much of that turmoil was over water. It still is.

Both the Tigris and the Euphrates rise in the moisture-rich and verdant mountain valleys of eastern Anatolia in Turkey. The Tigris flows southeast, winding through low mountain valleys and across the rolling Turkish plains, flirts with Syria (it's briefly the border between Syria and Turkey), and then heads southeast through Iraq, passing Baghdad and the Central Marshes before joining with the Euphrates to become the Shatt-al-Arab, which flows into the Persian Gulf. The Euphrates takes a more circuitous route. It starts nearer the Black Sea, makes a sweeping curve west and south, and wanders through a series of Turkish lakes and reservoirs before crossing into Syria and Lake Assad, the reservoir of Syria's Euphrates Dam. From there it, too, heads in a southeasterly direction, crosses into Iraq, and traverses the parched desert south of Baghdad before flowing through the Hammar Marsh – or the former Hammar Marsh, for it has now virtually disappeared, its wildlife and human inhabitants along with it – and so into the Tigris not far above the Iraqi city of Basra.

The Euphrates and its tributaries are Syria's major water source, on which rests the country's hopes for increasing food production to match its burgeoning population. Iraq, downstream from both Turkey and Syria, is dependent on both rivers. There are a few tributaries to the Tigris flowing in from Iran's Zagros Mountains, but they don't contribute anything substantial to the flow. Nearly 85 per cent of Iraq's population get all their water needs from the twin rivers. The claims and counterclaims of the three countries are in many ways mutually incompatible, complicated by ethnic conflicts and by historical memories. There is a long-standing suspicion of the Turks' perceived desire for regional domination, harking back to the expansionist aims of the Byzantine emperors. Turkey, as the upstream country, maintains it has absolute sovereignty over any water originating in its territory: "The water is as much ours as Iraq's oil is Iraq's" comes close to outlining Turkish policy. Downstream Iraq, in contrast, argues its case on the "natural course" of the rivers, and on the country's "historical rights" to waters used by the people of southern Mesopotamia since the dawn of civilization six thousand years ago, a prior-use doctrine that has been central to the struggle to develop international water law. Midstream Syria argues both ways: it cites the prior-use doctrine to Turkey but the sovereigntist one to Iraq, a classic case of speaking with forked tongue. This built-in rivalry is complicated by the fact of the Turkish Kurds, an ancient culture whose homeland straddles the headwaters of both rivers. The Kurds have been fighting a war of attrition against both Iraq and Turkey for more than two decades, determined to win territorial independence and prepared to use guerrilla actions against Turkish hydro installations as a weapon of policy.

There is hardly a hint of solution to these swirling controversies and disputes. Meanwhile, the amount of water available to all countries continues to shrink on a per capita basis; Turkey's development plans will make things worse for Syria and Iraq, and Syrian plans will imperil Iraq's supply even further. Taken together, the development plans of the three countries would consume half as much again as all the water the rivers currently hold. "Untenable" is the politest possible descriptor.

Turkish government statements are hardly reassuring. The secretary of state for water development, Kamran İnan, declared in a Turkish newspaper: "We give them [Syrians and the Iraqis] the water – they can't share it. Ninety per cent of the Arab population is dependent on water originating outside [their borders] and therefore it is normal that the tension in the region will escalate." Such sentiments are not reassuring to hear in Baghdad and Damascus, especially from a non-Arab.[1] On another occasion, İnan confided to a newspaper reporter that Turkey didn't feel "any international obligations" about the Tigris—Euphrates system, but softened his stance by referring to a "gentleman's agreement" not to harm neighbours. Nor is there any forum other than bilateral meetings in which these disputes can be resolved or even discussed. Turkey, whose political focus is primarily European, is nevertheless, at least for the moment, the dominant regional power both economically and militarily. As a consequence, Turkish water planners are paying little heed to the anxieties of downstream powers. The disputes in the region haven't come to war yet. But when the taps run dry, as they threaten to do, the Turks had better look to their borders.

Turkey is, by Middle East standards, water-rich, but in truth it doesn't have much water to spare. Its total available supply is estimated at just over 193 cubic kilometres annually, which works out at slightly less than the total flow of, say, the Danube. In 1995, divided among the country's 60 million citizens, this amount gives some 3,174 cubic metres per person per year, well above the water-stress level. In practice, things aren't quite so rosy. For one thing, the population is increasing alarmingly, and even a low projection has it at 78 million by 2025 (other projections put it closer to 90 million). This lower estimate would shrink per capita availability to somewhere between 2,100 and 2,400 cubic metres per year. Moreover, not all this water is easily reachable. Some of it is remote, and there are considerable difficulties inherent in damming and conveying it to where it will be needed. There are already shortages in the major cities, including Ankara, and along the Aegean coast, where tourism is gulping massive amounts of water that could otherwise be used for farming. In practice, therefore, most hydrologists assess the country's available water at 95 billion cubic metres annually, which would reduce the per capita availability to around 1,800 cubic metres – still adequate, but no longer rich, and around the same values as both Iraq and Syria, provided nothing changes.[2] But it is changing, and will change more.

İlter Turan, a political scientist at Istanbul University, has been tracking Turkey's water needs. "It is anticipated that consumption of water will rise rapidly in coming years. First . . . the government is pursuing a set of irrigation programs, the completion of which will increase demands on the country's water resources. Second, the population is growing. . . . Third, the country is urbanizing rapidly. . . . Finally, the rapidly expanding industrial base creates new and additional water requirements." By the time per capita income rises to $10,000 a year, Turan forecasts, the country will need all the available water except the amount it contractually supplies to downstream riparian states.[3]

What will change the supply most is Turkey's grandiose plan for the southeastern Anatolia region of the Euphrates, a massive $32-billion collection of twenty-two dams, some of them huge, and an attendant network of irrigation canals, weirs, and barrages on a multitude of rivers, large and small, that would irrigate 1.7 million hectares of currently non-irrigated land – 1.1 million in the Euphrates Basin and .6 million in the Tigris Basin. The project will affect 74,000 square kilometres and change the way nearly 5 million people live. The centrepiece of the GAP (Southeast Anatolia Project, but known by its Turkish acronym for Güneydogu Anadolu Projesi) is the Atatürk Dam, near the town of Şanlıurfa on the Euphrates (Fırat) River, the world's sixth- or ninth-largest dam, depending on whose statistics you believe. The dam was finished in 1990 and started to fill a year later. The water from its reservoir is carried to the Harran Plain in southern Turkey by the Şanlıurfa tunnel, a 26-kilometre behemoth that dwarfs even Colonel Gaddafi's Great Man-Made River (7.62 metres in diameter, to the GMMR's 4 metres). A second string to this massive bow is the construction of no fewer than nineteen hydroelectric generating plants that would produce more than 27 billion kilowatt hours of electricity, adding 40 per cent to Turkey's hydroelectric capacity, an amount equal to the country's entire power output in 1983. When completed – if completed, for money is short, and the World Bank has declined to write any cheques without Iraqi and Syrian agreement – the project would involve the massive urbanization of Turkey's relatively poor southeastern provinces, and forecasts suggest it would produce a 12-per-cent jump in national income. It would also, Turkish proponents point out, help regulate the alternating cycles of flood and famine, providing water in the lean years and preventing or minimizing the flooding in wet years.

The entire project was originally scheduled to be completed by the late 1990s. That was put off until 2005, and then, again, to 2010. Which was just as well: best estimates are that Syria, downstream from Turkey, will lose up to 40 per cent of its Euphrates water to the project. With the population growing at more than 3.5 per cent a year, Syria was facing shortages by century's end, even without GAP. With it, the crisis will turn to a catastrophe.

Internal Syrian sources agree. Mikhail Wakil of the University of Aleppo is more measured, but categorical: "Within the next two decades Syria will experience an ever-increasing water deficit that will have serious effects on its social and economic development. . . . A water crisis is emerging throughout the Middle East, creating the potential for future conflicts." Some time after 2010, Wakil's figures show, Syrian agriculture will need every drop of available water resources just to maintain the current per capita food production, leaving zero water for industry or domestic uses. And that's if the Euphrates supply does not diminish.[4]

Iraq, for its part, would lose somewhere between 80 and 90 per cent of its Euphrates allotment. Iraq still has the Tigris, but that, too, will be affected by projected impoundments. Neither conservation nor more intensive use of smaller rivers would be able to make up the shortfall.

Even without the GAP, consumption targets for the three countries exceed the total supply by 18.3 billion cubic metres per year on the Euphrates system alone, and 5.8 billion for the Tigris.

Both the Syrians and the Iraqis claim ancestral rights to the water. The Syrians argue that the Euphrates is an international watercourse and that disputes involving it should be negotiated in an international forum, something the Turks have adamantly opposed. In the 1970s, Damascus upped the ante by introducing the twin notions of race and religion, always sensitive topics in Middle East politics. Turkey, Syria asserted, was "stealing Arab water," and Arabs should pay it back in kind and discriminate against Turkish companies and citizens. Tellingly, Syria pointed out that Turkey had already reached amicable agreements with its European neighbours – Greece, Bulgaria, and Russia – about water. The fact that it hadn't struck a deal with any Arab country smacked of racism, of the worst kind of Turkish imperialism and hegemony. As the rhetoric notched upwards in stridency, the "real" Turkish aim was said to be to dominate the whole region by making its neighbours dependent on Turkish water supplies. And so on and so on – the diatribes were frequent and long. Then came the allegation that Ankara was planning to steal from the Arabs and supply water instead to Israel. This news came as a surprise to the Israelis, who had at that point no interest in becoming dependent on Ankara themselves. If they suddenly became water-rich, after all, Jordan and the Palestinians would have a much greater case for reassessing internal supplies and for demanding increased allotments.

In 1987, Turkey agreed to provide a steady 500 cubic metres a second at the border with Iraq, something it would never be able to do if the GAP project were ever completed. Syria and Iraq initialled the agreement, but continued to insist it wasn't nearly enough – 700 cubic metres per second, a total of two-thirds of the river's flow, would be more like it.

In January 1990, the Atatürk Dam was finally finished and needed to be filled. The 1987 agreement to supply 500 cubic metres per second notwithstanding, Turkey in effect stopped the river's flow for a full month, to furious-but-ineffective protests from both downstream countries. The Turks did issue advance warnings, and pointed out – correctly, if with over-heavy irony – that it's impossible to fill a dam if the full flow of water is allowed to pass it by. Also, by increasing the downstream allocation somewhat before the stoppage, they managed, with a little creative number-crunching, to prove to their own satisfaction that the "average flow" of 500 cubic metres per second had in fact been maintained. Three years later, with many of their dams at low levels after several seasons of mediocre rainfall, they did it again. This time they chose to turn off the tap in June during Üd al-Adha, the Muslim Feast of the Sacrifice, reducing the flow from 500 cubic metres per second to 170, to further howls of outrage and renewed cries of racism. The timing had been intended positively – the Turks had assumed that since all farming and industry stopped during the festival, Syria wouldn't need its regular allotment. Instead, the choice was perceived as a religious slight and further evidence of Turkish insensitivity, despite a visit from a delegation of Turkish water officials to assure dependent countries that Turkey had no wish to use water as a means of political pressure and wouldn't use its control of the Euphrates for political advantage.[5]

The low level of the Euphrates, aggravated by its poor quality due to agricultural pesticide runoff, chemical pollutants, and heavy doses of salts, has on more than one occasion forced the Syrians to curtail drinking water and hydro generation for the major cities of

Damascus and Aleppo. Damascus is frequently without water at night. The antiquated Syrian distribution system doesn't help: as much as 35 per cent of water simply disappears somewhere along the cracked and leaking network of conduits.

～

Disputes continued over that 500-cubic-metre allotment Turkey had promised. Both Iraq and Syria, still hoping to increase the amount, insist that the agreement in which it was stipulated expired with the completion of the Atatürk Dam. Their position had some merit. The Protocol of 1987 had put it this way: "During the filling up of the Atatürk Dam reservoir and until the final allocation of the waters of the Euphrates among the three riparian countries, the Turkish Side undertakes to release a yearly average of more than 500 m³s at the Turkish–Syrian border, and in cases where the monthly flow falls below the level of 500 m³s, the Turkish Side agrees to make up the difference during the following month." Since the agreement had mentioned the "filling" up of the Atatürk Dam, and the dam was now full, it might indeed be null and void. Therefore, Iraq argued, ancestral rights, which include prior use dating to Sumerian times and the existence of 1.9 million hectares of irrigated land, proved that Turkey should ante up. How much? Well, an allocation of 700 million cubic metres per second would do, it said, returning to earlier arguments. If the annual flow was 1,000 million, Turkey should keep one-third, with the remaining two-thirds to be shared by Iraq and Syria. This would be both equitable and reasonable. Syria, for its part, said the Euphrates should now be divvied up on a simple mathematical formula. Each side would present its demands, and if the total demands exceeded the total supply, deductions would be made from each country's declared requirements in the proportion they were first presented – a tidy formula with virtually irresistible built-in incentives for exaggeration and cheating.

Turkey's response was to reject out-of-hand the whole notion of acquired and inherited rights. As evidence for its position, it trotted out the eminent American water jurist, Stephen McCaffrey, who was for several years a *rapporteur* with the International Law Commission as it struggled to codify international water law. McCaffrey had written a piece for the *Denver Journal of International Law and Policy* in the spring of 1991 in which he suggested that "a downstream state that was the first to develop its water resources could not foreclose later development by an upstream state by demonstrating that the later development would cause it harm; under the doctrine of equitable utilization, the fact that a downstream state was the first to develop (and thus had made prior uses that would be adversely affected by new upstream uses) would be merely one of a number of factors to be taken into consideration in arriving at an equitable allocation of the uses and benefits of the watercourse." In effect, McCaffrey was saying that prior use was important, but so were a lot of other things. In this case, the Turks argued, the "other things" (unspecified) must prevail.

Turkey thereupon came up with what it called a Three-Stage Plan to work out this "equitable utilization." The plan would have entailed an international team of experts to make inventories of both water and irrigable land, and to evaluate needs that way. Iraq peremptorily rejected the plan. Syria ignored it.

In 1991, Turkey planned on hosting the Middle East Water Summit, a multilateral affair of twenty-two Middle East and North African countries, plus a scattering of ministers from Europe, the U.S.S.R., the United States, and Japan; even Israel said it would come. But the summit never happened. The Gulf War happened instead, the U.S.S.R. fractured into an ever-changing variety of ethnic nationalisms, and, to top it all, Turkey's President Turgut Özal suddenly dropped dead.

Sporadic negotiations continued. In 1993, a series of meetings was held between Turkey and Syria, but to such little avail that the two countries were unable even to issue a joint communiqué afterwards. Turkey then invited Iraq to join the process, but when Syria heard about the invitation, it failed to turn up for the next meeting, so the initiative came to naught. Another summit was assembled in Oman in 1994, with over forty delegations, including – a Middle East first – a group from Israel, which caused the absence of Lebanon and Syria. The summit was, in essence, results-free – nothing was agreed on, except that everyone agreed that co-operation was a good thing and that the meeting was useful. In a way, the meeting itself was its own result – the fact of its happening at all was progress. In the Middle East, you creep forward where you can.

Since then there has been only a series of ever-stiffer diplomatic notes, in 1994, 1995, and again in 1996. The last in the series was as blunt as diplomatic language ever gets, short of outright hostilities. Syria and Iraq accused Turkey of reducing and polluting the flow of water downstream, and insisted the Turks should stop. The Turkish reply was more measured: Iraq and Syria had their facts wrong. The accusations were neither technically nor ecologically sound. No reductions in supply had been made except three times, and briefly, for technical reasons, and then with prior warning. There was no pollution: that was a politically motivated accusation. There were no extra agricultural or other activities that would cause pollution. If there was increased pesticide use on the Harran Plain, that would pollute the Balık and Khabur rivers, which drain south into Syria, but not even Syria had alleged pollution there. Several studies had shown that it is possible to return "acceptably clean" water to river systems after "certain uses," and that Turkey could both use more than one-third of the water and, at the same time, release more than two-thirds. And if the dams did threaten obscure species like the hermit ibis, that was regrettable, but the losses were outweighed by the manifold benefits. Nothing, therefore, needed to be done.

⁓

Syria and Iraq are not much more cordial with each other than they are with Turkey. They signed an agreement in 1990 sharing the Euphrates – 52 per cent for Iraq and 48 per cent for Syria – but not before they had almost gone to war on several occasions.

In 1974, when Turkey and Syria were both filling their new dams, the flow to Iraq was reduced to one-quarter of the norm. After a series of escalating threats from both sides, and Iraqi claims that more than three million farmers were suffering water short-ages, Iraq moved troops to the Syrian border and threatened to bomb the Taqba Dam and its associated Assad Lake, constructed on the Euphrates with the aid of Russian engineers. In what might have been a put-up job, the Iraqis the following day arrested a Syrian soldier, who had apparently been planting explosives in the holy city of Kerbala. Syria denied everything, including lowering the river level, which was instead blamed on Turkey. Still, both sides mobilized and seemed on the brink of full-scale war before a team of Saudi and Russian negotiators helped cool things down. The dispute has never been settled, and a high level of fractious-ness is maintained by Syria's repeated insistence that what it does with its water is none of Iraq's concern.

Complicating everything were the shifting alliances in other regional conflicts: Syria against Iraq during the Iraq–Iran war; Syria and Turkey against Iraq in the Gulf War. In the Gulf War, Iraqi soldiers retreating from Kuwait set fire to dozens of oil wells and to Kuwait's entire desalination capacity. In return, the Allied forces intentionally bombed and destroyed Baghdad's already-antiquated water-distribution system. In the early days of the war, moreover, there were attempts to persuade Turkey to threaten to turn off the tap against Iraq by diverting the Tigris, though this direct assault on hapless civilians fortunately remained uttered but not implemented. There also remained the perplexing problem of the Kurds, who have been waging a violent and bloody revolt against both Turkey and Iraq for national independence. The Turks have frequently blamed the Syrians for giving refuge to the PKK, the militantly Marxist Kurdish guerrilla movement that has killed more than four thousand civilians in fifteen years and has promised to blow up the Atatürk Dam and the rest of the GAP as soon as it can. Turkey has more than once threatened to invoke its strangle-hold over the region's water in its struggle with the PKK. In 1989, Turgut Özal, then prime minister and later president until his abrupt and unexpected death, said he would cut off the Euphrates entirely unless Syria expelled the PKK.[6] How he was going to do that was never explained (where would he put it?), but the threat worked anyway.

In late 1998, Turkish–Syrian relations were at an all-time low. Turkish newspapers were speaking openly of air strikes against Damascus, and leaves for soldiers stationed on the frontier were cancelled. President Süleyman Demirel issued a terse statement to the Turkish news agency in which he said: "I am not only warning Syria, I am warning the world. This cannot continue." The Turkish chief of the general staff even declared, stridently and publicly, that there was a "state of undeclared war" with Syria. The situation was complicated by the PKK's public-relations successes in Western Europe – Italy allowed them to demonstrate publicly in Rome – and Turkey's irritation at being kept out of the European Union. For its part, Syria has never scrupled to use Lebanese guerrillas as bargaining chips against Israel, and has taken to doing the same thing with the PKK, even letting its leader, Abdullah Öcalan, keep an office in Damascus. Although it is officially denied, journalists have had no trouble finding him there.

Iraq, too, sheltered Turkish Kurds while waging a bloody war of attrition against its own minority Kurds, including, notoriously, the use of nerve gases and other chemical weapons. Israel, to stir the pot, cheerfully supplied arms to whichever Kurds would buy. After the First Gulf War, there was a mass exodus of Iraqi Kurds to Turkey, further complicating Ankara's already-difficult ethnic problem. Thousands of Kurds also fled to Germany as "guest workers," but increasing racial tensions there drove many of them home, and they escalated their clamour for an independent Kurdistan.

The Marsh Arabs in southern Iraq have also been beneficiaries of Saddam Hussein's ethnic attentions. Iraq's "river project," the 560-kilometre artificial Saddam River that cuts between the Tigris and the Euphrates, starting near Baghdad and ending near Basra, was originally designed to "wash" salinated land and reclaim it for irrigation, an apparently benign ecological scheme. But a large area of additional land was deliberately flooded while the marshes were being drained, pushing aside the Marsh Arabs and making them easy victims to Saddam's pacification schemes. It was against the Marsh Arabs that Saddam first tried out the chemical weapons he later turned on the Kurds. The word *genocide* was bandied about, with considerable justification.

As an added complication, Iraq and Syria were ruled by Ba'ath regimes, both of which claimed to be the authentic voice of Arab nationalism, though Saddam's violent assault on Kuwait in the First Gulf War devalued his claim. Syria is still ruled by the Alawites, a branch of Shi'a Islam, whereas the Iraqi Ba'ath party, militantly secular, was based on the Sunni sect. To further devalue Iraqi authenticity, Damascus was routinely referred to the Iraqi government as "the Takriti regime," Takrit being the home village of Saddam and most of his cronies.

As though that weren't enough, there are also territorial claims, some of them dating back to the fall of the Ottoman Empire. Syria has long claimed the disputed Hatay province of Turkey, while Turkey says it owns the Mosul district of Iraq, which just happens to be rich in oil, a commodity Turkey is otherwise without.

The Orontes (or Aşi) River, which rises in Lebanon and flows through Syria into Hatay, is the cause of yet another dispute between Syria and Turkey whose solution has been elusive. For years, Syria successfully bullied Lebanon into not damming the Orontes, wanting most of the water for itself. The river is effectively dry by the time it reaches Turkey, with a meagre flow of 25 million cubic metres out of the potential total of 1.2 billion, and most of that has already been "used" by Syria for one purpose or another. It has proved impossible even to start talks with the Turks – any agreement would imply Syrian recognition that Turkey owns Hatay. From Turkey's point of view, this stalemate only illuminates Syrian hypocrisy: Syria and Iraq both argue against Turkish appropriation of the Euphrates on the grounds of material harm to the downstream countries – themselves. But here, where Syria is the upstream country, it blithely does what it won't allow Turkey to do elsewhere, and takes all the water for itself.

With this antagonism as background, Özal's surprise offer, while on a trip to the United States, of a "Peace Pipeline" to carry Turkish water to thirsty Middle East countries fell predictably on deaf ears. So has the more-outré scheme to tow water from the Manavgat River, which otherwise flows harmlessly into the Mediterranean, to Israel or anyone else who wants it, in so-called Medusa Bags.

Still, the Peace Pipeline is nothing if not ambitious. There would be two branches, a total of 6,500 kilometres long, distributing more than 2 billion cubic metres of water a year from Turkey's Ceyhan and Seyhan rivers throughout the Middle East and Persian Gulf. The western branch would take water to Syria, Jordan, and western Saudi Arabia, eventually reaching Jeddah and Mecca; the Gulf branch would serve Kuwait, eastern Saudi Arabia, Bahrain, Qatar, the Emirates, and Oman. Both pipes would be buried two metres underground and would pierce mountain ranges in special tunnels. A study by an American consulting firm estimated it would take fifteen years to build and cost around $20 billion in 1998 dollars, thus rivalling for cost and for degree of political improbability other Middle East water schemes – such as an aqueduct from the Euphrates in Iraq to Jordan, a canal from the Nile across the northern Sinai to Gaza, and a diversion of water from Lebanon's Litani River via Israel to Palestine and Jordan.

None of the downstream countries particularly want to be beholden to Turkey for their water, however, no matter the protestations from the Turks that they would never turn off the tap. Everyone in the region has been obdurate in water negotiations with everyone else, and importing water is easily perceived to weaken other national aspirations. The Arabs don't want to rely on the Turks. The Israelis, ever mindful of the ease of sabotage of such a pipeline, don't want to be dependent on anyone, and don't want the notion of imported water to weaken their strongly held position that they are entitled to control all their domestic supply. Jordan, too, is fearful that accepting Turkish water will diminish its entitlement to local supplies. Kuwait most certainly doesn't want to depend on water coming from Iraq, nor does Jordan want to depend on the Syrians. Imported water is no one's first choice. The Turks, for their part, while occasionally speculating idly about the lucrative water market the notion opens up, protest their innocence. As İlter Turan puts it, a touch indignantly, it's not as though Turkey doesn't need the water for itself. There are plans afoot for both the Seyhan and the Ceyhan rivers, and the idea "must be taken as a major gesture of goodwill towards [other countries in the region]."

Indeed, shoot an arrow of peace into the air and get a quiverful of suspicions and paranoia in return. So much for water peace in Old Mesopotamia.

12 | THE NILE

With Egypt adding another million people every nine months, demand is already in critical conflict with supply.

One evening not so long ago I was leaning on the railing of a creaking passenger vessel bound for Yemen and Mogadishu and beyond, staring at the waters of the Rosetta Branch of the Nile Delta below. Off to the right was Alexandria, now a grim place, shabby and mean, but once an illustrious seat of learning, a beacon of civilization. There is nothing much left except a chamber of commerce that talks earnestly of glories restored, as though talking would make it true. Far off to the east was the other branch of the delta, at Damieta, near the mouth of the canal that cuts through to the Red Sea. There used to be more branches – the delta was once a maze of shifting channels – but they have gone, as much history as the Library of Alexandria. The shifting silt that made them used to come down from Ethiopia, but now it stopped at Aswan High Dam or some other barrage, and the sea is reclaiming its own. I looked down at the water. There was a blood-red moon, but the water was warm, sluggish, and black in the Egyptian night. For a brief moment, a pair of dolphins splashed, a silvery glitter that was gone in an instant. I could hear the sound of a rough-hewn diesel, *pukkapukka*, across the harbour. There were low murmurs from the quay, where a group of Arab men in djellabas smoked and gossiped. The dolphins splashed again. They had swum in from Homer's wine-dark sea, here where it meets the Emerald River, flowing down from the Mountains of the Moon. . . . Is there any more romantic spot on earth than this? Did not Mohammed, the Prophet of God himself, say that the "Nile comes out of the Garden of Paradise, and if you were to examine it when it comes out, you would find in it leaves of Paradise"?[1]

The Nile, the *nakhal* or river-valley in Old Semitic, is the longest river on the planet, wandering over 6,800 kilometres through thirty-five degrees of latitude. If you backtrack up the Nile from its delta, you'll pass Al-Qahira, the upstart city Cairo, only a few millennia old, and traverse the Eastern Desert, passing all the grand places of Egyptian history, Deir Mahwas, Qena, Qus, Luxor, Aswan, and Abu Simbel. From there you'll travel up the massive wound of the Great Rift, to the African heartland. In Sudan you'll pass the places where the old empires of Kush and Meroë flourished; they governed Egypt at times, and in other periods Egypt governed them. Past Khartoum you'll have to choose, the White Nile or the Blue. The White and its many branches will take you through the Sudd Marshes, where the Nuer fishermen stand and wait, patient as storks, and on, south to the Great Lakes, to Jinja on Lake Victoria, to Burundi (where the Nile is really born), and to the Ruwenzori, the Mountains of the Moon. The Blue Nile, on the other hand, will take you into Ethiopia, where legend says the Ark of the Covenant was hidden (and is still kept) after its disappearance from the Temple in Jerusalem – Ethiopia, whose culture goes back to the time of Solomon the Wise, king of the Jews. It is the annual drenching of the Ethiopian highlands that spills over into the Nile, and the resulting deposit of silt in the delta that enabled the fecundity of Egyptian civilization.

When Nero ordered his centurions to find the source of the great river, they got no further than the impenetrable marshes of the Sudd. Stanley, on his travels through Africa, recounts many legends of the Nile, most of them borrowed, in his magpie journalistic way, from early Arab manuscripts: "As for the Nile, it starts from the Mountains of Gumr. . . . Some say that word ought to be pronounced Kamar, which means the moon, but the traveller, Ti Tarshi, says it was called by that name because the eye is dazzled by the great brightness."[2]

Veru, in his *Antiquities*, says Isis, Queen of Many Names, had travelled down the Nile from Ethiopia to Egypt. The Ethiopians, whenever they want to remind the Egyptians of the source of their bounty, are fond of quoting this one: most recently it was trotted out by Dr. Mesfin Abebe, Ethiopia's minister of natural resources and environmental protection, in a speech in Cairo to the eighth World Congress on Water Resources. The speech had as its theme the need for regional co-operation, but he spent a good deal of time being proprietary about the Nile, reminding the assembled delegates that almost two-thirds of the lower Nile's water comes from the Blue Nile and from a tributary called the Atbara, both of which originate in Ethiopia.

Which is a reminder that there are more practical matters in Egypt than religious esoterica, however romantic, and the Egyptians were (and still are) an immensely practical people. The Nile produced the first agronomists, the first engineers of the farm. Achmed, son of Ti Farshi, in his book on the description of the Nile, says historians relate that Adam bequeathed the Nile unto Seth, his son, and it remained in the possession of these children of prophecy and of religion. They came down to Egypt (or Cairo), when it was called Lul, and they dwelt upon the mountains. After them came a son Kinaan, then his son Mahaleel, and his son Yaoud, and his son Hamu, and his son Hermes – that is, Idrisi the prophet (Enoch). Idrisi began to reduce the land to law and order: he was the first man to regulate the flow of the Nile to Egypt. Idrisi gathered the people of Egypt and went with them to the first stream of the Nile, and there adjusted the levelling of the land and the water by lowering the high land, raising the low land, and other things according to the science of astronomy and surveying. Idrisi was the first person who spoke and wrote books on those sciences. It is said that in the days of Am Kaam, one of the kings of Egypt, Idrisi was taken up to Heaven, and he prophesied the coming of the flood. He remained on the other side of the equator and there built a palace on the slopes of Mount Gumr. He built it of copper and made eighty-five statues of copper. The waters of the Nile flowed out through the mouths of these statues and then into a great lake, and thence to Egypt.[3]

Legends, lost in the suburban pages of history, but in a way true enough. The Nile is the Father of Rivers, the Seed of Civilization. Modern engineers, at the urging of the pan-Arabist Gamal Abdel Nasser, didn't do nearly as well as Idrisi of old. They put a stop to the Nile Delta's infusions of silt by building the Aswan High Dam. No one yet knows the full consequences.

The Nile is important – critical – to Egypt, which gets no usable rain and has no other water but a few rapidly diminishing aquifers under the desert. Only 2 per cent of Egypt is not desert, and water stress is rising every month. Egypt's 70 million people, rising to 85 million in a decade, are entirely dependent on the river. The Nile flows through eight other countries before reaching Egypt, the last in line. According to a 1959 agreement with Sudan, the Egyptians are entitled to 55 billion cubic metres of Nile water a year, while Sudan gets 18.5 billion. In 1990, according to Sandra Postel, total water availability in Egypt was 63.5 billion cubic metres, but in 1998 demand was already passing 68 billion and rising steadily. USAid predicted in the mid-1990s that Egypt would experience a 16- to 30-per-cent water deficit by the end of the century. By 1998 that alarmist prediction was not yet true, but the Egyptians were closing in on it. The targets for recycling water were not quite being met, despite more or less heroic efforts, and drought was reducing the Nile's overall flow.

Where is the extra water to come from? Egypt's water minister at the time, Dr. Mohammed Abdel Hadi Radi, maintained in 1994 that Egypt was already efficient at water reuse: "every drop is reused at least twice, and water efficiency is estimated at 75 per cent," almost as high as Israel's. Still, a study by Dr. Hussam Fahmy for the National Water Research Centre suggested that Egypt's supply could be improved by capturing another 2.5 billion metres from the Nile by completing the Jonglei Canal with Sudan and by squeezing another 4 or 5 billion cubic metres from more intensive reuse of drainage water. Egypt has to some degree privatized agriculture, allowing farmers to sell directly to consumers instead of through the government, and is considering charges for irrigation water to offset the revenue losses. This trend is expected to improve efficiencies even further, since wasteful use of water hits the individual farmer directly. In the Nile Delta, efforts are being made to reclaim land. New college graduates of agriculture are given two hectares of land and a few hundred dollars a year to farm the land and feed themselves with tightly managed (and recycled) water supplies. The model? The kibbutzim.

But even this development would simply take supply up to meet demand, with no margin for error. In many other countries in the Nile Basin, water availability is on a collision course with irreducible demand. Tanzania, Burundi, Rwanda, Kenya, and Ethiopia are already critically short of water. So, indeed, are other North African countries far out of reach of the Nile: Libya, Tunisia, Algeria, and even Morocco, though the Moroccans manage to feed themselves and still export food.

Egypt has said in the past it is willing to go to war to prevent anyone upstream from tampering with its water flow, and there is no reason to suspect that anything has changed. In the mid-1980s it was on the point of ordering air attacks against Khartoum for perceived water threats, until cooler heads prevailed. What has saved Egypt until now is that both the Sudan and Ethiopia were too poor and wracked by civil war to contemplate any large-scale water diversions. This was why the settlement of Ethiopia's internal war and the destruction of its Derg tyrants raised such alarm in Egypt. And the reason why the cynical Egyptians aren't particularly sorry that Ethiopia seems to be settling back into its grim military fugue as the millennium nears its end.

Current tensions between Egypt, Ethiopia, and Sudan are in one way a continuation of a two-thousand-year-old struggle over who will control the water. Often in their history, Egyptian rulers have sought to unify the Nile Valley under their rule by conquering Sudan. In poor years, when the rains failed in Ethiopia and the annual flooding didn't come, it was common knowledge that human mischief was afoot. More than once an Egyptian sultan sent his ambassadors to the Ethiopian kings to plead with them not to obstruct the waters. The Scottish explorer, James Bruce, who spent years in Ethiopia and at the Gonder court in the eighteenth century, recalled that the king had sent a letter to the pasha in 1704, threatening to cut off the water. The current anxieties have only sharpened the potential for conflict. It's not just Egypt whose demand is increasing. Development, that Holy Grail pursued by all nations, will inevitably up the demand all along the river. The supply, however, cannot be changed.

The British, who controlled Egypt and Sudan on and off for good parts of the nineteenth and twentieth centuries, were the first to try to impose a basin-wide plan. To improve navigation, they sent sappers in to hack away at the series of natural dams and impediments that had built up over the years in the Sudan's Sudd marshes, impeding all shipping in the region. Like most of the developers who followed, the English were careless of the effects this would have on the marshes and the local climate, or on the local people, whose living was made from fishing. Before they could get very far, however, they had to leave, for Mohammed Ali, pasha of Egypt, sent mercenaries upriver to set up his own fiefdom at Khartoum. After Ali's death, control of northern Sudan became disorganized and corrupt. There were more and more European intrusions; the British in particular, who were once again meddling in Egypt, returned to the Sudan in a typically imperial mix of profit-taking, high-mindedness, and an eye for the main chance: control of the Nile. With Egypt falling into bankruptcy, matters became even more chaotic, and in 1881 the Sudanese rebelled in a fervent religious and political uprising called Mahdism, led by Mahdi Mohammed-Ahmed, who preached a version of Sufism. Among his victories was the massacre of the army of British general "Chinese" Gordon at Khartoum in 1885. The Mahdi ruled until 1898, when he was defeated and overthrown by an Anglo-Egyptian army under Lord Kitchener.

Shortly afterwards, the sappers returned to the marshes in full force. By 1904, they had hacked a narrow path all the way through the marshes, and army engineers began drawing up plans to control and regulate the Nile's full flow. These plans were complicated by the fact that Britain dominated Egypt and Sudan, but not Ethiopia. Although it went against the imperial grain, the British were obliged to sign an agreement in 1902 with the Ethiopians that neither side would tamper with the Blue Nile. The British also had to bully the French and the Italians into going along with their plans; both European powers had asserted rights in the basin. By the time they agreed, however, the Egyptian government was baulking. It wasn't until 1929 that Britain was able to sponsor the Nile Water Agreement, which regulated the flow of the Nile and apportioned its use.

The Second World War postponed these useful endeavours. After it was over, the British, still in their own minds the political masters of the Nile Basin, commissioned what was to be the first thorough hydrological study of the Nile. Alas, once again it wasn't completed. The Ethiopians, restored to independence and jealous of their rewon prerogatives, declined to participate.

The study was finally released in 1958 as the Report on the Nile Valley Plan. It was in substance uncontroversial, but it did suggest a number of ways to increase the amount of water that reached Egypt. The most critical was the construction of the Jonglei Canal from Bor to Malakal, finishing what the British had started seventy-five years earlier – to push a channel through the Sudd marshes and straighten out a "wasteful" U-turn in the course of the White Nile proper. The point now was not so much navigation – where, after all, would the boats go? To Uganda? The point was pushing the waters of the Nile faster through the marshes. The canal would eliminate the enormous evaporation in the tropical sun during the water's sluggish passages through the marshes, freeing up what the report suggested would be four billion cubic metres a year for downstream users. Again, what this would do to the local inhabitants or to the ecosystem of the marshes was left unexplored. In the 1950s, marshes didn't have ecosystems, or if they did they were not perceived to benefit anyone. That climate is dependent on circulating moisture and evaporation patterns was still unknown; climate didn't seem so easily tampered with in those innocent days.

The plan, put out with great fanfare in Khartoum, went nowhere. Prominent in its preamble was the notion that the Nile Basin must be treated, hydrologically speaking, as a single entity, and that the newly independent states would have to look beyond their own narrow interests and work together for everyone's benefit. This was a bit thick, coming as it did from a former colonial power, and only a few years after French and British paratroopers had dropped uninvited into Suez. Egypt, in any case, had somewhat lost interest in the Jonglei. It had developed plans of its own for building a dam at Aswan, and was busy playing the Russians off against the Americans to generate funds to do so.

In the early 1960s, Julius Nyerere, then the leader of independent Tanganyika (not yet Tanzania), declared that "former colonial countries had no role in the formulation and conclusion of treaties done during the colonial era, and therefore they must not be assumed to automatically succeed to those treaties." Nyerere bluntly told Britain and the other Nile countries that he had no intention of adhering to the 1929 Nile agreement. Kenya and Uganda agreed with Nyerere. Egypt – which benefited most from the now rejected agreement – did not, but there was nothing it could do about it.

In Khartoum, the Egyptian plans for Aswan received a cool reception. A few years earlier, when the Sudanese had proposed a dam of their own on the Blue Nile at Roseires to generate power for Khartoum, the Egyptians had been furious. Any reservoir above Khartoum would slow the Nile down, increase evaporation, and lose the Egyptians water. The Egyptians blustered and bullied, and the rhetoric became ever more extravagant, until the climate was right for the Egyptian army to draw up battle plans. The Sudanese plan for the Blue Nile came to naught, however, and the crisis evaporated. Now it was the Egyptian turn to build a dam – and it would flood vast quantities of northern Sudan, force the resettlement of more than fifty thousand hapless Nubians, and change life on the banks of the Nile. Where was compensation to be found?

After a few years of simmering hostility, the two nations finally got together in late 1958. Once around the table, it looked a little easier, and in early 1959 they initialled an agreement for "full utilization of the Nile waters." Sudan was placated by upping its allotment of Nile water from the 4 billion cubic metres stipulated in the 1929 agreement to 18.5 billion cubic metres. It would now be allowed to build the Roseires Dam, and was encouraged to go ahead with the Jonglei Canal. In exchange, Egypt would be allowed to proceed at Aswan without interference. The treaty set up a joint committee to supervise all development projects that might affect the flow of the river.

There was one major drawback to this newfound cordiality: the agreement divvied up the Nile's water without consulting Ethiopia, from which most of it came. Ethiopia, in retaliation, declared it would reserve the right to use Nile water in whatever way it chose, which once again sent Egypt's military planners to their map books.

The hapless Nubians, up to one hundred thousand of them by now, were duly resettled. Unfortunately, some had to be moved a second time, when construction started on the Roseires Dam. The Sudanese government promised compensation, but thirty years later no money had been forthcoming, and the Nubians weren't holding their breath; many of them joined the insurgents in the south.

The Aswan High Dam was finished in 1970, and, behind it, Lake Nasser – six hundred kilometres long and fifty kilometres wide in some places – started to fill. The same year, Sudan and Egypt began joint construction of the Jonglei Canal, and were pushing it steadily southward, when construction crews were driven out by guerrilla raids by South Sudanese rebels of the Sudanese People's Liberation Army. They have never returned, and nothing has happened in the marshes since 1983. Sudan is still mired in a debilitating civil war. The two countries had spent more than $100 million of donor-country money and had nothing to show for it. Egypt needs the water the canal is supposed to yield, has paid real money for it, and is getting nothing. In the last few years, the go-it-alone growls from Cairo were getting louder. Rumours – no more than rumours – say that another spasm of Egypt's military musculature occurred in 1994: Sudan can't get itself together, we need the water, so let's do it. The story is denied, of course, by all sides. But the following year there was an attempted assassination against President Hasni Mubarak of Egypt; Egyptian security officials didn't blame Sudan, exactly, but didn't exonerate it either. Border clashes between the two countries flared up again in summer 1998. Tensions remain high, and at this point the chances of Egypt getting water from the Jonglei are more-or-less zero.

~

For the Egyptians, Ethiopia is even trickier than Sudan. For one thing, it is further away, and thus inherently less controllable. And, at least, the Sudanese are fellow Muslims. You never know with the Ethiopians. About half of them are Christians, when they're not being Marxist. They've always gone their own way, and still do. Prickly, too. A month after Sudanese independence in 1956, and again during the Suez crisis, Ethiopia issued diplomatic notes highly undiplomatic in tone. Ethiopia, they declared, "reserved the right to utilize the waters of the Nile for the benefit of its peoples, whatever might be the measure of utilization of such waters sought by riparian states." In other words, Buzz off. The Nile is ours to use as we see fit.

For a while, it looked as though they meant it. In the 1960s, the Ethiopian government, at the personal urging of the Lion of Judah himself, the emperor Haile Selassie, hired the U.S. Interior Department, in the form of the Bureau of Reclamation, to develop a master plan for the Ethiopian Blue Nile and its tributaries. This was like honey to a bear: bureau engineers had spent half a century damming everything they could find in the United States, in legendary competition with their prime enemy, the Army Corps of Engineers, and the notion of a tabula rasa for dam-building – all those pristine, dam-free rivers, flowing uselessly through to the sea – must have been irresistible. They went ahead with sublime indifference to declared State Department policy, which was not to hassle the Egyptians in the aftermath of the Aswan debacle, in which the Soviets were thought to have scored precious diplomatic brownie points among Third World countries. (The dam's evil side-effects were not yet known, though Egypt's own engineers were already grumbling about the dam, *sotto voce*.)

In any case, the bureau's planners came up with a scheme that involved the building of twenty-nine irrigation and hydroelectric projects, varying in size from modest to enormous. The bureau's own figures show that, had all the schemes been implemented, the annual flooding of the Blue Nile – with its precious deposits of silt – would have been eliminated altogether, and the total flow would have been reduced by some 8.5 per cent. Talk about provocative!

In any case, only one project was started, and even that was never finished: Egypt blocked the necessary financial approval of a loan by the African Development Bank. And in 1974 Haile Selassie was overthrown by the Derg, a cabal of junior military officers with radical Marxist and dictatorial leanings, under the sinister influence of Mengistu Haile Mariam.

Mengistu brought his book-learned Marxism to bear on the Nile water problem. To develop an irrigated agriculture in the Nile Basin, he devised resettlement schemes along the tributaries of the Blue Nile and along the Baro–Sobat, a tributary of the White Nile. At the height of the most notorious human-induced famine in recent history, in 1984, the Derg pronounced that it would move more than 1 million people into these new settlements, and the first 15,000 were rounded up and bundled off to their new "homes." In the next year and a half, 1 million more were arbitrarily uprooted and moved, a piece of social engineering that caused immense suffering and thousands of deaths, and succeeded in irrigating enough land to feed about 100 people.

It didn't work – so of course Mengistu did more of it. If you didn't have to confront what actually happened on the ground, resettlement satisfied so many policy initiatives. It relieved population pressures in the cities, helped famine victims, settled down nomads, and provided land to the landless. So Mengistu announced he would move ten times as many people as before.

The policy was an organizational and economic failure, and a human disaster. Fortunately, Mengistu's control was never complete. The Eritrean guerrillas continued their war, and were soon joined by the Tigrai People's Liberation Front. In May 1991, with rebels only a few kilometres from the capital, Mengistu fled the country and took refuge in Zimbabwe. Eritrea gained (or regained, depending on your historical perspective) its hard-won independence, and a coalition of rebel groups under Meles Zenawi took control of a country with an empty treasury, a moribund economy, and a ruined agricultural sector. Despite this, they announced multi-party elections and a program of economic reform.

In 1993, a study by J.A. Dyer and his colleagues at Ottawa's Carleton University found significant potential for agricultural development in Ethiopia. True, parts of the country were arid – the Ogaden, for example – but "regions such as Kefa and Nyala have very rich land resources, capable of year-round production of food and cash crops." Other regions had potential, but would demand substantial irrigation. Irrigation would mean dams. And dams would alert the ever-vigilant Egyptians.

In 1998, however, fighting once again broke out between Addis Ababa and the Eritreans.

So far, nothing has been done in Ethiopia to damage Egypt's interests. But the Egyptians are always watchful.

In 1990, while Mengistu was still in charge, there were widespread reports that Israeli water engineers were examining Lake Tana, the source of the Blue Nile, with a view to new irrigation schemes there. This notion was plausible: Israeli policy had always been to support the Ethiopians as a countervailing force to the dominant Muslim cultures of the region, and there were well-founded rumours that the Israelis were secretly helping the South Sudan guerrilla rebels against Khartoum.

In 1993, a general agreement was reached between the new Ethiopian government and Egypt. It was vague, but it did contain a clause in which each country agreed not to do anything to the Nile that might harm the other, and, perhaps more important, they agreed that "future water resource cooperation would be grounded in international law," without specifying what law, exactly.

But it is not so much what the Ethiopians have done, as Swedish anthropologist Jan Hultin said in a survey for Leif Ohlsson's *Hydropolitics*, but rather what they might be doing. Former UN secretary-general Boutros Boutros Ghali, when he was Egypt's foreign minister in 1990, pointed out that "the national security of Egypt, which is based on the water of the Nile, is in the hands of other African countries." Later, he was even more blunt: "The next war in our region will be over the waters of the Nile," he said, echoing a comment made by Anwar al-Sadat fifteen years earlier.

The potential for war is very real. But, with some political give and take, so is the prospect for regional co-operation. The best way for Egypt to "store" Nile water would be in massive reservoirs in the Ethiopian highlands, where evaporation rates would be far lower than they are at Lake Nasser, in the middle of the desert. It loses two metres every year to the sun, and in the 1979–88 drought reduced the water level to a degree that threatened hydro-power generation. Some studies have shown that enough water could be saved in these reservoirs to quadruple Ethiopia's irrigated areas without affecting the downstream countries.[4] This solution would take a level of trust that none of the countries of the region have hitherto shown. But Egypt's leaders have not been strangers to vision and they know the need is critical. No one really wants a war. If it comes, it will come only as a failure of political imagination.

In 1997, construction was supposed to start on an Egyptian scheme to deal with the fact that all the country's people live on a fraction of the land, a narrow cultivable strip along the Nile. The notion, first concocted by Sadat in the 1970s, was to construct a brand-new "valley of the Nile" – a new, self-sustaining river that would flow through the Western Desert. A canal, called the New Valley Canal, would connect a series of oases south to Al-Kharga and Dakhleh, taking advantage of the excellent-but-waterless soil that exists in many parts of the country. A large number of people would be resettled there – not as the Ethiopians had tried to do, by arbitrary decree, but by land-holding incentives and subsidies. Early surveys said there was enough groundwater in the deep aquifers to make the scheme sustainable; the aquifers were thought to be recharged by outflows from the Tibesti and Ennedi mountains. But by the 1990s it was far from clear that the aquifers were being recharged enough, or at all, and even the optimists were saying that the scheme would, at best, last a hundred years. After all, the wells in the two southern oases had been no more than 150 metres deep just a few decades ago, but are now more than 1,000 metres deep, and still dropping. Testing has shown that the drop is directly proportional to the wells' output – the water is therefore being unsustainably mined. At the same time, the population of the Dakhleh oasis has grown from 45,000 to 75,000 people over twenty-five years.

These dismal numbers pushed the notion onto the bureaucratic shelves marked "inaction." But in 1997, Hosni Mubarak hauled out the plans again and dusted them off. He called his version the New Delta Project, or sometimes the Toshka Project, and said it would draw off some of the Nile water from Lake Nasser – 5.5 billion cubic metres, about a tenth of the country's demand. The world's largest pumping station would then raise the water 50 metres and drive it through a canal that leads to the Toshka Depression, 180 kilometres south of Aswan. Mubarak tactfully called this brand-new canal the Sheikh Zayad Canal, after the president of the United Arab Emirates, who stumped up the money. The first water flowed through the canal in 2001, and by 2002 the first irrigated crops were already being harvested. Plans are to extend the canal to Kharga and Dakhleh by 2017, since it is obvious that the groundwater there will be depleted in about twenty years.

The president's stated intention was to "finally get Egypt out of the narrow confines of the Nile Valley" by developing a new zone of three hundred thousand hectares for farming, with an additional hundred thousand hectares around the oases. Some six million Egyptians would be resettled in the new zone.

But the whole scheme depends on a number of risky and entirely speculative assumptions. No account has been taken of the massive evaporation that will occur in the canal itself. The slow-moving water will be liable to schistosomiasis, just as the canals below Lake Nasser are; in addition, the poor drainage in the destination area will result in salination and further desertification. The project was said by the country's lenders to be sucking the lifeblood out of the economy – of the $4 billion or so Egypt owed in 2000, more than half was destined for foreign contractors building the Toshka scheme. In any case, where would this "extra" Nile water come from? The only way Egypt could get more water was to renegotiate the agreement with Sudan. However, Sudan said No – it, too, needed more water. Both Sudan and Ethiopia have fertile soil, easily irrigable, and in theory could develop vigorous and productive agricultural industries based on irrigation. Of course, by doing so, they would be reducing Egypt's supply.[5]

Ethiopia's reaction to the Toshka scheme was predictably hostile. It immediately demanded an amendment to the 1959 treaty between Sudan and Egypt (which had left Ethiopia on the outside looking in) and said it wanted an additional share of Nile water, unspecified in amount, to allow it to implement its own projects. The tame Ethiopian press duly carried strongly worded statements from Ethiopian officials. The chief engineer in the Ethiopian ministry of water resources was quoted as saying that there would be "no co-operation among the Nile countries until a clear commitment to the principle of fair and equitable use was established." The statement was ignored in Cairo. Agreement seemed as unlikely as ever.[6]

The only short-term solution available, apart from intensifying recycling and conservation, is to divert more water from farming – to turn more farmers into city dwellers, if the economy can absorb them. In the medium term, limiting population growth is essential; in the long term, so is regional co-operation. The same Cairo meeting to which Ethiopia's Dr. Abebe delivered his keynote address ("The Nile – Source of Regional Co-operation or Conflict?") failed to come up with any solutions, only pious platitudes. Abebe called for the appointment, at least, of a panel of experts to study such things as irrigation, hydro-power generation, flood control, flow augmentation, fisheries, and the sound management of the environment. No such panel was appointed. In fact, despite Abebe's speech, Ethiopia was widely criticized at the meeting for obstructionism. It was noted, ironically, that increased silt loads in Sudan's dams could be blamed directly on deforestation in Ethiopia and its consequent erosion. Ethiopia flatly refused to discuss the issue, or any other issue it decided was "purely domestic."

In 2003, that's where the matter rested. Everyone wanted more Nile water, and everyone was planning schemes to use more of "their" allotments. Demand remained in critical conflict with supply.

13 | SOUTHERN AFRICA

At the other end of Africa, politically induced famine overlapped a catastrophic natural drought, and millions were at risk of starvation. But some countries were getting water policy more or less right.

F ew places better illustrate the intricate intersection of water and politics than Southern Africa, which was in 2003 faced with the appalling prospect of famine among nearly thirty million people – a famine brought on by prolonged droughts, but also by gross mismanagement and political bossism verging on genocide.

There were also bright spots among the gloom. On the political side, South Africa itself was getting its water policy right, despite opposition from all sides; Namibia, for its part, had more rain in 2002 than it had for years, and the Namib desert, normally one of the driest places on earth, was wearing a thin skin of green for the first time in decades.

In the fall of 2002, a flurry of stories hostile to the South African government suddenly started appearing in the reliable organs of the Green movement – those served by the ENS (or Environmental News Service) and in magazines such as America's *Mother Jones*, an indefatigable opponent of free-marketism. Curiously, all these stories had the same basic structure. The reporter would find a poor family somewhere driven to desperate straits, and use their plight to beat the African National Congress government of Thabo Mbeki over the head.

Entirely typical was a piece for ENS by Mohammed Allie, titled "South Africa Flounders in Its Search for Free Water." It was date-lined Cape Town, March 26, 2002. Here are the first few paragraphs:

> David Shezi did not necessarily think he was doing anything criminal when he stole some water one day in January. He was doing no more than any other man would do to help his family survive in conditions of extreme poverty.
>
> With an income of only R100 (US$8.65) a month and with eight children to support, Shezi could not afford to pay for the water supplied to his home via a water meter – and used a pipe instead to illegally siphon it off in his village in KwaZulu–Natal.
>
> Despite a government ruling that all South Africans should have free access to a basic amount of clean and safe water, Shezi was arrested for stealing water.

After that, the piece was entirely predictable. It briefly acknowl-edged the government's success is bringing clean water to seven million citizens who had hitherto had no access (a grudging tribute that was published adjacent to a photograph with the caption "Street children in Johannesburg have uncertain access to clean drinking water"), but then found a person identified as a "researcher" who could reliably fulminate about privatization, in this case from the University of Leeds, in England.

Nevertheless, when the ANC inherited the country from the last apartheid government, 12 million of a total population of 38 million had no access to clean water. By 2002, that number was down to 7 million of a total population of 42 million. The ghastly shantytowns outside the major cities were ghastly still, but there were piped water to public standpipes and basic latrines for the first time. It may not have looked like much to outsiders, but it *was* progress.

The government has achieved this entirely remarkable result by a combination of political will and the willingness to experiment, devising policies that initially met opposition from both the development banks and NGO lobby groups.

The twin underpinnings of the policy were a series of regional public-private partnerships, and, more controversially, a determination that water would have to be priced sufficiently high to both deter its waste and underwrite the maintenance of its infrastructure.

Ronnie Kasrils, the water minister, understood perfectly well that millions of the poorest citizens couldn't afford to pay for their water. He determined, therefore, that each family should receive the first six thousand litres of water per month free of charge; after that, they would have to pay, on a rising scale according to consumption. This meant that the wealthiest South Africans – generally still white suburbanites – would pay a much steeper rate if, for example, they wanted to water extensive gardens or fill a swimming pool. The chosen system was a variant of the electricity metering card already in widespread use in South Africa

By the end of 2002, some local water authorities were still trying to work out how to put these measures into practice. Many, like David Shezi's local authority in KwaZulu–Natal, involved private companies in water-management systems. David McDonald, co-director of the Municipal Services Project, which has been monitoring the privatization of water services in South Africa, says privatization "is most often seen as part necessity, part political choice because of fiscal constraints," such as reducing public borrowing, taxes, and outlays.

There is, clearly, a long way to go. There are inequalities between the rural and urban populations and, of the 12 per cent of South Africa's water consumed by households, more than half goes into gardens and swimming pools in the wealthier (white) areas, and less than a tenth is consumed by black households.

Still, in terms of national supply, there was adequate water if properly managed. The southern area of the country had good winter rains for the first two years of the millennium; there was drought to the north, but the dams of Lesotho's controversial Central Highlands Project were now on-stream, and the supply to the central part of the country, the industrial heartland around Johannesburg, seemed at least assured.

Things were not going so swimmingly on South Africa's borders. To the northeast, Mozambique, still recovering from twenty years of civil war and a series of years in which massive rains caused widespread flooding, was suddenly facing drought-caused water shortages. The main problem was, however, that the drought was not local – Mozambique's rivers all originate outside its borders, in South Africa, Zimbabwe, or Swaziland, and by the time they reach Mozambique, rivers like the Limpopo and Incomati are substantially drained or polluted or both.

The Incomati, which flow through the Kruger National Park game reserve along the border with Mozambique, is the most contentious of these, because the South Africans have stepped up agricultural withdrawals from Incomati tributaries. Many Mozambican farmers, already struggling with a ruined infrastructure, poor roads, and uncertain access to markets, suddenly found themselves without reliable water, and sometimes with no water at all.

A regulatory treaty, signed between South Africa, Swaziland, and Mozambique as long ago as the 1980s, allocated Mozambique a minimum flow of two cubic metres a second of the Incomati, a modest enough number. South Africa professed itself satisfied,

Mozambique did not. In 1998, the southern Mozambican province of Maputo accused South Africa of violating the agreement over water use. Farmers in Maputo pointed out, with considerable justification, that, at the slightest hint of drought, their South African counterparts impounded as much Incomati water as they could, and on occasion the river was little more than a drainage ditch by the time it reached the border. There were echoes of another dispute far away, over the Colorado River, which the Americans had effectively stolen from the Mexicans.

By late 2002, the issue was still unresolved, despite otherwise cordial relations between the two countries. South Africa's department of water affairs dealt with the allegations largely by ignoring them; Mozambique, for its part, threatened to take South Africa to the International Court of Justice – a piece of political theatre designed to force South Africa's hand on the Incomati by damaging its standing in the Organisation of African Unity (OAU), which at the time was still mutating into the African Union.

To South Africa's northwest, the real problem was between Namibia and Botswana, two of the driest countries on earth. Namibia has no perennial rivers of its own, only those on its southern and northern borders – the Orange, which it shares with South Africa, the Kunene, which it shares with Angola, and the Okavango, which it shares with Angola and Botswana. Botswana isn't much better off.

The Okavango has been – and still is – the locus of an environment struggle between two well-meaning states, in which need is trumping neighbourliness. The third-largest river in southern Africa, it rises in the moist tropical hills of Angola, where it is known as the Cubango, and flows for about 400 kilometres through Namibia and into Botswana; there it soaks into the flat plains of the Kalahari and spreads out in a dazzling array of channels that make up Africa's largest oasis and the world's most spectacular inland delta. This vast marshy zone, known in local tourist lingo as "the jewel of the Kalahari," is 175 kilometres long and 180 wide, a luxuriant emerald-green paradise of almost 16,000 square kilometres. The last free-roaming herds of Cape buffalo are found here, along with a host of other wild animals and some 350 species of birds, including the jacana, the bee-eater, the malachite kingfisher, and the African fish eagle.

Once the Okavango Delta was a lake, but a geologic era ago the region began to dry up. The Okavango River continued to pour down across what is now the Caprivi Strip from the moist hills of Angola's Benguela Plateau as it always had, but the lake disappeared. The river became tangled in thickets of reeds, giant papyrus, and mud, and then just wandered into the desert and vanished. *Riviersonderend*, the Afrikaners called it, River Without End. There are many such rivers in the wild and desolate north, the Great Thirstland, but the Okavango was – and is – the grandest, most fertile, and most beautiful of all.

In good years the waters still overflow the marshes into the Boteti River and reach the parched, arid surface of the Makgadikgadi Pans to the southeast. In the best years of all they seep northeastward through the Selinda Spillway into the even more remote and mysterious Linyanti Lagoons. This marsh is one of the grandest places on earth. So many legends, so entangled with mundane and exotic fact! The *munu*, the black tree baboons that folklore says stalk Okavango women, longing for the day they might be men again. Lions hunting at night. Lagoons boiling with hippos. Endless forests of thorn acacia and mopane trees, home to leopards, cheetahs, wild dogs, Cape buffalo, giraffes, a shopping cart full of monkey species, the greatest migrations of zebras on earth, and huge herds of elephants. These sixty thousand elephants are the subject of bitter controversy. The Botswanan government prohibited all elephant hunting in 1983, to universal acclaim, but since then the herds have proliferated, causing massive habitat destruction. And in the late 1990s they were being culled; some fifteen thousand elephants had to be shot, here and in Zimbabwe and Angola, to end their own suicidal eating binge.[1]

The Okavango swamp has other enemies: drought, for one. In the 1990s, the Kalahari region and Angola suffered from the worst drought of the century; El Niño was blamed, but when the disturbance ended and the rains still did not come, human-induced climate change became the cause of choice. Mostly, the inhabitants were reduced to praying for rain, but the water levels continued to shrink. And from the south, from Botswana's orderly and well-managed cities, comes an enemy even worse than drought – growing human populations and their demands for more, more, more. And from neighbouring Namibia too. Namibia is a desert country and always parched. But its population is growing, and in 1996, with the country's reservoirs standing at 9 per cent of capacity, Namibian authorities turned their eyes eastward, to Botswana's Okavango Delta.

By 1997, there were already threats of a "water war" between Namibia, which wanted to take twenty million cubic metres a year from the Okavango system, and Botswana, whose own dams and reservoirs were critically low after ten years without rain. In the end, this talk of war was mere hyperbole. Still, there were threats of sabotage if a pipeline proposed by Namibia were ever to be built. Late in 1997, residents of Maun, Botswana, walked out of a conciliatory meeting called by the Namibian government. And both countries went to the International Court of Justice at The Hague to dispute a minor boundary issue, a sure sign of international fractiousness.

～

Maun, on the southern fringes of the Okavango, is Botswana's safari capital, a dusty, chaotic African village grown into a town. There are still clusters of traditional huts within its boundaries: small, neatly painted concrete-block houses with corrugated iron roofs. Commercial buildings are randomly scattered about. The whole place is as complex and organic as the Okavango Delta itself, which lies just to the north. It's no wonder that ecologists like Maun, while engineers despise it. Tourism in the Okavango keeps Maun alive. It is also the third-largest foreign-currency earner for Botswana, and the government is looking to tourism to replace the ailing minerals sector.

The Botswanan government, sensitive to water and ecological issues, is fully aware that sustainable development involves wise management of water resources. Early in the 1990s, it designed the Southern Okavango Integrated Water Development Project to increase food production and to provide water for both Maun and a cluster of diamond mines located in the desert 280 kilometres to the southeast. The project involved dredging 42 kilometres of waterways in the southern part of the delta and creating two large reservoirs that would have submerged good agricultural land in order to create new irrigable land elsewhere. The local people were hostile and, at a seven-hour *kgotla*, or meeting, seven hundred villagers asked the government to drop the idea. Greenpeace International, meanwhile, had been alerted to the situation, and, according to rumour, had plans for a worldwide diamond boycott under the slogan "Diamonds Are for Death." When the diamond interests concerned distanced themselves from the project, the government suspended work pending further evaluation. A team from the International Union for the Conservation of Nature came to Botswana and suggested numerous smaller-scale alternatives to address water shortages in the Maun area, but in 1992 the government shelved the whole project.[2]

In 1996, when the annual floods failed, Maun was put immediately at risk – not just tourism, but the villagers, too, who needed the river for washing, for fish, and for water-lily roots and reeds to build houses. The town's drinking water, drawn from boreholes, was also drying up. Botswana's hydrologists scrambled to find out just how rapidly the water table was dropping. Alarmist stories were everywhere. Namibia's pipeline would mean permanent drought, residents were told. Wells and boreholes would go dry. The district health department report for 1997 showed that water in seven of Maun's twenty-two standpipes, or communal water dispensers, was contaminated by human or animal feces.[3] Tawana Moremi, the senior chief of the Bayei, was indignant. "I don't like this pipeline very much," he told everyone who would listen. "We should buy more planes and bomb it."[4] In 1997, peaceable Botswana tried to buy tanks from Europe, but was rebuffed – no donor country would give it the money, mostly (or so Botswanans believed) because of a pitch from the Namibians.

Namibia, in turn, had its anxieties. Few places on earth are drier than Namibia. Of the meagre rain that does fall, four-fifths evaporates immediately. Only 1 per cent recharges groundwater tables. Worse, Namibia has no perennial rivers, only seasonally flowing ones that are reduced to a trickle for several months and dry up completely in others.

In the old days, this fluctuation wasn't a problem; the local people would migrate, along with the game, from one water source to another during the dry season. But now there are too many people and they have become more urban: city people are by definition less mobile than hunter-gatherers. To augment the water supply, Namibia has been tinkering with other options, including desalination and pumping groundwater from its fossil aquifers. But desalination, expensive enough anywhere, is still prohibitive in Namibia, where the major population centres are inland. The few desalination plants in operation require huge amounts of energy. As well, overpumping of groundwater has caused dangerous increases in water and soil salinity, as well as the rapid depletion of the aquifers themselves.

This chronic water shortage prodded Namibia to launch a planning process to extend its already massive network of supply pipelines and aqueducts, the Eastern National Water Carrier, to the Okavango River, which runs throughout the year along its northeastern border with Angola. The first phases of the plan would divert an estimated twenty-million cubic metres of water annually (seven hundred litres a second) from a point on the Okavango River near Rundu in northern Namibia – well before the river gets to Botswana – and pump it uphill through a 250-kilometre pipeline to Windhoek, the capital. Namibia saw the pipeline as the only feasible solution to keep pace with the water demands of its growing urban centres, despite their exemplary record in water management, and Windhoek's extraordinary recycling program, which saw raw sewage turned into potable water. In 1996, hydrologists were predicting that Windhoek would run short of water in just a few years. Clearly, the need for both governments to negotiate a long-term solution was urgent.[5]

Not surprisingly, the Namibians defended their plans and maintained that the amount they'd be taking was negligible, even though they intended to quadruple or quintuple it to somewhere around a hundred million cubic metres a year. Even that, they maintained, would draw off only 1 per cent of the water flowing through the Okavango system. Richard Fry, Namibia's deputy secretary of water affairs, was blunt. "The severity of Namibia's water crisis leaves us with little option," he said after the Okavango floods failed in 1996. The dams supplying Windhoek and the central areas of Namibia were at an all-time low, and were expected to run completely dry in a year or two. Namibia's senior water engineer, a bluff Afrikaner called Piet Heyns, was blunter still: "If we don't build the pipeline and the rains fail again . . . we'll be in the shit," he told the press.[6] The corpses of sixty thousand cattle, dead of thirst, littered the landscape.

An immediate crisis was averted when engineers started pumping water from a disused mine near the northern town of Grootfontein. This expedient was expected to see Windhoek through for a further twelve months, even if there was no rain and the dams ran dry. And there were still unexploited fossil aquifers in the area that could be "mined" for water in the short term.[7]

It was hard, in the terrible droughts, to see any alternatives. It was not that Namibia wasn't trying. If desalination is too expensive, and pumping water from disused mines was only temporary, what was left? Namibians are not profligate users of water. In the past decade, Windhoek's residents cut back water use, achieving a 30-per-cent saving. The city's annual consumption of seventeen million cubic metres has not increased significantly since independence eight years ago, despite a growth in population from 130,000 to 220,000, and it is now only one-third of the water consumed per capita in another desert city, Las Vegas. Significant increases in the efficiency of use – what Sandra Postel has called poetically "the last oasis" – are therefore unlikely, and, despite conservation, the country's water needs are expected to double in twenty years. Where is this water to come from? Namibia has agreed to an extensive environmental-impact study before spending more than half a billion dollars to dip its pipeline into the Okavango. But what happens if the study says the delta would be irreparably harmed? What happens if Botswana furiously objects? Richard Fry believes that Botswana will see Namibia's water crisis in the light of "humanitarian need" and will ultimately respond sympathetically to the pipeline project. But if it doesn't, what are the Namibians to do? And if the Namibians go ahead despite objections, what are the Botswanans to do?

The nagging questions remain: Is it safe to interfere with the delta's only supply of water? The Okavango is robust enough to survive anything except the water being turned off. If that happens, a Garden of Eden would return to Kalahari dust, the wildlife would migrate or die, and one hundred thousand humans would be reduced to slum dwellers in cities already unable to cope. If it isn't safe to divert the water, is it necessary? Where, in the balance of competing interests, does natural justice lie? At what point does man's stewardship of the planet and its resources collide with man's own needs? What is the ethical position? Has it come to this, a stark choice between human misery and the destruction of one of the planet's most magnificent jewels?

And perhaps the most difficult question of all: In any ecology, beyond a critical point within a finite space, freedom diminishes as numbers increase. This is as true of humans in the finite space of a planetary ecosystem as it is of the Okavango elephants or of gas molecules in a sealed flask. The human question is not how many can possibly survive within the system, but what kind of existence is possible for those who do survive.[8]

In late 2001 and early 2002 the unexpected happened: it rained, copiously and often, especially around the capital, and within a few months the aquifers had been recharged and the reservoirs were full – in March 2002, large and festive crowds would march down to the city's two main reservoirs to gawk at the miraculous overflowing of the dams. The pressure was off the Okavango – for the moment.

Nevertheless, Namibia still faces massive problems, despite all its best efforts. The coastal towns of Swakopmund and Walvis Bay together consume about 9.25 million cubic metres of water a year. Of this amount, 7.25 million is extracted from the Kuiseb River system and its groundwater aquifer. However, the sustainable yield of the sporadically flowing Kuiseb is only 5.25 million cubic metres. This means that coastal Namibia is "in overdraft" to the tune of about 2 million cubic metres a year. The water they are drawing on will soon run out, and a replacement supply is urgently being sought, because everyone knows the rains won't last. In this arid place, one thing is always sure: drought will return.

In 2002, the politicians and engineers were turning their attention, once again, to a grandiose scheme rejected as impractical only a decade before. This was to bring water in through a pipeline from the Congo River, more than a thousand miles to the north, on the far side of desperately poor Angola, which is still struggling to emerge from its decades-long civil war. In its scale (and its cost) the project was similar to one espoused by Sese Seko Mobutu, Zaire's unlamented late dictator, to harness the waters of the Congo River in a series of massive hydroelectric schemes "to supply power to most of Africa," but the Congo–Namibia aqueduct was taken seriously by smitten bureaucrats. "It's not out-of-the-world," mused Namibia's deputy water secretary, pointing out that Botswana itself would benefit, since not all the diversion would be used by Namibia.[9] Botswana's water affairs minister, Boometswe Mokgothu, said in 2002, "We need a lot of water in our country. Agriculture will come into force, wildlife will come up, and tourism will be revived" if water is diverted into Kalahari Desert rivers such as the Nossob and Molopo, which form part of Botswana's border with Namibia and South Africa, respectively. Sam Nujoma, Namibia's otherwise sensible president, has allocated $50,000 U.S. to study the possibility, and is reportedly in negotiations with the Democratic Republic of the Congo for access to Congo River water.

It is a project challenging enough to deter even the Californians or Chinese, the planet's pre-eminent water engineers. The project would require pumping water through 1,000 kilometres of pipeline to the Angolan Highlands, raising it a full 1,500 metres above the Congo River. The water could then be channelled into any of a number of rivers that originate in Angola, including the Okavango, which flows to Namibia and Botswana. For the Namibians, clearly, even that wouldn't do. The Okavango cannot deliver the water to areas where it is needed most, and Namibia would have to pump the water again through a 250-kilometre pipeline that climbs 500 metres to an existing canal that could then deliver it to Windhoek. Botswana, for its part, would have to pump the water at least 400 kilometres to the nearest agricultural area, or 700 kilometres to the capital.

Whether the pipeline ever gets built, even if the reservoirs once again run dry, is moot. The cost is estimated at somewhere around $1.5 billion U.S., with annual operating costs in the $175-million range, which would make the water deliverable at about $2 a ton, more than four times existing rates and comparable to the cost of desalinating sea water and pumping it uphill from the coast to the capital. In addition, no one has been able to estimate the biological and ecological hazards of transferring tropical wildlife from the Congo into the pristine Okavango ecosystem. The most persuasive argument against the project, however, was the nature of the territory through which the water would have to pass: neither Congo nor Angola are exactly the most stable of entities, and none of the southern countries particularly want to be hostage to their erratic governments.

Directly to South Africa's north lies Zimbabwe, and then Zambia. The dreadful droughts of the 1990s that so ravaged Namibia and Botswana seem not to have disappeared, but only migrated, for by 2002 both these countries were suffering – for once, the word is accurate, not just a careless cliché – from a lack of water that dropped crop yields more than 60 per cent, and millions of people were facing starvation.

In Zimbabwe, the effects were exacerbated by politics, as Zimbabwe's erratic dictator, in his desperate attempt to cling to power over mounting opposition, unleashed a furious assault on the country's productive white farmers (no doubt to the temporary relief of the country's homosexuals, who had until then been the main targets of Robert Mugabe's unstable paranoia). The net effect was entirely predictable: the country changed rapidly from an exporter of food into an international beggar. But as millions of Zimbabweans faced starvation in late 2002, donor countries were reluctant to ship more food aid, since it was known that Mugabe was diverting the food from those who opposed him to those who didn't – a deliberate policy of starving political opponents that met the classical definitions of genocide. As the year closed, NGOs on the ground were reduced to pleading that it was better to save some, even if they did support Mugabe, than none at all.

Zambia was in better shape politically, but there the drought was even worse. In parts of the country, crop yields were reduced to zero, and millions of Zambians, too, faced starvation. All of which made it even more bizarre that food aid was being turned away at the borders on the grounds that the crops from which it was made had been genetically modified. The country's president, Levy Mwanawasa, was adamant, however. He didn't want GM goods in his country. In years to come, he said, the county's farmers would thank him that they hadn't been contaminated. The European Union, he pointed out, was considering a ban on importing GM foods, and allowing them into Zambia now would preclude future exports to the country's largest export market. This curious calculus – starving citizens sacrificed to potential future exports – was vociferously protested by donor countries and by those NGOs on the ground that were not themselves prone to paranoia about genetic modification, but to no avail, and the bags of GM corn sat on rail sidings, unopened and unwanted.

At the start of 2003, there was no sign of rain.

14 | THE UNITED STATES AND ITS NEIGHBOURS

Whose water is it, anyway, and how are they using (and abusing) it?

When two hundred Texas farmers drove their tractors onto the Los Indios international bridge over the Rio Grande near Harlingen, Texas, in May 2002, it was clear that the simmering environmental dispute over Mexico's appropriation of the river's water was turning dangerously political. Which was, in a sense, deeply ironic, because far to the west, the Colorado River has been reduced to a drainage ditch by the time it crosses the Mexican border, and for years the Mexicans have complained, to no avail, that the Americans have stolen the river away. Diplomatic Notes over the Rio Grande had been shuttling back forth between Washington and Mexico City for several years, increasing in asperity as time went on without a solution, though the media, preoccupied in 2002 with "Confronting Iraq," more or less missed the whole thing. Far to the north, meanwhile, in the councils of power in Ottawa, policy-makers were largely ignoring the real problems associated with the parlous state of the country's waterways, and were still obsessing over water's political implications, mostly through the deeply irrelevant twin lenses of privatization and bulk water exports.

In the south, it was unusual politics. In the north, alas, just politics as usual.

～

The state of the continent's water, meanwhile, just continued to get worse. In the middle of 2002 there was drought over wide areas of the continent, with consequent water rationing – and not just in Texas and Arizona and the other Plains states fringing the desert (where Texas and Arizona had even forbidden householders to collect rainwater for their own use). All up the East Coast, from Delaware to Maine, water shortages reached crisis proportions in the spring, with reservoirs in some places down to 30 per cent of capacity, due to a lack of winter snow reaching back, in some places, five years. Spring rains took the urgency off the situation, but cities like New York nevertheless maintained their ordinances forbidding restaurants from providing diners with unasked-for glasses of water, and there were new thousand-dollar fines to prevent the time-honoured hot-weather trick of using fire hydrants to cool down. The Ogallala Aquifer, which underlies many of the Great Plains states, was exhibiting puzzling characteristics. It was dropping much more rapidly than expected in some places, while in others the rate of shrinkage slowed to a few inches a year. It was still an hydrologist's truism, however, that the aquifer would be exhausted in many places in ten to fifteen years. In Arizona and Colorado the drought led to devastating fires in the summer of 2002, while Congress dithered over whether to pass the National Drought Awareness Act, which called for a national drought council, better forecasting, and more and better reservoirs and wells. Best guesses were that if the act did pass, it wouldn't be for another five years. In Canada, a four-year drought meant that the Prairie provinces were facing severe water shortages and farmers were forced to sell off their livestock at a fraction of their worth; in June, on the contrary, some parts of the West experienced more rain in a few days that they'd had in three years, and floods were causing widespread destruction. Alas, the flooding did little good – the water ran off, and when the rains were over, the drought simply resumed as though the rains had been a cruel mirage. Alberta was wrestling with new water policies, striving to share sparse water equitably between farmers and the oil patch. In southern Ontario, in the booming quasi-dormitory towns to the north of Toronto, water tables were dropping, and nearby farmers were forced to deepen their wells. Some aquifers in Quebec dried up altogether – in at least one case because of the withdrawals of a new bottled-water company. In the Annapolis Valley, Nova Scotia's Napa-like farming community, farmers were facing the fourth year of drought, despite decent spring rains. Even in Vancouver, with its reputation as the Sodden West, they were contemplating water shortages.

The sense of looming crisis was compounded by the dismal state of much of the water that did exist. An International Rivers Commission report found that while there have been minor improvements in some places (including the notoriously polluted Hudson River), not a single one of America's major rivers is any longer fit to drink in its "wild" state. There was a massive dead zone in the Gulf of Mexico that was killing the gulf shrimping industry; it was blamed on phosphorus runoff from Midwest farms a thousand miles to the north. In several major cities, such as Atlanta, residents were faced with periodic intrusions of debris and mysterious sludge of unknown origin into their domestic water. More than nine hundred deaths were reported in the United States from contaminated water; in Canada, no one seemed to be keeping similar statistics. The Great Lakes were said to be cleaner than they had been, but again there was a massive dead zone in Lake Erie, and the International Joint Commission acknowledged that there were unacceptable levels of toxins in the lakes' wild creatures. The IJC also reported rather discouragingly that scientists could only identify about 30 per cent of the chemicals they found in fish, and had no idea what the others were or what effect they would have on humans. It also became clear that lowered funding meant fewer pollution monitors on the lakes, and an inability on the part of the IJC to implement its own remedial policies. Foreign vessels using the St. Lawrence Seaway were routinely dumping toxins into the lakes with impunity, mostly because of an inadequate bureaucratic definition of what constituted garbage. In Canada, more than forty communities in Newfoundland were under boil-water advisories, a term that was so current in the popular imagination it no longer needed explanation. Agricultural runoff was contaminating the groundwater and killing fish in streams in Prince Edward Island; pesticides and herbicides have killed the salmon in some of New Brunswick's most glorious rivers; in dozens of towns like Bridgewater, Nova Scotia, natural leaf moulds have been found to combine with chlorine to cause complex carcinogens. In Ontario, in the aftermath of the Walkerton tainted-water tragedy, officials were still found to be leaving reports of laboratory malfeasance for dangerously long periods; and some communities still seemed to treat water through a combination of finger-crossing and folk remedies. Environmental scientist David Schindler, a man of measured words and not readily given to hyperbole, took to employing apocalyptic words such as "catastrophe" to describe what was going on.

El Mayor, on the Colorado Delta in Mexico, is a scattered collection of ramshackle plywood and concrete-block shanties on a couple of unpaved streets, rutted and pitted over time, dusty in the dry years, and filled with mucky water when the rare rains come. Children play, as they do everywhere in places like this, paddling in the stagnant and dung-filled water, their playthings the shabby castoffs of industrial civilization – rusting auto parts, scraps of old tires, broken bicycles and baby carriages. El Mayor is too small to appear on any map, and its "mayor's office" doesn't have a phone, but it is still the principal town of the Cocopa people, who once fished and hunted around here, innocent of the Gringo scourge to come.

Or maybe not so innocent. When the Spanish explorer Francisco Vásquez de Coronado first passed through in 1540, an aide recorded the Cocopas' first response to seeing the European interlopers: "They ran towards us, gesturing with fierce threats to go back where we came from," he wrote. Would that they had been more successful. The Colorado River that watered this place has gone now, and the delta, rich fishing grounds for over two thousand years, is a sad thing, desiccated, salt-poisoned, and polluted, nothing more than cracked mud flats, a rural slum. Aldo Leopold, the American naturalist, canoed here in the 1920s and called it "a milk and honey wilderness" filled with lagoons and dense forests.[1] There were thousands of the Cocopa, then. Today, there are fewer than 1,500, living in straggling communities along the "Colorado" and the little Hardy River, now no more than a drainage basin for farm runoff.

The water that does come down the Colorado from the United States is regulated in volume and quality, but has often been below the required volume and more saline than allowed. Some 90 per cent of it has been drained away by U.S. cities and farms, or to fill swimming pools in Los Angeles and generate electricity for the Vegas strip. By the time it reaches Mexico, there's very little left, and the towns and farmlands of the border country take most of that, concrete channels distributing it to nearly two hundred thousand hectares in the Mexican desert. Mexico duplicates what the Americans have already done, creating farmland in an area where there should be none, crops growing in a region with less than eight centimetres of rain a year. The valley along the river south of Mexicali produces one-tenth of Mexico's wheat, and important quantities of sorghum, alfalfa, and food crops.

By the time the water gets to the delta, and even when there are heavy rains upstream, a few steel culverts under a gravel road easily take care of what was once referred to as an "American Nile." In dry years, or when the Americans are recharging their dams, there is no water at all.[2]

The Cocopa fish in the Gulf of California now, where there are still fish, but many fewer than when the Colorado River flowed unchecked to the sea, bringing rich nutrients along with it. What it brings now is pollution.

The Mexicans are trying. A ditch to divert salty runoff from Arizona land has been used to recharge the Ciénaga de Santa Clara wetlands area, which had been drained, irrigated, and poisoned. A small reservoir is planned to recreate wetlands along the Hardy, and treated sewage is supposed to be diverted into the area. But there is no money for any of this development. Soon the last few Cocopa will be gone.

~

American high-handedness with "Mexican" water goes back to the last century, to the years following open warfare between the two nations. At the end of the century, skirmishing began again, this time over the Rio Grande, the border watercourse between Texas and Mexico from El Paso to the Gulf. In an ironic prelude to what was to come – American accusations about Mexican water theft – the Americans had been diverting water from the river to irrigate West Texas and New Mexico rangelands without agreement or permission, and the Mexicans filed a protest with Washington. The State Department, which handled border disputes, asked the U.S. attorney general, Judson Harmon, for a legal opinion.

The Rio Grande, said Harmon, rose in the territory of the United States. The diversions were being made with American water. The United States had absolute sovereignty over its natural resources, of which water was one, no different in kind from gold or iron. No international law existed compelling the United States to share its water, and it would not allow itself to be bound by such a law. The Harmon Doctrine, discredited now even in the land of its birth, is still occasionally trotted out by recalcitrant countries trying to squeeze more water from transnational water systems – at the expense of downstream neighbours.

In the years that followed, the State Department devised more flexible rules, without ever actually repudiating Harmon. It reached a tentative agreement with Mexico in 1906, and a more thorough one in 1944 that incorporated not just the Rio Grande but also the Tijuana and the crucial Colorado. That agreement was carefully worded by State Department lawyers. Mexico was allocated specified quantities of water, 1.5 million acre-feet in the case of the Colorado. But there were interesting nuances. The U.S. side viewed the allocations as unilateral – that is, they were made by the United States as though there were no border between the two countries. That meant the Mexicans were granted water, but not as a right. Power will out.[3]

The Mexicans didn't argue. At least they had treaty rights to water in reasonable quantities.

What the treaty didn't mention, however – no one thought at the time it would be a problem – was the issue of water quality. Water was water. You could irrigate with it, grow crops with it, drink it. Sure, you could poison it, but no one anticipated that.

In the 1950s, things began to change. Rapid industrialization of the West, and the American desire to irrigate ever more unproductive land, began to change the turbidity of river water in the boundary region. The Rio Grande had long since ceased being a Texas legend. Since the 1970s, it has been one of the most polluted waterways in America, and was declared in 1993 to be America's most endangered river, containing almost as much untreated sewage, pesticides, and industrial waste as water, the unwanted and unforeseen product of an unrelenting economic boom. Pesticides like DDT and chlordane, long banned in the United States, are present in the river. In the Nogales region, water-related problems have been identified as the primary health concern for people on both sides of the border.

~

Those Texas farmers who blockaded traffic on the international bridge in the middle of 2002 had a curious case to argue. Whose water is the Rio Grande's, anyway?

Once, the river lived up to its name. It is, after all, the fifth-longest river in North America, and among the twenty longest rivers in the world; from its source high in the Rockies, it travels a little more than 1,800 miles, and drops 2 miles before emptying into the Gulf of Mexico. It passes through three states in the U.S. (Colorado, New Mexico, and Texas) and five states in Mexico (Chihuahua, Durango, Coahuila, Nuevo Léon, and Tamaulipas). From El Paso to the Gulf of Mexico – a distance of about 1,250 miles – the river is the international frontier. The river's watershed is massive, about 335,000 square miles, about 11 per cent of the area of the continental United States. But for a good part of its run, it passes through one of America's largest deserts, the Chihuahuan, where there are no tributaries to replenish it, and intense demand has over the years depleted it, until by 2000 the river no longer flowed all the way. American cities and users have extracted every drop from the river by the time it reaches El Paso, Texas, and the river doesn't resume flowing until the confluence with the Río Conchos, 250 miles downstream. The Conchos and its tributaries rise in Mexico, and so the lower reaches of the Rio Grande, the source of the growing dispute between Mexico and the United States, consists almost entirely of Mexican water.

"The Rio Grande," Will Rogers once said, "is the only river I know in need of irrigating." Which is barely an exaggeration. Diversions take nearly 95 per cent of the river's water already, and

claims considerably exceed the actual supply. The cause of the
deficit is partly natural. There have been years of drought and a
curious lack of a phenomenon that in better years is more feared
than yearned for – hurricanes. But humans are having their effect
too. The population of the border regions has increased dramatically since NAFTA was signed (1,055,000 in 1945; 4,162,000 in 1990;
6,900,000 now); and Mexican farmers have been shifting towards
crops such as pecans, which are more profitable but thirstier than
corn. In addition, the aquifer on which the city of El Paso depends
for its domestic water, the Hueco Bolsón, will run out of recover-
able water by 2025, which will add to El Paso's burden.

The laws governing the river are arcane even by western-
American standards. The *Cascadia Times*, an online news and
information service on water and environmental issues, points out
that its governing laws "reflect the influence of aboriginal rules and
customs, Spanish and Mexican laws from before the 1848 treaty of
Guadalupe Hidalgo, international treaties, an interstate compact,
the federal government's trust responsibilities for Indian tribes, and
its public-interest trust responsibilities as stewards of the nation's
resources, and the unique laws and institutions of three states."[4]

Complicated or not, there is another and simpler way of
looking at it: the Americans unilaterally extract all the water from
the Rio Grande before El Paso, seeing as it runs through their territory. None of that water reaches Mexico. Texas farmers, on the
other hand, are demanding that Mexico yield up more water from
the lower Rio Grande (which is from the Río Conchos and its
tributaries), which the Americans can't extract because they don't
control it . . . Hmmm.

Not that the argument gets very far with Texas farmers.
Southern Texas is one of the poorest areas of the United States,
and drought brought losses of $1 billion between 1996 and 2000,
according to a Texas A & M University study. By June 2002, the
Texans had already used up all their annual quota, and had to leave

their crops to desiccate and die. Bumper stickers appeared along the border bearing the slogan "Mexico sucks."

According to the treaties, by the end of 2002 Mexico "owed" the United States 1.5 million acre-feet of irrigation water. (Part of the 1944 agreement, which reserved 1.5 million acre-feet of Colorado water for Mexico, was Mexico's promise to reserve for Texas 350,000 acre-feet a year from the six rivers flowing into the Río Bravo del Norte, as the Mexicans call the Grande.) A potential lever against the Mexicans, or so the Texans argued, was "to shut down the pipe on the Colorado."

~

The Colorado's problems start a long way from the border. Irrigation, as we have seen, tends to make runoff saltier, and so do storage dams. The Colorado's rush to the sea is first stopped at Glen Canyon Dam, near the Utah–Colorado border, by which time it has already received the agricultural runoff from the irrigated lands along the Green, White, and Yampa rivers. The river is stopped again at Lake Mead at Hoover Dam, between Nevada and Arizona, again at Lake Mojave, and once again at Havasu Lake; at each stop, the saline levels increase. Finally, it takes in the Gila River, the last tributary flowing into the Colorado before it crosses the frontier into Mexico. There was irrigation along the Gila as early as the sixteenth century, when the Spaniards planted extensive gardens and feed grains in the valley. But the soil was clay, and the water-logged soil was soon so salty that the crops died and the basin was abandoned. Then, in the 1940s, the United States built the Wellton–Mohawk Irrigation and Drainage District of Arizona, which included a huge drain to carry away the salty residues. This drain is dumped into the Colorado just before the border.[5]

In November 1961, Mexico had finally had enough. The government protested to the United States that "the water received is not suitable for agricultural uses, and agricultural production in the Mexicali Valley is being adversely affected." It insisted that the Americans were breaking international law by violating the 1944 treaty. This diplomatic language masked a growing popular fury.

But instead of getting better, matters got worse. Two years after the protest, the Glen Canyon Dam was completed and, while it filled, the delivery of water to Mexico dropped sharply. The salinity of the Colorado rose to more than 1,500 parts per million from an annual average of less than half that, salty enough to poison most agricultural crops. The resultant Mexican uproar fell on deaf ears in the United States. Hell, what were they worrying about? The treaty hadn't said anything about potable water, had it?

But by the middle of the decade, even the Americans began to be embarrassed, and in 1965 they undertook to do something about it. Mexico and the United States signed a five-year agreement promising "various measures," mostly unspecified, to deal with the salinity problem. Predictably, most of the state politicians and congressmen mistook the agreement itself for actual results and, having agreed to do something, no longer felt the need to do anything. A few steps were taken on the local level, the sum total of which was to release some stored water when Mexico ran really low. This had the effect of lowering the salinity levels somewhat, but only down to around 1,245 from 1,500, still much too high. The long-simmering anger among Mexicans bubbled to the surface, and the 1972 presidential campaign of Luís Echeverría was a Yankee-bashing affair based on the wretched state of Mexican agriculture. He won by a large majority. When the new president threatened to take the United States to the International Court of Justice at The Hague for treaty violation, President Nixon, eager to think of a sound way to do nothing, established a task force to study the problem, the politician's classic delaying tactic. He turned the issue over to the U.S.–Mexico International Boundary Waters Commission. Their job was to look at the Colorado Basin states

(California, Arizona, Nevada, Colorado, Utah, and, through tributaries, Wyoming and New Mexico), as well as the Tijuana and Rio Grande rivers.

The task force hemmed and hawed, and came up with a variety of options in decreasing rank of political probability. In the end, it agreed that, in all equity, the United States should provide Mexico with decent water. But how? The options were simple, but all of them expensive. The Americans, for example, could continue to pour salts into the river, in which case they would either have to treat the water before it got to the Mexican border or substitute quality water from somewhere else, source unspecified. Or the demand could be made that the salination of the river cease. The only way of making that happen was to take farmland out of production to eliminate the source of the highest salinity.

The two governments resumed desultory negotiations based on the task force's report. Probably nothing much would have been done had not 1973 been the year of the OPEC oil crisis and the consequent shock of seeing Americans in long lineups for gasoline. It came just at the time that seismic soundings showed Mexico to be, at least potentially, a major exporter of oil. The Nixon administration was quick to pay attention.

In 1974, the two countries announced they had found a "permanent and definitive solution" to the salinity problem. Mexico was allotted the same amount of water as in the 1944 agreement, 1.5 million acre-feet of the Colorado River flow, and the United States agreed to meet specified standards of water quality – essentially the same standards as those offered American farmers. This solution was arrived at not as a treaty, but in the form of a minute of the International Boundary Waters Commission. This dodge avoided the need for ratification by either side, and neatly sidestepped the prospect of clamorous and acrimonious congressional debates, either in Mexico or in the United States. However, Congress was still very much involved. Congress, after all, had to pay for it.

Unerringly, of course, Congress opted for the most expensive of the options. To improve the quality of the water delivered across the border, the United States would build a series of drains and other engineering works on long stretches of the Colorado, and – the linchpin of the scheme – build a massive desalination plant to process the water from the Wellton–Mohawk Diversion in Arizona. This facility would "scrub" the water back to legal levels just before it entered Mexico. In addition, Congress would build at its expense a drainage canal to carry the now-massively saline effluent from the treatment plant directly to the Santa Clara slough in the Gulf of California, bypassing the Colorado entirely; and it would spend money to help Mexico rehabilitate and then "improve" the Mexicali Valley. It was expensive, Basin State congressmen acknowledged, but it had many advantages. They wouldn't have to take farmland out of production by buying out the farmers, they would minimize loss of water, they wouldn't have to discourage development and expansion in their own states, and the project would surely lead to important progress in desalination techniques. The total cost was estimated at somewhere around $1 billion. This way, as Marc Reisner put it, they would provide water to upstream irrigators at $3.50 per acre-foot, and then scrub the same water free of the resulting irrigation salts at $300 per acre-foot. The economics made no sense whatever.

The desalination plant at Yuma, Arizona, used a reverse-osmosis membrane filtration process that was supposed to obtain a 70-per-cent recovery capability of water with salts in the 250-parts-per-million range, certainly well within treaty levels. Engineers optimistically suggested that, with proper management and later improvements, efficiencies as high as 90 per cent might be encountered, despite that target never having been reached anywhere but in experimental facilities.

And when all the salt had been removed, where was it to go? The minute had opted in the end for dumping it into the ocean through a diversion canal. The only other options were to evaporate it in large ponds, in which case mountains of raw salts would remain that were likely to blow about on the winds, polluting land hundreds of kilometres away; or to use deep-well injection, an untried and untested technique that would force what is essentially brine into aquifers far underground, with unknown consequences. At the time this seemed like a possibility, for the United States was still storing nuclear waste in "entirely safe" shafts deep underground.

～

The Colorado River basin covers 632,000 square kilometres and provides water for 30 million people. This is the most watched, argued-about, measured, and contested water in the United States. And the most manipulated, for water policy in the Colorado basin has amounted to what Marc Reisner calls

a uniquely productive, creative vandalism. Agricultural paradises were formed out of seas of sand and humps of rock. Sprawling cities sprouted out of nowhere, grew at mad rates. . . . It was a spectacular achievement and its most implacable critics have to acknowledge its positive side. The economy was no doubt enriched. . . . Land that had been dry-farmed and horribly abused was stabilized and saved from the drought winds. . . . "Wasting" resources – rivers and aquifers – were put to productive use. The cost of all this, however, was a vandalization both of our natural heritage and our economic future, and the reckoning has not even begun. Thus far, nature has paid the highest price. . . . [But] who is going to pay to rescue the salt-poisoned land? To dredge trillions of tons of silt out of expiring reservoirs? To bring more water to whole regions, whole states, dependent on aquifers that have been recklessly mined?[6]

The Colorado has an average flow of around 14 million acre-feet a year, which occasionally increases to 18 million in good years and decreases to around 12 million in bad. By treaty, Mexico must get 1.5 million of those acre-feet. By a complicated series of regulations, laws, and agreements, some of them of debatable legality, California has acquired the "right" to a minimum 4.4 million acre-feet, rain or drought. Evaporation from the river and its multiplicity of storage dams varies from year to year, but averages around 2 million acre-feet. The upper basin, or "producer," states, which contribute to the river's flow, have an accumulated withdrawal allotment of around 3.6 million acre-feet. The Central Arizona Project, a 540-kilometre $4 billion canal, the largest and costliest water transfer system ever built, has the capacity to withdraw more than 2 million acre-feet, though it has been operating only at three-quarters capacity and is drawing just shy of 1.5 million acre-feet. In bad years, that pretty well sucks the whole system dry. State-by-state allotments are 4.4 million acre-feet for California, 3.9 million for Colorado, 2.8 million for Arizona, 1.7 million for Utah, 1.5 million for Mexico, 1 million for Wyoming, 843,750 for New Mexico, and 300,000 for Nevada.

The lower basin states – California, Arizona, and Nevada – have themselves been arguing amongst themselves about allotments for years. California has, by compact, used all the surplus water of the other two states, plus that of Wyoming, Utah, Colorado, and New Mexico, to water its cities and farms. But the rules are changing. The Western Water Policy Advisory Commission, in a report in 1998, speculated that, of the ten American states that will grow fastest between now and 2025, five of them are in the Colorado Basin. The report uses typically extravagant Western language: "It may be fair to say," it says, "that the major period of settlement in the West did not occur in 1850: it is just now taking place." Nevada, especially thirsty Las Vegas, already a city of 1.5 million people, wants its full allotment back from California, and then some.

As part of the re-examination that continues in all water-stressed states, California finally got around to redressing a problem of its own making, the damage it did to its own Owens Valley through the great water theft engineered by Los Angeles eighty-five years earlier, the subject of Roman Polanski's movie *Chinatown*. Decades of bitter conflict culminated in a deal between the Owens Valley and Los Angeles in July 1998, which was intended to bring about an end to the massive dust storms that had plagued the valley for generations. The Los Angeles Department of Water and Power agreed to begin work at the long-dry Owens Lake by 2001, and set a daring date – eight years – by which the people of the Owens Lake area would once again be "breathing air that met federal health standards." Owens Lake – where the actual lake has long been gone – is by far the single largest source of air-particle pollution in the United States, and, for decades, people living around the Sierra Nevada have been exposed to the clouds of saline dust contaminated with toxic pesticide residues that swirl off the "lake" surface, little different in volume and substance from those that blow off the Aral Sea. The solution came when Los Angeles, for the first time in that thirsty city's history, surrendered water without a lengthy court fight. The lake will not be refilled, but the city agreed to keep enough water there to cover the bed with a thin film of moisture, while gradually treating the surface by planting salt-resistant vegetation and laying down gravel. It is the last major hurdle to repairing the damage Los Angeles inflicted when it opened its aqueduct and drained the valley of its mountain-fed water.

In late 1998, in an interesting case of life imitating art, water was about to flow from the water-rich but cash-poor farmers of the Imperial Valley in southeastern California, to arid San Diego County, one of the most rapidly growing areas of southern California. The valley's water authority, the Imperial Irrigation District, had agreed to sell two hundred thousand acre-feet of "surplus" Colorado River water to San Diego at $245 per acre-foot, more than twenty times the price it pays for the water in the first place. The deal began when the billionaire Bass brothers of Fort Worth, Texas, quietly assembled more than two hundred thousand hectares of Imperial Valley land in the expectation of acquiring water rights, which they could then sell to San Diego. The deal foundered when they discovered that the authority owned the rights, not the farmers. Inspired by this attempted chicanery (or, depending on your point of view, free-marketism), the authority decided to play the game of marked-up prices itself. It first had to do a deal with the Metropolitan Water District, the Los Angeles behemoth, which owned the only aqueducts, but soon found a typically Californian solution: it persuaded the state legislature to pay off the Met through general tax revenues, thus giving the thirsty citizens of San Diego and the property owners of the Imperial Valley a massive subsidy, another neat way of benefiting the few while charging the many. Conservationists were left shaking their heads. Sure, San Diego's water supply was ensured, but if the precedent sets off a rush of similar deals by other water-scarce California communities, the last barrier to un-restrained urban growth will be trashed. The Imperial Valley farmers, according to the agreement, were to use the money they received to install conservation devices on their land. Conservationists pointed out gloomily that enforcement mechanisms were nowhere to be found.

Another hiccup in the process was the contention of the Coachella Water District that it had been unfairly deprived of water by Imperial – since 1934. Finally, however, in late 1998, in a speech to the seven states that depend on the Colorado, Bill Clinton's interior secretary, Bruce Babbitt – who for a while took to referring to himself as the River Master – announced the set-tlement of the Imperial–Coachella dispute too. Well, sort of. "I am not going away, I am continuing the pressure," he scolded the almost-partners. He threatened among other things to reduce the amount of water California takes from the Colorado unless the state learns to be more efficient, an assessment that was indignantly received by the profligate Californians. But the secretary was soothing: "I am very impressed," he said, "that Coachella and Imperial have at last discovered their fraternal bonds." He called it "a minor miracle."

Indeed, in August 1999, negotiators for the Imperial Water District, the Metropolitan Water District, and the Coachella Valley Water District finally signed a comprehensive agreement. The impetus was not just Babbitt's threats, but the realization that they were jointly facing a 37-per-cent increase in demand over the next two decades, with a far-from-assured supply. The declared long-term goal was to show the federal government and the other six Colorado River states that California was able to stop its internal feuding and could learn to use its Colorado allocation frugally. In return, they hoped aloud that their newly self-defined status as conservators would persuade the federal government to "liberal-ize" (a nice word) its definition of what constituted a "surplus" in Colorado water, and thereby up California's allocation beyond its 4.4 million acre-feet. Reaction from the other states was unenthu-siastic. Arizona, for one, declared itself unimpressed by these meagre signs of Californian self-restraint.

There has been no such miracle in the northern part of the state, not at all. In the San Joaquin Delta, the state's largest watershed, which provides water for 22 million Californians, the ecosystem was close to collapse. Babbitt had hoped for another "minor miracle" there, but it was not forthcoming. Agribusiness, environmentalists, and the urban water agencies could hardly agree even on an agenda for meetings, and Babbitt was reduced to calling them all "intransigent." The water agencies and the farmers wanted more water transfers; the environmentalists wanted con-servation. Peter Gleick asserted halfway through the process that the state's department of water resources "grossly underestimated potential water savings." Gary Bobker, another environmentalist, explained it this way: "There is an engineering logic to the idea that you can build a lot more flexibility into the system if you can bank more water in more pots. What is not proven is whether you can take more water out of the system and bank it without doing harm to the environment.[7]

Delayed for what was said to be "more study" was the most divisive issue facing CalFed, a federal-state "superagency" that was supposed to come up with a thirty-year, $10-billion plan for the area: whether to build a forty-two-mile canal to divert Sacramento River water around the Delta and so more directly to southern California. The absence of this canal is an intense irritant to southern water districts, who see the stubborn north as awash in water it doesn't need.

For some time, Babbitt had tried diligently to make sense of the maze of regulations, entitlements, laws, and agreements, without much result. In the end, he authorized a form of water banking that allows exchanges of Colorado water between lower basin states. In practice this scheme benefited mostly Arizona, which had been taking but not using its full allotment, prudently storing the surplus water in underground aquifers, where it neither evaporates nor runs away. As of 1998, it could sell that water to Nevada – but not directly, of course. The rules allow Arizona to bank the water in aquifers for Nevada. For Nevada to "buy" the water, the state dips it from the Central Arizona Project and issues a credit to Arizona. In return, an Arizona water user, perhaps the city of Tucson (which otherwise relies on rapidly depleting deep wells), pumps the same amount of water from the aquifer underlying Arizona, with Nevada paying all the costs.[8] Nevada, in turn, subsidizes its water to its users. Water traded in this way costs about one-third of what it does in Santa Barbara, California – which until recently used its "own" water, without joining the California distribution system, and saw its rates skyrocket when it finally signed up.

Of course, the system encourages ever-greater use. Nevada can now get more water than its entitlement, and can attract more industry and ever more people – which means it will need even more water. Nevada had some warnings for California: "There are seven sovereign states on the river, not one sovereign state and six lesser partners," said Patricia Mulroy, general manager of the Nevada Water Authority. "California needs to remember that."

It is entirely possible, though, that the Colorado River will be in overdraft well before the scheme is implemented. There are still half a dozen major diversion projects on the drawing boards of seven states. If only half of them are built, the river will finally be overdrawn, and extra water will have to be found from . . . somewhere. But where?

Water planners are clearly worried. In 2004, California will for the first time incorporate climate change into its five-year water-management plan. Jonas Minton, the deputy director of the California Department of Water Resources, pointed out that, over the past fifty years, winter precipitation in the Sierra Nevadas has been falling more and more in the form of rain, increasing flood risks, instead of as snow, which supplies farmers and taps alike as it melts in the spring. He blamed a warming climate.[9]

Mexico hasn't got any to spare, though its 1.5 million acre-feet of Colorado entitlement is vulnerable: after all, by the start of 2003 it owed the Americans 1.5 million acre-feet from the Rio Grande. Some northern-California rivers are still running free to the ocean, but the long history of animosity between the north and the south of that state makes any more diversions from that source implausible.

Which means making more water (desalination) or using less (conservation), or importation (either from the Mississippi – though why anyone would want the toxin-laced water of the Mississippi is anyone's guess – or from the north). If the provisions of the North American Free Trade Agreement are read in a certain way, if water can be certified as a commodity under the trading rules, well then, Canada has lots of water, doesn't it?

~

On the continent's East Coast, matters were not nearly so dire in 2002, but there were still considerable causes for concern. Even in Florida, a state that gets more than an adequate annual rainfall, water was causing problems. Efforts have been made to preserve the Everglades, but in many parts of the state – given its chaotic and personality-driven politics – that familiar unholy coalition of developers and petty politicians was still draining "swamps" and building new subdivisions. In a number of cases, these subdivisions began sinking into the muck a few years after they were built, when the aquifers that had been keeping the soil dense enough were drained. In others, wells drilled into the underlying aquifers began depleting the water for neighbouring communities, and in 2002 the lower courts were becoming clogged with angry water litigants.

A case in point was the situation on the state's west coast.

The Floridan Aquifer, a massive body that rivals the Ogallala in volume, still contained plenty of water, but some time in the early 1990s, Pinellas County, a spit of land nearly surrounded by the salt water of Tampa Bay and the Gulf of Mexico, exhausted its supply of potable water, allowing sea water to encroach and turn the west/central edge of the Floridan Aquifer brackish. The area affected was substantial: it included the Tampa Bay area, Pinellas, Hillsborough and Pasco counties, and the major cities of St. Petersburg, Clearwater, and Tampa itself.

Desperate for new supplies, Pinellas and the city of St. Petersburg solved the local problem by buying large tracts of land in Pasco County to the north and in Hillsborough County to the east, and began pumping water from under their neighbours to supply their own requirements for irrigation, car washing, cooking, and drinking.

The results were familiar: overpumping meant drawing down the aquifers faster than they could be replenished. Wetlands around the wellfields began to dry up. Cypress trees died and toppled. Private wells dried up. Homes sagged and collapsed as the porous ground beneath began to cave in on itself.

Emotions ran so high in neighbouring counties that, when the local governments collectively hired a facilitator, she soon threw up her hands and declared that, if she couldn't even get them to agree on the date of the next meeting, she wasn't going to get them to agree on anything else. She departed, and did not return.

And the point of this small but lamentable history? Water is politics, and politics can fix as well as fail, given the proper will and the proper conjunction of personalities. St. Petersburg elected a new mayor, David Fischer, who promptly made a dramatic suggestion: if the west coast could organize a true regional water utility, he would be willing to sell all of the city's wellfields, treatment plants, and delivery systems to the new organization if his neighbours would do the same. It took two more years of meetings, often rancorous and always running twelve to fourteen hours, to hammer out the agreement, but it went into effect in 1998.

In 2001, the new utility, Tampa Bay Water, signed a contract for a 25-million-gallon-a-day desalination plant. Another 60 million gallons of new supplies will come from a reservoir to be filled by skimming high water from area rivers. More desal is in the works.[10]

At the same time, further to the north, a wet fall and early winter snows relieved the pressure on the reservoirs after a four-year drought, but the forecasters were gloomy, predicting more droughts to come. The main problem in cities like New York and Boston was a creaking infrastructure, and a lack of redundancy – if the main pipeline to New York had to be taken out of service for repairs, how would the water get to the city? The reserves were perilously thin, and, in the newly heightened consciousness that terrorist acts had induced, vulnerable to attack.

In Boston, too, there were problems, despite the city's success in reducing consumption.

No one in Boston thought the city would ever have a problem of supply. In the 1930s the city had "solved" its water problems once and for all, by the neat trick (lately more attributed to the Chinese) of flooding a valley and driving out its inhabitants, without compensation and with only modest warning, thereby creating the Quabbin Reservoir. Even without any rain, the Quabbin holds enough water for two years, thus ensuring that the three million customers of the Massachusetts Water Resources Authority have a secure supply.

The water issues that remain include contamination, mostly by a massive influx of seagulls, whose guano is causing complex cleanup problems. Thousands of homes in the area still have lead pipes; the main aqueduct had no backup that would allow it to be closed for repairs (another issue for the analysts of the newly created Department of Homeland Security), and ancient water mains in many of the forty-one cities and towns served by the authority are leaking billions of litres a year.

At the start of the millennium, work had already begun on a 17.6-mile underground aqueduct that will finally allow the MWRA to repair its main pipeline. New corrosion controls have cut lead levels in tap water by 30 per cent since 1996, though the authority acknowledged in 2001 that many households with old lead pipes still have dangerously elevated readings.

The U.S. Environmental Protection Agency thinks the authority isn't doing nearly enough. In a scathing indictment, the EPA declared that Massachusetts was relying too much on "good fortune, shocks of chlorine, and the water's initial cleanliness to prevent outbreaks of giardiasis and other waterborne diseases." This stinging assessment was contained in documents produced in court after the EPA sued the MWRA in an attempt to force it to add a $180-million filtration system to the $261-million ozone treatment plant it is supposed to open in 2004.

Still, there are problems, and then there are other problems. As the *Boston Globe* put it in a piece in 2000: "water-poor communities in eastern Massachusetts would gladly trade problems with the MWRA. Brockton, a city of 90,000, is so short of water that it has considered building a costly desalinization plant to make sea water potable. And limits on outdoor water use are a ubiquitous part of summer for many communities north and south of Boston. Not surprisingly, some of these towns petition to join the MWRA practically every year."[11]

At the continent's heart, the greatest concern in 2002 was over the parlous state of the Great Lakes.

In October, environmental reporter Martin Mittelstaedt took two full pages in the Toronto *Globe and Mail* to declare that "there was a problem with Lake Erie's return from the brink: it didn't last." His main point was that the famous "dead zone" that had appeared in Lake Erie in the 1960s, and which had led to a multibillion-dollar pollution-abatement program in dozens of lakeside cities as far upstream as Chicago, had reappeared, bigger and deader than ever.

"The dead zone is growing again, and the floating mats of smelly algae are back, along with a blue-green variety that is toxic. As in the 1960s, a vicious strain of avian botulism is claiming thousands of the birds that feed on Erie's fish, and chemical toxins are building up in the fish and fowl." No one, Mittelstaedt found, knew why this was happening. The causes of choice included global warming and the notion that alien invasive species have so altered the ecology of the lakes that the old rules no longer applied.

Lake Erie is critical to the health of the Great Lakes as a body, because it shows negative effects earlier. This is because Erie has more aquatic life and far less water than the other lakes. "In fact, the renewed algae blooms plaguing Lake Erie have already begun to show up elsewhere – on lakes Huron, Michigan and Ontario." Ralph Smith, a biologist from the University of Waterloo, was quoted as saying his research vessel often stalls ten kilometres from shore when algae clogs its propellers.

The IJC, the intergovernmental body charged with responsibility for monitoring all boundary waters, was predictably more measured than the *Globe and Mail* in its assessments. In its 2002 report on the state of the lakes (a report that began with a ringing endorsement of its own existence), the commission admitted that its "RAPs" (Remedial Action Plans) for the lakes have been stalled by recent staff reductions and budget cutbacks in federal, state, and provincial agencies. In Lake Superior, for instance, the IJC's major cleanup plan envisaged monitoring the major dischargers of effluent into the lake. Since there were only forty-one of them, it was supposed to be easy. But there was not enough money, and the information has not been forthcoming. The report also acknowledged that, even without money – and therefore without any human-devised RAP – natural recovery is possible once the source of pollution has been removed. This, however, is a leisurely process. It can take as little as five years (to remove, say, the toxic insecticide kepone from the James River) but as long as ten thousand years (to eliminate radioactive wastes).

After this more or less discouraging preamble, the report cut to the chase: "despite significant improvements in water quality during the past two decades, current concentrations of polychlorinated byphenyls (PCBs) in samples of Great Lakes water are still 100 times the water quality criteria [envisaged] under the Great Lakes Initiative. Without sediment cleanup, injury to the health of humans, fish and wildlife will continue." There has been some success in controlling point-source discharges of toxic chemicals into the lakes, although the effluent from large-scale hog and beef

production was an ongoing worry, and so was increasing urbanization, with its attendant burden of pathogens, industrial chemicals,
and pesticides. The real problem was, no progress had been made in
cleaning up what it called "the historic burden of toxic chemicals
in contaminated sediment"; nor had there been much abatement of
airborne toxins, which had increased if anything.

The problem of contaminated sediment will likely get worse,
not better. The U.S. Army Corps of Engineers, the agency that
drained the Everglades and re-engineered the lower Mississippi,
has announced plans to dredge the St. Lawrence Seaway to allow
access for the new generation of supertankers, which need massively deeper channels than their predecessors. The Corps wants to
enlarge locks and deepen channels from Montreal to Minnesota.
Their own background papers admit that the dredging would
disturb hundreds of millions of cubic metres of contaminated
sediment containing PCBs, thus resuspending it in the shipping
channels. "Concerns about threats to municipal water supplies by
pollutants released during dredging operations may be warranted,"
the Corps acknowledged.[12]

In 2002, fish caught on the Great Lakes still contained many
neurotoxins, including PCBs and methyl mercury, and many studies
have indicated that anyone consuming such fish was vulnerable to
a shopping-list of negative side effects. Anglers are particularly
advised not to serve fish to children or pregnant women. The
problem is that scientists have only a hazy notion of what chemicals these fish actually contain. Researchers have been able to
identify only about a third of the substances they find in fish.
"Several halogenated compounds as well as antibiotic and other
pharmaceutical residues in Great Lakes samples remain unidentified.
The presence of brominated diphenyl ethers, chlorinated paraffins
and naphthalenes, and PCB metabolites in the tissue of a variety of
species ranging from snapping turtles and herring gulls to polar
bears and humans, remains a mystery." Worse, because they can't

identify the chemicals themselves, there is no way to guess at what chemical mixtures might do to animal tissue. The chemical soup ingested by fish, and thus by humans, has unknown properties. "Some of these chemicals interact with each other, but how mixtures affect the biota remains unknown."

Towards the end of the long report, the authors raised another issue, the discharge of garbage from ocean-going vessels. This, it turns out, is a bureaucratic problem. The discharge of garbage is, in fact, forbidden to all vessels operating in the Great Lakes, and the potential fines are substantial. On the other hand, the "discharge of cargo residues" is not prohibited by any rule. On the contrary, areas of the lakes have been designated for just such discharge, despite the fact that the International Maritime Organization thinks the one is just as deadly as the other, and includes "cargo waste" in its definition of garbage. And so vessels that want to dump cargo waste have merely to wait until they are far from the sea and safely up the St. Lawrence Seaway, and then they can legally dump the stuff into the lakes. The report recommends further study, as well it might.

The rules can get even more arcane, and more ridiculous. Vessels are not allowed to dump ballast water in the lakes, for fear of what are called "alien invasive species." But if they declare themselves to be NOBOB ("no ballast on board"), they are allowed to rinse out their ballast tanks, thereby flushing the residue into the lakes – residue that has, indeed, been found to contain both active and dormant alien species.

The national and local press on both sides of the international frontier has been silent on these issues. The cities, meanwhile, continue to draw their drinking water from the lakes, to an apparent complete lack of public concern.

If water is essential to life – and if that life-giving fluid is taken to consumers through a finite and known network of conduits – how safe is the water infrastructure to terrorist attack?

The answer is: not very. But some.

The most obvious apparent danger, that terrorists would dump a poison or a virus into a major storage reservoir and kills millions of people, is essentially a fantasy. Most reservoirs hold between 15 million and 150 million litres of water, which would significantly dilute any poison to the point that terrorists would have to release enormous quantities to do serious damage. In any case, poisons would be destroyed when the water is purified at a treatment plant. Chlorination, used in virtually every municipal system, kills or inactivates viruses as well as bacteria like *E. coli* and salmonella. Some plants also treat water with ozone, which is more effective in killing protozoa. In most facilities the water is also filtered. Removing particles larger than one micron in size will eliminate threats from anthrax and botulism spores.

Still, that pathogens in water supplies can kill, we know only too well – in Canada, Walkerton and North Battleford provide the grim evidence. In the United States, the largest such contamination was in the city of Milwaukee in the spring of 1993, when cryptosporidium, a protozoan, passed undetected through two water-treatment plants and, once it reached customers' taps, caused more than four hundred thousand illnesses and somewhere between fifty and a hundred deaths out of some eight hundred thousand customers who drank the water. Cryptosporidium is often present in mammal feces, and so might have come from a nearby sewage plant; authorities never figured out what happened.

There are about 168,000 public water systems in the United States, some serving 8 million people and some serving as few as 25. To thwart a deliberate repeat of the Milwaukee poisoning, security has been stepped up around the country's water supplies. Since the events of September 11, for example, the Coast Guard has been patrolling the area of Lake Michigan where Chicago's water intake is located. New York City has increased the number of daily samples it takes at 900 sites from 2,060 to nearly 2,500, has blocked off some roads that traverse reservoirs, and has stationed armed guards at critical sites.

Water-delivery systems, of course, are vulnerable to more than poison or pathogen attacks. Tom Curtis, an executive of the American Water Works Association, whose member utilities supply water to 80 per cent of the U.S. population, pointed out in the *Boston Globe* in late 2002 that a single truck bomb set off beneath a pumping station could disrupt urban supplies for millions of people for months. At least one U.S. city, he said, without being specific, "has six giant pumps, and they're all in one building. If you crashed an airplane into that building or blew it up, it would cause half a million people to lose their water supply almost instantly. Pumps of this size must be custom-built and can take as long as eighteen months to replace."

And while we're dealing with paranoid fantasies, there is also the easily managed possibility that ordinary household taps or fire hydrants could be used as reverse siphons to dump pathogens into a localized water supply system. A simple bicycle pump, or even a vacuum cleaner, would be enough to do the trick.

In July 2002, as though to underline these concerns, the FBI issued another of its pre-emptive warnings. This one spoke of possible terrorist attacks against water-pumping stations and pipes that serve its cities and suburbs. The bulletin said that documents had been discovered in Afghanistan that indicated an Al-Qaeda interest "in ways to disrupt the U.S. water supply on a massive scale."[13]

North of the U.S., the concerns were rather different. When I published the first edition of this book, and for a year or so before that, the hot-button issue in Canada was not water quality,

however, or even water supply, but the notion of bulk water
exports of "our" water to "them" – meaning, of course, the
United States. It was an issue tailor-made for the nationalists and
the anti-free-traders, and it has gotten worse in the interim, as
Canada-friendly Clinton gave way to couldn't-care-less Bush.

The issue was kick-started by a permit granted by the newly
laissez-faire province of Ontario to a small company, based in the
border town of Sault Ste. Marie, to export six hundred million
litres of Great Lakes water via cargo ship to Asia. The very notion,
of course, was improbable – water is precious, but not yet valuable,
and it is heavy. The company concerned, Nova Group, declined to
say how it would profit. Where would these tanker ships go? You
can't sail from the Great Lakes to Asia without first travelling down
the St. Lawrence River, along the east coast of North America,
and through the Panama Canal or around the Horn. What all that
distance would cost in freight charges alone never came up. And
who would buy the stuff? Who in Asia was that short of water?
Nova didn't say – it hadn't yet looked for customers. It didn't have
time: the mere notion brought the anti-free-traders and the cultural nationalists rushing to the barricades to protect Our water
from Their greed . . . "their," in this case, for once, not meaning
the Americans. Except indirectly.

Running through the argument were consistent errors. The
notion that Canada has "two-thirds of the world's freshwater
resources" became a commonplace in news stories, though it is
entirely wrong (the real number is around 6 per cent of annual
global runoff). There were even reports that said the Great Lakes
themselves contained half the world's fresh water, which was not
only wrong but ludicrous – all the Great Lakes combined contain
about as much water as Lake Baikal in Russia. Still, water is
copious in Canada. By some measures, Canada already "exports"
about 79,000 cubic metres of water per second as it flows into the
sea or across the border, and the Nova Group proposal would have
taken, each year, a volume equal to about ninety minutes' worth of

the average annual outflow of the Great Lakes Basin. Put another way, the 600 million litres is the amount of water that flows into the St. Lawrence River every minute. Lake Huron alone is the size of the entire Ogallala Aquifer.

A year after the first edition of *Water* came out, a group of Newfoundland entrepreneurs bought a lake called Gisborne, and declared their intention to export it. The Newfoundland government approved the notion, and when I gave a speech in St. John's not long afterwards, in which I expressed skepticism about the idea, I was met by a wave of hostility, as though I was taking jobs away from deserving islanders. But I had made the same point over and over to journalists, who called me in the years after my book came out, seeking comment on whether we should protect ourselves from American rapacity by forbidding them access to our pristine water. My point was pretty simple: you can't sell something without a buyer – so who are the customers? I urged each journalist who called to ask the "owners" of Gisborne Lake, as well as Nova Corp. and Sunbelt Water Inc., who were then in the British Columbia courts, who their customers were. Each journalist said they'd asked, but had not received an answer. The given reason was commercial confidentiality, but the real reason was that there were no customers.

When more journalists called, I urged them to examine the business plan. How did these companies propose to make a profit? The only plans I saw proposed taking water in tankers to . . . well, to somewhere . . . or, perhaps more practically, to pipe or conduit it from the British Columbia coast to thirsty southern California. Never mind the thorny politics of traversing ecologically friendly regions, such as Oregon and northern California – what would it cost?

Except for bottled waters, the exporting of water, in reality, makes little business sense. The only real customers for Canadian water are the southern Californians and the arid states of the

Colorado basin down to the High Plains of Texas; the most optimistic business projections for delivering water to customers came
in at about $2 a ton, and that figure depended on a number of
more-or-less implausible assumptions. Set that against the fact that,
in August 1999, Tampa Bay Water in Florida announced it had
hired a consortium to build a new reverse-osmosis desal plant that
would produce 25 million gallons of fresh water a day, at a declared
cost of 45 cents a ton. And consider that, in 2002, the Israelis
announced a plan to build a massive desal plant at Ashkelon that
would produce water at 10 cents a ton. It simply reinforced what I
had been telling interviewers for the better part of a year: if I had
any money to invest (a vain hope) I'd sink it into desalination
research and not into the illusory business of exporting water. The
cost of desal was bound to drop rapidly – and in fact was dropping
rapidly, even before I began looking at it again. Why would anyone
buy Canadian water when they could make it more cheaply themselves? Even Nevada, which was a long way from the sea, could use
desal: it could build a plant for California and in return keep some
of California's allocation of the Colorado River. In sum, if Canada
did end up exporting bulk water, it would be a short-term industry.

Nevertheless, there were scientists who looked at the ecological
consequences of bulk water diversions, should they occur.

"You'd think there wouldn't be any problem," said Pierre
Boland, a scientist and one of three Canadian commissioners on
the IJC. "Maybe there isn't now. But, as with many other things,
you start taking a little here and a little there and eventually you
find out that you're taking a lot of water. If you take enough,
you end up lowering the level of the Great Lakes."

A point not often understood is that only about 1 per cent of
the water in the Great Lakes is replaced every year through the

natural water cycle, such as feeder rivers and rainfall. The other 99 per cent is fossil water, from the melting of glaciers about twelve thousand years ago. If you start to affect the flow-through, you start to mine the system, and the whole system starts to change.

But what does it take to affect the flow-through? The numbers are always tricky, and forecasts trickier still. In 1977, the Canadian Environmental Law Association (CELA) issued an analysis of the Great Lakes system that forecast a drop of 25 per cent in the flow-through in less than forty years. Climate change and "outrageous overuse" of water, mostly by agriculture, were blamed. I had already heard from Eugene Stakhiv, the Army Corps of Engineers researcher, who had noted that Canadian estimates of changes to the lakes were notoriously inflated. Of course, he was talking about climate-change forecasts, not about overuse. But Canada's federal department of the environment met the CELA report with equanimity, pointing out that, in 1977, the Great Lakes were "at record levels, excessively high." No need to worry; come back and see us in forty years. And indeed, halfway through the forty years, nothing much had changed.

An editorial in the *Globe and Mail* compared the flap over Nova's application to General Jack D. Ripper in Kubrick's *Dr. Strangelove*, who seemed an eerily Canadian figure in his obsession with foreign conspiracies "to sap and impurify our precious bodily fluids." The *Globe* pointed out, with some justification, that Canadian steward-ship of its water resources was hardly impressive. Canadians were water hogs, second in the world only to the United States in per capita consumption of water, taking three times more each day than the average German, for example. The numbers are 326 litres per day per person in Canada, 400 litres in the United States. Canadian water costs somewhere around 40 cents for 1,000 litres, a little lower than American prices. By comparison, the English pay more than $4 for 1,000 litres, and consume less than half as much. More than half of Canadian households are not metered, and can use as much water as they like for a low flat fee. A third of Canadians are not even served by waste-treatment plants.

Canadians have been slow to acknowledge, moreover, that the U.S. per capita consumption of water has been dropping steadily – by more than 20 per cent since 1985 – whereas Canada's hasn't. Said the *Globe*:

> With respect to water, Canadians and Americans suffer from the same disease: We say that it is priceless, but act as if it were absurdly cheap. Most North Americans pay far less for their water than even just the cost of supplying it, cleaning it up and returning it to the environment. Yet subsidizing water use is economically and ecologically disastrous. In fact, heavy subsidization of water in the U.S. is the cause of any water "shortages" that may exist there. In California, for example, agriculture consumes 80 per cent of the state's water, often growing low-value and water-intensive crops in the desert. If everyone paid the true value of the water they used, they would use less, freeing more for those who need it. The so-called shortages would then stand revealed as water-misallocation problems that would rapidly disappear. But why would Americans endure that pain if Canadians are willing to treat their water like so many other natural resources, selling it at less than its true value as an artificial prop to growth and employment? If Canada's water is priced at its full value, U.S. demand will be moderate, because the huge diversions and other projects that so frighten many Canadians would be wildly uneconomic compared with more sensible alternatives, such as conservation, economization and desalination.[14]

They might have pointed out to their readers that, in any case, the Great Lakes are not "ours." The border runs somewhere down the middle of the lakes, and "their" half is therefore their water, and what is to stop them simply pumping it away? If they are as bad as we say, what is to stop them taking as much as they like? What are Canadians going to do? Put up a fence?

Of course, the Americans have no intention of doing anything of the sort. Nowhere in all this debate was another fact unpalatable to Canadian nationalists even hinted at, which is that the American border states, some of them part-owners of the Great Lakes, were by and large better at conservation and pollution abatement than their Canadian counterparts. Environmental protection laws were considerably stricter in, say, Wisconsin than in Ontario. Canadians prefer to adhere to their fond belief that "Americans" are worse than they are.

The Nova issue came to naught, in the end, when the company, chastened at the public uproar it had triggered, withdrew its application, on the condition that Ottawa work with the provinces to ban all water exports. Quebec's separatist Parti Québécois government, in its turn, announced that it would declare all water under its territory as the property of the state, which would until recently have been a very French solution – except that France began privatizing its own water a few years ago. In another largely ignored snub to the federal government, a provincial communiqué said that Quebec reserved the right to sell its water wherever it wanted.

Early in 1999, the federal government, its nationalist hackles now up, promised legislation to ban water exports via "supertankers, pipelines or man-made trenches." The environment minister, Christine Stewart, was clearly on the side of virtue. "We consider this an urgent issue," she said.

Nevertheless, the actual policy statement she issued was more measured than her press releases. Her notion was to call for a moratorium on water exports until a national accord could be worked out. The governing idea of the policy was to protect Canadian watersheds, and in this context (an environmental context), the whole notion of NAFTA was neatly sidestepped.

The International Joint Commission itself entered the debate with a report in August 1999, calling for a six-month moratorium on exporting water from the Great Lakes. They were responding to low water levels in the summer of 1999 and declared, essentially, that there was no surplus water to spare. The Canadian government responded by calling the report "too permissive."

And nowhere in any of this did the notion ever emerge that water was not – or should not be, strictly – anyone's property. Water is not "ours" or "theirs," but the planet's. We use water, and it passes on, and then it comes back to us. But it is not, surely, something we should either hoard or prevent others from using.

And so it was left to a letter writer to the *Globe and Mail*, Shirley Conover, to set the ecological record straight and remind the politicians, the editorial writers, the corporations, and the citizens what water really is. "Water is not a renewable resource," she wrote. It only seems renewable because it keeps falling from the sky. But that is an ecologically primitive way of looking at things. It may be common sense, but, as so often happens, common sense can be uncommonly ignorant. "Renewable resources can reproduce themselves, that is, living things such as trees, cows and people. Water cannot reproduce itself. Water is recycled by means of the hydrological cycle: evaporation plus transpiration by plants, to cloud formation, to rain and snow, back to plants, rivers and ground water, to the oceans and cycling around again by means of evaporation, transpiration and precipitation. The hydrological cycle is an ecosystem service, a life-support system for all living things, including humans." By removing water from one basin to the next, the basin being the hydrological cycle's recycling unit, you are tampering with this life-support system, with uncertain consequences. The Aral Sea, as we have seen, is a grotesque example, but there are many others.

Along with water-export the other hot-button issue in Canada was privatization. The Canadian Environmental Law Association, which may have been over-hasty on the Great Lakes, nevertheless has a finely tuned sense of what is happening in resource industries, and believes strongly that the newly privatized water companies are gathering for the kill. The British and French water companies, behemoths in world water management, are prowling the world looking for customers. Like any corporation hungry for business, they are seldom scrupulous about their partners. In 1998, the British company Thames Water and the French firm Lyonnaise des Eaux had their contracts to manage Jakarta's water supply suspended after their contacts with cronies of the dictator, Suharto, were revealed. Lyonnaise des Eaux became the largest water company in the world after merging with another French company, Compagnie de Suez, in 1997. Along with Thames Water and Vivendi, Lyonnaise now distributes water to 230 million people in thirty countries, and has contracts with half a dozen Canadian municipalities, including Montreal (a subsidiary, Janin, built the Confederation Bridge linking the provinces of Prince Edward Island to New Brunswick). The company has been actively pushing the trendy notion of public-private partnerships, which the cynical interpret as a way for private companies to get access to all that wonderful tax money.

A Canadian company, Philip Services Corp., which runs the water system in the Ontario city of Hamilton, successfully negotiated a contract whereby the politicians, and not the company, would be held liable for "any problems rising from the management of the facility," a neat way of having your cake, eating it too, and never having to pay for it. The American electricity and gas company Enron has taken over one of the British privatized water companies, Wessex Water, and is looking actively at the Canadian market – or it was, until its ignominious collapse. Other American companies such as American Filter (which started its life as a treatment facility) have already signed contracts to "manage" local water companies in Canada, and in at least one case have filed for status as a municipality under Canadian law, to bypass the tax code. Jim Southworth, president of another company, called Consumer Utilities, has expressed his dismay that the Canadian tax code seems, inexplicably to him, to favour public projects. He had hoped, he told the industry newsletter *Global Water Report* in 1997, "for a more level playing field on the 7-per-cent Goods and Services Tax, which is levied on private- but not public-sector services." Many other water companies have expressed their dismay at the "slow pace" of privatization in Canada. CH2M Waterworks, for example, an arm of Colorado's MH2M Hill, points to the more than two years it took to get regulatory approval to build a water-treatment plant in Halifax, Nova Scotia.

Canadian companies are getting into the water business too. "Bottled water," one of them told the Canadian Broadcasting Corporation approvingly, "sells for a dollar a litre. Gasoline sells for half that." His company has been diligently mining an aquifer in Quebec, depleting it so much that local farmers are now obliged to treat the water they give their cattle. A proposal for another water-bottling plant in the little Quebec community of Ways Mills set off a citizen uproar that sent the chastened city fathers back to the bylaw drafting table to keep "predators" out.

~

Still, despite the politicians' perverse lack of interest, in the last few years the notion that there is a water crisis has bubbled up into a public consciousness that had hitherto been more or less impervious. That it did so in Canada, was due to a concatenation of events, some of them deadly in nature, if, at least in hindsight, entirely predictable.

The most obvious "event," and the most serious in the short term, were the half-dozen deaths in the bucolic little town of Walkerton, Ontario, where a strain of the *E. coli* bacterium somehow infiltrated the civic water supply and infected the town's most vulnerable citizens – the elderly, the very young, and the already ill. This was followed by a similar "event" in the Saskatchewan town of North Battleford. The most depressing result was the outpouring of recriminations and finger-pointing between the provincial governments and their ideological opponents, both sides using tragedy as a political weapon, and leaving observers shaking their heads, unsure whether the blame-laying was the result of cynicism or stupidity.

The most encouraging result, on the other hand, was the way the tragedy made many of the citizens of Canada suspicious of how their water got to their taps, thereby focusing public attention both on the origins of the water and the process of its delivery in a way that hadn't been possible before.

Another positive result of the bad news was the wave of entrepreneurial energy it set off all over the country, as inventive people sought remedies for the newly perceived problems. These "remedies" ranged from massive and capital-intensive schemes to home-brew remedies. Even in my own area, I saw the results, as two inventive neighbours, John McGee and George Palmer, set up a company to deliver potable water for residential and commercial use, employing a shopping list of off-the-shelf technologies, such as inline chlorination, ultraviolet and ozone purification, and multi-media filtration systems. Their new company, The Water Clinic, even offered new techniques for producing drinking-quality water from the air's humidity, aimed at cottagers and boat- and RV-owners.

The other result was that a wave of alarums went out from suddenly thorough municipal waterworks, and dozens of communities issued boil-your-water-just-in-case bulletins. This was when little Prince Edward Island suddenly woke up to its transformation into a fiefdom of agribusinesses. Island fishing associations have been horrified at the results. "The river I deal with is as thick as chocolate milk," said one of them. "We can't go on like this."[15]

That the country's attention has been slowly turning to the quality of its water is also due to the energy and tirelessness of people like David Schindler, a scientist with impeccable credentials, who seems to have developed a knack for the quotable phrase. His assessment is stark: pollution, habitat destruction, and climate warming will compromise Canada's freshwater supplies so dramatically in the next fifty years, he has said, that freshwater fisheries could disappear altogether and drinking supplies will be in a state of crisis. In a paper in the *Canadian Journal of Fisheries and Aquatic Sciences*, in fall 2000, he laid into Canada's political masters for their "tiresome, juvenile turf war about who controls what."

In fact, he says, a continued warming trend is not just a theoretical danger for Canada – it will cause many of the Prairie lakes to disappear in thirty years. Evaporation rates, he points out, have increased 30 per cent in recent years, with no end in sight. Summer flows in the South Saskatchewan River are now only 20 per cent of values measured early in the twentieth century. And pollution problems are creeping northward. "Recent data show that Lake Winnipeg is now in about the same state as Lake Erie was in the late 1960s. This is the result of massive increases in agriculture and decreased renewal resulting from the decreased flow of the Saskatchewan River."[16]

Canadian carelessness with water is hardly new. There are still dozens of Canadian communities in which water use isn't even measured, let alone metered. And virtually everywhere in Canada, users were paying far below the true cost of delivering clean water – water was one of the most heavily subsidized commodities in the economy, another factor that encouraged waste. Until the 1970s, almost a third of Canadian municipalities had no wastewater treatment at all, but simply dumped their effluent into the most convenient river (of course, as we have seen, we still have two major cities, both provincial capitals, that dump raw sewage into waterways without treatment). Even in "Super, Natural British Columbia," to use the province's own marketing slogan, things had been going awry. In and around the city of Kelowna, the parasite called cryptosporidium, familiar from the Milwaukee outbreak in 1993, was rampant in municipal water supplies; as many as fifteen thousand people were infected. No one was entirely sure how it got there, but the most likely culprit was the runoff of cattle waste from farms in the Okanagan Valley, British Columbia's most valuable farmland. In addition, individual Canadians now dump about 300 million litres of used motor oil into various Canadian waterways each year – more than seven times the oil spilled during the *Exxon Valdez* disaster.

Dan Braniff, who now heads a southern Ontario citizens' group, Residents of West Grey, who came together when a gravel company was given permission to undermine the local aquifer, remembers a long history of dealing with careless pollution. He had his first encounter with water contamination at the age of nine, when his family went camping on the banks of the Grand River north of Kitchener, which was contaminated with raw sewage and cattle manure. Even so, the kids happily swam in the polluted water. "We fished but fortunately never caught any," he says. In 1945 he summered at Meaford, Ontario, swam in the harbour every day, simply ignoring the signs fastened to the old bridge: *Warning: Water Infected with Polio.* "Our society was already taking water pollution for granted."

In the spring of 2002 Braniff sent me a piece he'd written for a local paper:

In the early fifties while working as a Bell engineer I got an insider's glimpse of community water stewardship. My work required coordination with local utilities. At a luncheon meeting in Palmerston Ontario the PUC superintendent abruptly stopped me from lifting my glass whispering, *Don't drink the water.*

Afterwards he explained that the town's water might make me sick but *it would do no good to unnecessarily alarm the residents.*

A few months later he gave me a personal tour of the municipal reservoir proudly demonstrating how the town had remedied the problem. After tossing a shovel full of pellets into the reservoir the surface exploded into a bubbling turbulence. Mystified, I questioned "What is that stuff?" "Dog food," he replied. "Thanks to these magnificent trout you can again, without hesitation, safely drink our water." As a bonus he bragged, "Palmerston now has its own thriving municipal trout farm."

He invited me several times for a trout dinner but I had no stomach for it. I understood that dog food often contained meat rendered from sick cattle. Somehow the image of seeing the dead carcasses lying along the rural roadside waiting in the heat for pick up was enough to quash my appetite for fresh gourmet trout. It also kept me from drinking water in small Ontario towns for a long time.[17]

Not all the water news in Canada is bad, however and the IJC, despite its occasionally self-congratulatory tone, is part of the good news.

Among the good news is that Americans and Canadians have, after all, a long history of more-or-less amiable relations on water matters, partly because the politicians of both countries saw eye-to-eye on the values of exploiting water and controlling recalcitrant rivers, and partly because Canada's sheer abundance – or perceived abundance – of water made it difficult to generate much political heat over water matters.

In 1909, the U.S. and British governments signed the Boundary Waters Treaty, which was designed to minimize what frictions did exist, and created, almost as an afterthought, the International Joint Commission, which was given the right of approval or refusal of any transboundary water issues. A shopping list of other treaties followed, including the Lake of the Woods Convention, the Skagit River Treaty, the Columbia River Treaty, and the St. Lawrence Seaway Project.

In the almost ninety years since its creation, the IJC has become a model for managing cross-border resources: its careful balance and perceived objectivity have meant that its decisions are seldom challenged by either government. Its amiable American chief commissioner, Tom Baldini, has explained its success in admirably non-bureaucratic language: "What can happen is that one country controls the source and can pretty well screw the other guys," he said, after delegations from the Aral Sea area and the Lake Titicaca dispute in South America had trooped to the IJC offices for advice. "What we have, and what we tell the others, is that there has to be a willingness on the part of all participants to reach a fair and equitable agreement. And it's difficult – water is such a precious commodity."

The IJC has helped this willingness along by maintaining a rigorous objectivity, and by ensuring that the asymmetry in the U.S.–Canadian power balance is not reflected on its boards. The chief Canadian commissioner, Len Legault, maintains that objectivity is built into the IJC's very existence. "We take a solemn affirmation of objectivity," he says. "We are an independent body, and we maintain strict impartiality. Commissioners represent the commission, not the governments who appoint them." Not unexpectedly, there have been ripples in these placid waters from time to time. In 1996 the former Canadian chief, Adèle Hurley, nearly broke up the organization with her fury at its refusal to take a more activist stance on acid rain. After she departed, to the relief of both national delegations, the IJC reasserted its image as impartial scientific problem-solver.

Late in 1998, the two national governments expanded the IJC's responsibilities. It would establish a chain of local transborder watershed boards with a much broader mandate to manage, to make decisions, than is currently granted to local water-control and pollution boards. The first such board has already been set up on the St. Croix River that straddles the Maine–New Brunswick border. Another dozen are planned, from the Atlantic coast to British Columbia rivers that flow into Alaska. A second millennium project for the IJC involves rethinking emergency-response procedures for cross-border disasters, such as the Red River flooding of 1997.

There have been rumblings from Ottawa that the IJC should be handed responsibility for the hottest U.S.–Canada issue after logging: the dispute over the Pacific salmon. In response, Baldini would only say, "That's something that has crossed people's minds, but . . . [considering what else we have on our plates] that would be an incredible diversion."[18]

In fact, the IJC might have to turn its attention instead to updating the Columbia River Treaty, a more-or-less successful international-management treaty which is now showing serious signs of strain. In essence the treaty, signed in 1961, was designed for flood control and hydroelectric generation, and permitted the building of three new dams (the Duncan, the Keenleyside, and the Mica) along the Columbia on the Canadian side of the border, with certain American entitlements to the power generated in return for much of the capital used to build them. In addition, the Americans were allowed to build the Libby Dam on the Kootenai River in Montana, which backed up about seventy kilometres into Canada. A series of codicils compensated Canada for lands flooded by the Libby, outlined the water entitlements of both countries, and provided for a cross-border payment schedule for power used when generated in the other country. For the first thirty years, this arrangement worked well enough, but at the halfway point in its sixty-year lifespan, disagreements are emerging. Under the treaty, Canada consented to sell its full entitlements to downstream power generation at a fixed price to the Americans. In 1995 this agreement expired, and the Bonneville Power Administration, which runs the power-generation schemes on the American side of the border, broke off negotiations with British Columbia, maintaining that the province's new demands and price structures were "unrealistic" – a nice euphemism for "gouging." The IJC, on its record, will probably solve this dispute fairly easily.

More seriously, the original treaty was driven by the power industry and left no room for environmental issues, such as the West Coast salmon fishery. An article in the *American Review of Canadian Studies* in 1997 put it this way:

> Nowhere in the treaty are there formal provisions for the integration of fisheries and other environmental concerns with existing power and flood control priorities. Although adjustments can be made . . . to accommodate multiple-use concerns, such changes must be compensated for if they reduce power production on the other side of the border. Unfortunately this narrow approach to decision-making [has] created a climate in which management decisions, which ultimately are taken by hydroelectric and reservoir operation entities in both countries, are viewed as "nonpower vs. power" trade-offs. On both sides of the border, this faces the primary river managers with a most controversial issue: the inherent conflict between their obligations under domestic environmental legislation, and obligations under the Columbia River Treaty.

In effect, the treaty is between the classic rock and a hard place: the rock is the Endangered Species Act in the United States and the Fisheries Act in Canada; the hard place is the way the system has come to depend on managed water for electricity.[19]

The treaty's framers were innocent of environmental concerns – in 1961, rivers were, by definition, to be managed. That's what they were for. As the power-generation industry will tell you, good management is what made us all prosperous. But, as the Canadian Environmental Law Association (CELA) will tell you, it's that's attitude that got us into this mess in the first place.

~

A number of provincial governments have, at least on the surface, been paying attention to drinking-water problems. In June 2002, the British Columbia government, despite wholesale carnage in the civil service and drastic budget cuts, allocated $15 million for cleaning up the water and hired "drinking water protection officers" for each of the province's health regions; these officers would report to the health department, not to Resources or Agriculture, where their reports would more likely be buried. There was some skepticism – the government rejected a recommendation for a lead agency at a Cabinet level responsible for all water issues, opting instead for a nebulous inter-ministerial committee of dubious effectiveness – but the initiative was welcomed by most of the stakeholders, including NGOs and political opponents. The most positive sign was that the plan eschewed grandstanding for data collection. Residents of other provinces were left hoping it might be catching.

~

British Columbia's initiative was, in point of fact, a refreshing change from the work of earlier generations of water planners, who had dreamed very big dreams for Canadian water, some of them grand on a scale that would impress even the engineers who have covered the American West with its web of pipes, canals, ditches, siphons, dams, and pumps, and who are used to shifting a trillion litres across a landscape where it is naturally absent.

One of these plans started with an ecological catastrophe in the making, the infamous Love Canal, more a dumpsite for chemicals than a true canal. In the 1970s, hydrologists began examining the ooze that had entered the canal and found it a toxic brew of PCBs and other chemicals too numerous to catalogue. Some of this pollution had already entered the groundwater and was seeping steadily towards the gorge of the Niagara River. There were alarming eyewitness reports, and for a while television crews were stationed along the lip, watching for breakthrough. What a great story! You could actually see the seepage along the escarpment wall. It would take only a widening fissure to create a clean-water emergency of massive proportions. The cities of Toronto and Montreal, Canada's two largest, drew their water from the Great Lakes and St. Lawrence system, downstream from Niagara.

A plan was floated by engineering consultants, which took on, albeit briefly, a virtual existence in the media and the minds of provincial politicians. The notion was to connect the pristine waters of Georgian Bay, a deep and protected adjunct to Lake Superior, to the communities of the St. Lawrence valley. Georgian Bay had the advantage of being upstream of Niagara, as well as upstream from most pollution sources, except for a few struggling communities north of Superior, such as Thunder Bay. The plan's proponents thought it probable that they would be able to supply Toronto with potable water in its hour of need – and charge a stiff price for the privilege.

After a while, when the seepage stopped and the scientists began sucking the muck from the canal upstream, the TV crews lost interest. So did the potential investors: without an emergency, there was no profit. Toronto continued drawing its drinking water from Lake Ontario and, when the pollution levels reached a medically scary level, it solved the problem by pushing the intake pipes deeper and further. In 1998, impressed with the cold temperature of the water they were drawing up, Toronto proposed extending a pipe even deeper, to the really cold water, kept frigid since the ice age, and "mining" that water – not for the water itself but for its temperature. Why not use the cold water to air-condition all of downtown Toronto through heat-exchangers, and export the resulting warm water downstream to Montreal? Let them worry about temperature gradients. No one thought to question how this would affect the life cycle of the lake itself.

The end of the Georgian Bay Canal wasn't the end of the Big Idea. In fact, it was only the start. The province of Quebec had already been building a massive series of dams and reservoirs to construct the $35-billion James Bay Hydroelectric Project, which would export huge amounts of power to the United States. The project was developed at ruinous cost to the indigenous culture of the Cree Indians, many of whose ancestral lands were submerged. Robert Bourassa, premier of Quebec on and off for several decades, began in 1985 to push for an even-more-massive project. The acronym his planners gave it was GRAND, for Great Replenishment and Northern Development Canal Concept. It would turn Ontario's James Bay port of Moosonee back into a freshwater terminus. A gigantic dike would be constructed across the mouth of Hudson Bay; the rivers pumping millions of tons of water annually into the bay would gradually transform it into a freshwater reservoir. What would happen to all that new fresh water depended on who you were talking to – or on who, as a newspaper report at the time put it, had been smoking what. An aqueduct to the Great Lakes seemed the most popular version, with perhaps a canal from there to the Midwest or High Plains states.

All these schemes paled by comparison with the North American Water and Power Alliance (NAWAPA), an even-more-grandiose plan, dreamed up by Californians, to dam most of British Columbia's dozens of rivers and turn them backwards into the Rocky Mountain trench, a new storage reservoir about eight hundred kilometres long. The province contains the third-, fourth-, seventh-, and eighth-largest rivers in North America, and could easily spare enough water to slake at least ten times California's most extravagant desires. Some of the water would irrigate all the High Plains, other bits would go to the Colorado and the Great Lakes and even the Mississippi system. This project never got very far, not merely because of its cost – no one got around to working that out, though it would probably climb to somewhere around half a trillion dollars – but because the ecological consequences would be, almost literally, ungraspable. It would do as much damage to the environment as all the water diversions in America combined.

No one now thinks these projects, or any like them, will be built. The day of the massive eco-engineering scheme seems to be over. There is now much more understanding of the damage such projects do. In 1998, as a signal of changing sensibilities, the forest giant MacMillan Bloedel agreed to end clear-cutting once and for all, the given reason being "customer demand." And in the year the West Coast salmon fishery collapsed because of overfishing, pollution, and habitat destruction, there was more talk of work to re-engineer projects back to the way they were than of building new diversions.

Throughout North America there is a now a good deal of pressure to repair the damage some of the grandiosity of the past has caused. The U.S. government is now "studying" – a nice evasion – the breaching of the three lowest dams on the Snake River, in Idaho and Washington, to the delight of the environmentalists and the fury of the farm lobby, who in early 1999 were planning what they were calling a "massive pro-dam rally." A farmers' meeting in Pasco, Washington, was addressed by anti-environmentalist gadfly James Buchal, who believes, in the teeth of the evidence, that the "salmon crisis" in the Columbia River basin has been concocted by federal bureaucrats intent on "gaining control of the Columbia River." The steam went out of the plan and the opposition both, with the election of George W. Bush. But even so, the 215-metre Glen Canyon Dam on the Colorado, and the 290-kilometre Lake Powell above it, may also soon be history, if a campaign started by the Sierra Club gathers enough political clout. The campaign also represents something of a *mea culpa* for the Sierra Club, whose director, David Brower, agreed to let the dam be built in the 1950s in exchange for the scrapping of two other dams he considered worse. Lake Powell drowned several thousand Native American ruins and obliterated the rapids in Cataract Canyon and some of the most startling riverbank scenery in America. All that can be restored, the club's analysis says, and the notion that the reservoir provides enough water for Salt Lake City and Las Vegas is false. It loses almost as much to evaporation as it gives to those cities, and more water could be provided more cheaply through simple canals. Not everyone agrees, however, and among the dissidents are some eight thousand people living in the town of Page, Arizona, which exists only because of the dam. A Page resident was quoted in 1997 as saying, "they ought to line up those environmentalists and shoot 'em, every last one of 'em."[20] Nor does everyone necessarily understand the Sierra Club's desire for pristine wilderness. A Scottsdale, Arizona, company, for example, has proposed something it calls, with a sublime disregard for topographic logic, Canyon Forest Village, a 2.8-square-kilometre development near the entrance to the Grand Canyon which would contain more than 5,000 hotel rooms and 46,000 square metres in retail space, and which would draw its water from wells struck into the deep aquifers that feed some of the canyon's most beautiful creeks and falls. There is opposition, mostly from business people in neighbouring towns and from environmentalists, who rightly fear the unleashing of further hordes on the canyon's fragile ecosystem, but in this case the project will likely get built.

As we move further into the new millennium, we find the Mexicans hunkering down, hanging onto every drop they can; the Canadians hunkering down, determined to lose no more water; and the Americans in the middle, trying to make the best use of what they have. The Americans retain their trademark combination of reckless use (egged on by a corporate system that will, if necessary, pave over the wilderness it wants to protect so that ever more people can see it) and creative energy (using that same corporate system to fine-tune demand and supply, and conserve by the logic of efficiency). Denver's Chamber of Commerce has already pointed out that, while irrigated agriculture is an industry pulling in somewhere around $200 million a year, tourism pulled in twenty-five times that, more than $5 billion. For the first time, conservationists have an argument more compelling than using the Endangered Species Act to save the snail darter or another obscure warbler: money. In some places, at least, conservation is demonstrably good for business. It will help raise tax money. It will help politicians get elected. It will, everyone agrees, help the land look better. Most certainly, if conservationism can make everyone a dollar, conservationism has a future.

15 | SOUTH ASIA AND THE INDIAN SUBCONTINENT

Some water issues are intractable, and some that look intractable have, to everyone's astonishment, been amicably solved.

Ll over Asia, in the early years of the millennium, regional banks and development agencies were beginning to issue dire warnings about the worsening state of the region's water. The language of these reports still was generally measured, but there was, increasingly, an undertone of anxiety that was not yet panic. Typical was a study by the Asian Development Bank, completed early in 2002, that found water resources throughout the region under "severe stress" from widespread overuse and pollution caused at one and the same time by quickly expanding populations, poverty, and bad management. "The pressures are rapidly becoming acute," the report said. "For the first time, water security is a live issue in the region."

South Asia generally, where more than a sixth of the world's population lives, is in particularly bad shape, and getting worse. The bank estimated that availability of fresh water per capita in the region has dropped a discouraging 70 per cent since 1950.

Indonesia's crowded main island of Java, with 60 per cent of the population, is facing acute water depletion and pollution problems.

At the same time, rampant deforestation is causing worsening floods and increasing soil erosion.

A major part of the bank's concern was focused on the Mekong River, which flows for 4,400 kilometres, from the Tibetan plateau through six countries – China, Myanmar, Laos, Cambodia, Thailand, and Vietnam – to the South China Sea. The Mekong basin has a population of 60 million, which will reach 100 million by 2025. None of the riparian countries want to surrender sovereignty over their portion of the river to an international body, but, on the other hand, complaints about upstream rapacity are commonplace. China is the most frequent target, for the Chinese are building a series of increasingly large dams along its section of the river, which, they have assured everyone who will listen, are for flood-control purposes. Downstream countries, for their part, remain dubious. They fear that China's growing demand for water will persuade the Chinese to divert too much of the impounded water for their own use.

Severe as the problems are elsewhere, however, India and its neighbours perhaps illustrate best the complexity and intricacy of managing interstate relations in regions where major rivers cross national frontiers and where demand is outstripping supply.

The news is not all bad, either.

~

There are 19 million new Indians every year – a Canada every two years or so. In 1993, the last time anyone counted, fewer than half of urban Indians (42.9 per cent) had any access to sanitation facilities, and a puny 3.5 per cent of rural Indians. Some 16 per cent of city people and 20 per cent of rural people had no clean drinking water. Farmers currently take 85 per cent of all India's renewable water, and – just to feed the new people – demand will rise 50 per cent by 2025, by which time another 400 million new Indians will be seeking sustenance. Industrial centres, most of them in the same areas as the most fertile land, will triple their demand for water by the same year. As an FAO study put it, "[the results do] not identify India as suffering from severe water stress, [but] India's percentage is well above [worse than] the solvable water level." By 2025, however, nearly three-quarters of all Indians will be living in regions with less than 1,000 cubic metres of water per year, too little to sustain development. The Indian government is not unaware of the extent of the problems. In 2002, water resources minister Arjun Charan Sethi took to using the hitherto-forbidden word *crisis*, and was quoted as saying that 22 per cent of the country was facing "absolute water scarcity."[1]

On the other hand, the ramshackle, sprawling, chaotic, ethnically complex more-or-less democracy that is India was widely expected to be an international pauper and an agricultural basket case by the 1960s, and instead it has become self-sufficient in food production and a net exporter of food grains.

On the *other* other hand, the Green Revolution and the grand scheme of large-scale water-development projects launched by Jawaharlal Nehru after independence have now run their course. India's ambitious use, or overuse, of its rivers served its purpose in its time. The Green Revolution helped knit the unwieldy country together, and food production rose from a measly .3-per-cent growth a year to well over 3 per cent, a phenomenal increase. But they have also caused irritation and antagonism in downstream countries. India's quarrels with Pakistan and Bangladesh are not just religious and ethnic; they are also resource-based. The Green Revolution depended not only on more productive seed grains but on irrigation and pesticides. Both have run out their string, maximized their potential. There is no more water to be had, irrigation is increasing the amount of saline soil, and pesticides have poisoned an unsettling amount of what land is left. And the inexorable grind of population is now negating what few gains there are still to be made.

The river Ganges, the holiest of rivers to the Hindus, the gift of God to civilized mankind, giver of blessings and cleanser of sins, rises on the southern slopes of the Himalayas in India, 2,510 kilometres away from its mouth in the Bay of Bengal. Well, that's what the Indian texts say, but it's not quite so clear-cut: a look at any map will show that much of the Ganges water comes from China and Nepal before flowing into India. About 40 per cent of the total flow, and nearly three-quarters of the dry-season flow, come from outside the country, facts that both the Chinese and Nepalese have impressed upon India more than once in the last few decades. The river wanders through the Ganges flood plain, briefly forms the border between India and Bangladesh, where it takes the name Bhadma, and then joins the Jamuna–Brahmaputra, which itself rises in China and flows in a great circle around Bhutan before passing through India into Bangladesh. It then empties into the Bay of Bengal.

It is a grim irony that one of the highest concentrations of very poor people in the world live on the fertile plains of one of the world's largest river basins. The Ganges and Brahmaputra river systems take up less than .2 per cent of the world's land mass, but their basins are home to 10 per cent of the world's population. The whole area is prone to floods, droughts, and cyclones; at times it desperately needs water, at others it seems the whole place is drowning. For more than forty years, national governments in the area have wrestled with allocating and managing the available water. The environmental issues are hopelessly complicated in a techno-political tangle, while demand and supply are caught up in a difficult swirl of nationalist and ethnic political issues.

The Ganges disputes go back almost to independence in 1947, and at least to 1951, when Bangladesh was still the eastern province of the Federation of Pakistan. In that year, India's new government announced it would create a barrage across the Ganges at a place called Farakka, in the province of Bengal close to the frontier with Pakistan. Its purpose would be to resuscitate "the dead and dying spillways" of the river in the area and divert water through a canal to the Hooghly River and the Indian port of Calcutta, which was in serious danger of silting up. The Indians needed enough water to flush the silt from the harbour; but taking that much water would seriously damage East Pakistan, which was entirely dependent on Ganges and Brahmaputra water.

Just as happened in the Tigris–Euphrates quarrel, the downstream country (in this case Pakistan) squawked, but to little avail. The Pakistanis argued strenuously for an international solution; India said the resource was hers to dispose of. Pakistan argued for historic rights; India said it would make equitable distribution, taking into account the population, the land, and the demonstrated need. If any jurisdiction felt it was harmed, it would be up to it to prove its case. Any country can build a dam anywhere on its land, India asserted, and the water the dam stores is that country's to dispose of.

There were a number of inconclusive meetings between the two countries, but in the end India went ahead without an agreement and started construction in 1962. Negotiations resumed after that, but the Indo-Pakistan war of 1965 cut them short even before an agenda had been worked out. Afterwards, negotiations were poisoned not only by residual hostilities but by the separatist movement in East Pakistan. Pakistan tried to stall, hoping the Bengalis in their eastern province would see that India was attempting to steal their water. India, for its part, tried to stir the separatist pot by implying that West Pakistan didn't much care whether the East got any water or not.

In 1971, East Pakistan duly broke away from the West, constituting itself as the new country of Bangladesh. Since India had been of considerable help in their independence struggle, observers assumed that an agreement over the Ganges would follow quickly.

And indeed, an Indo-Bangladesh Joint River Commission was duly formed.

But again, domestic politics intruded. The Bangladeshi leader, Mujibur Rahman, knew how emotional his water-dependent people could become about Ganges water, and he couldn't allow himself to be seen as an Indian stooge. At the same time, he knew how precarious his position was, and how dependent he was on India. Instead of acting, therefore, he dithered, and it wasn't until 1974 that an "interim agreement" was reached. It allowed the barrage to proceed, and let India release between 310 and 450 cubic metres per second in the dry season into the Hooghly and towards Calcutta. The Indian government boasted that the agreement was an outstanding example of mutual understanding and accommodation, but it was nothing of the kind. The Indian opposition was angry that India didn't just take the 1,100 cubic metres per second that Calcutta's siltation problem demanded; and the Bangladeshi opposition said the whole agreement had been a craven bowing to Indian power. Shortly afterwards, Rahman was assassinated by the army. The water agreement was blamed.

A few years later, when Bangladesh's new leaders began receiving moral and material support from the Islamic nations in the Middle East, and anti-Hindu rumblings were echoing across Bangladesh, India's prime minister, Indira Gandhi, began to take a tougher line. In January 1976, at the start of the dry season, India went ahead and diverted the water at Farakka without consulting anyone. The charismatic Maulana Bhasani promptly led a "long march" to the Indian border; more than a hundred thousand people crowded against the frontier barricades and chanted anti-Indian slogans, protesting that the Indians had stolen their water.

And so it went. There was a brief interlude when Gandhi's Congress Party unexpectedly lost the Indian elections to the Janata

Party in 1977, and the Janata leaders made overtures to India's neighbours, pledging co-operation on the water front. They signed an accord guaranteeing minimum water (63 per cent of the river flow) to Bangladesh in the dry season, in return for Bangladeshi agreement to build a new channel between the Brahmaputra and Ganges to augment that flow.

The channel was never built, and the accord was widely denounced. Nevertheless, when the Congress Party returned to power in 1980, Indira Gandhi signed a Memorandum of Understanding with her Bangladeshi counterpart, extending the 1977 agreement. The extension expired after eighteen months, and nothing had been done about it when Gandhi was assassinated by her Sikh bodyguards in 1984.

Of the classical elements – air, water, earth, and fire – only one seems to have any relevance to Bangladesh, and "Bangladesh often seems to be a land upon water or a watery sheet upon the land."[2] Almost one-third of the country is water, even in the dry season, and when it rains, more than two-thirds is under water. When it floods, it floods catastrophically, and never more so than it did in the wet season of August and September 1988. Three-quarters of the country was underwater, and the water remained for over two months. More than 35 million people were dispersed, and tens of thousands of wells were submerged, rendering their water unfit to drink. The Bangladeshi president, General Ershad, appealed for international help, but refused to take aid from India, which, he declared, caused the flooding by its inhuman unwillingness to manage the Ganges. Later that year, when the dry season approached, India once again began diverting the Ganges Water unilaterally.

The political situation continued to deteriorate. Bangladesh converted itself constitutionally into an Islamic republic, and militant

Hindu fundamentalism was on the rise in the border areas of India. Neither side wanted to "sell out" to the other for domestic reasons, and when the two governments met in 1992, no agreement was reached. Only their disparity in military might seemed able to prevent a conflagration. Bangladesh was much too small to take on the Indians, but terrorism and guerrilla raids seemed inevitable.

In 1996, Bangladesh complained once again that India's diversions had led to serious water shortages and environmental degradation in the southwestern part of the country. "The reduced dry season flow has produced adverse impacts in almost every sector of life in the Ganges-dependent area in Bangladesh," a government report said, maintaining that India was diverting much more water than it was entitled to. The minimum flow at Hardinge Bridge, downstream from Farakka, was 362 cubic metres per second in 1995 and 350 in 1996, compared with 1,740 before diversions began. The decreased flow had dried up the river Gorai, the most important source of fresh water for the southwest of Bangladesh, and the general lack of water increased the salinity of rivers, groundwater, and soil, as tides from the Bay of Bengal crept further up the estuary. A pulp-and-paper plant 150 kilometres inland was forced to close when its water became too saline. Higher salt concentrations were killing trees in the coastal Sundarbans, the largest mangrove forest in the world. "Water has become a test case," a Bangladeshi official said in 1996. "If we can't get this out of our hair, there is a real danger relations will bog down."[3]

It was to everyone's astonishment, therefore, that a formal Ganges Water Sharing Treaty was signed between the two countries in 1997, signalling an end to what Bangladeshi journalists had taken to calling "our Farakka jeremiad." At the heart of the treaty were the water-sharing clauses and their inbuilt formulae. When the flow was 1,785 cubic metres per second, India would get 952, Bangladesh the rest. When the flow was higher than that, between 1,785 and 2,122 cubic metres, Bangladesh would receive 833 cubic metres and India the rest. At lower levels, between 1,190 and 1,785 cubic metres, the two countries would share the water equally. If the water availability at Farakka were ever to fall below 1,190, "the two sides will meet immediately to decide the shares." The critical point was to ensure at least 833 cubic metres per second to Bangladesh in alternate ten-day periods during the driest period (March to May). However, this "guarantee" never made it into the final draft of the treaty.

In the treaty's preamble there is a self-congratulatory paragraph that recommends the Ganges agreement as the basis for sharing the waters of the other fifty-three common rivers between India and its neighbours. Hydrological opinion is that the self-congratulation is, for once, thoroughly deserved. If equitable agreement can be found here, among the shrill ideological conflicts and ancient ethnic feuds, it can surely be found anywhere.

It was in the northwest of India, in the Indus Basin and its tributaries, that the Green Revolution achieved its most dramatic results. The Indus, which rises in Tibet and flows through India and Pakistan to its delta in the Arabian Gulf, is the largest and one of the oldest irrigation systems anywhere on the planet, its flow three times the size of the Tigris–Euphrates system and the equivalent of two Niles.[4] At the same time, the Indus is a source of friction and conflict, both within and without the Indian federation.

Unexpectedly, the international dispute with Pakistan was more easily settled than the domestic one between Hindus and Sikhs. This despite the fact that some of the major tributaries of the Indus – the Sutlej, the Beas, and the Ravi – flow from India into Pakistan, and that two of them form the national frontier for several hundred kilometres, and despite also that Pakistan is dependent on irrigation water for 80 per cent of its food. Moreover, India and Pakistan have been at war more than once (and were recently mired in a mine-is-bigger-than-yours series of nuclear explosion tests, accompanied by much national chest-thumping and braggadocio).

~~~

The British-induced partition of the subcontinent left part of the Indus Basin in Pakistan and part in India; Pakistan got most of the canals and irrigated lands, but India, being the upstream riparian, retained control. It was a solution pregnant with possible rancour.

It wasn't long in coming. In 1948, only a year after partition, the Indian state of East Punjab, without any warning, cut the water flowing to West Punjab in Pakistan. The Pakistanis were furious, and prime minister Liaquat Ali Khan fired off a telegram to his counterpart in India, Jawaharlal Nehru: "Such stoppage is a most serious matter and . . . will cause distress to millions and will result in calamitous reduction in the production of food grains, etc."[5]

The Indians declared they wouldn't restore the water until the Pakistanis admitted they had no right to it. The Pakistanis said they were entitled to the water by prior right and natural equity. The Indians said . . . the Pakistanis said . . . and a series of telegrams flew back and forth, predictably escalating in rhetoric. But good sense intervened, and the two prime ministers agreed to an international conference to discuss the issue. As a gesture of goodwill, the water started flowing again on the conference eve.

It was left to the World Bank to broker an agreement called the Indus Waters Treaty of 1960, which essentially apportioned the Indus waters between the two countries by allocating the Sutlej, Beas, and Ravi (the eastern rivers of the basin) to India and the western rivers (the Indus itself, the Jhelum, and the Chenab) to Pakistan. At the same time, the bank sweetened the deal by financing a series of water-management projects that had the net effect of increasing the flow to both countries.

This treaty – despite the political and ethnic tensions between the two countries – is still in force.

~

The other major dispute over the waters of the Indus has been, and is, an internal Indian one. Militant Sikhs have been demanding control of these waters as part of their ongoing campaign for independence; the acts of violence that the campaign has generated include the previously mentioned assassination of the Indian prime minister, Indira Gandhi.

The trouble started in 1955, when the new federal capital of New Delhi, an enclave carved from the Punjab, allocated the waters of the Sutlej, Beas, and Ravi among the desert-bound state of Rajahstan (8 million acre-feet), Punjab (7.2 million acre-feet), and Kashmir (a paltry .65 million acre-feet). Bound up with the allocations was a parcel of development projects, dams, and canals, including a major canal into Hindu-dominated Rajahstan from the Sikh portions of Punjab. The first conflict was over who should control the flows. Rajahstan, which had no water of its own, wanted New Delhi to have the major say. The other players believed water rights should be controlled at the state level. There were already anti-Delhi stirrings in the Punjab, left over from the 1947 act of partition.

But the pot really started to boil when Punjab was itself partitioned in 1966 into Sikh Punjab and Hindu Haryana. Partition was driven by linguistic and religious considerations, and water rights were never mentioned; consequently, the Sikh enclave got all the rivers and effective control of all the canals. Haryana, citing "need and the principle of fundamental equity," demanded 4.8 million acre-feet out of the total allocation of 7.2 million. Punjab refused. The rivers run only through Punjab and are ours, they maintained, and we will control what we do with them. In a cynical display of manipulative politics, they used the ethnic card to whip up anger against New Delhi and its Hindu creatures, Rajahstan and Haryana. Haryana appealed to New Delhi for help.

The politics were, as always, complicated. A central government cannot tolerate complete disregard for its policies; on the other hand, enforcing those policies encouraged the very separatism that caused the recalcitrance in the first place. If they cracked down, the ethnic separatists would win popular support. If they failed to crack down, the separatists would win anyway. Internal dynastic difficulties within the cumbersome Congress Party only made things worse. Indira Gandhi's real power had gone into eclipse after 1975, when she had declared a State of Emergency, and her controversial son Sanjay was effectively running India with his cronies. This wouldn't have been troublesome had Sanjay been a good administrator, but cronyism hardly makes good government, and his impulsive actions were seldom well-thought-out. Punjabi separatism – irritating enough to more calm-headed politicians – prodded him into an ill-advised executive order, unilaterally assigning a minimum of 3.5 million acre-feet to Haryana, .2 to the capital itself, and the balance (but never more than 3.5 million acre-feet) to Punjab. Not unexpectedly, there were howls of anger from Chandigarh, the Punjabi capital it shared, inconveniently, with Haryana. The Akali Dal regional party routed the Congress Party in the subsequent 1977 election. They promptly took the 1975 executive order to the Indian Supreme Court and demanded it be reopened.

In the interim, the Congress Party began covertly financing a violent Sikh fundamentalist movement, a campaign that met with some success. In the 1980 election, Indira Gandhi, once again ascendant, came back to power in Delhi and, to everyone's surprise, Congress succeeded in taking over the government of the Punjab as well. One of the first things the leaders did was to remove the Punjab water case from the Supreme Court docket – you couldn't have a Congress state government fighting a Congress central government in court, could you? – to a predictable outcry from the Sikhs. The executive order of 1976 was replaced with an "agreement" imposed on the three states in 1981, with a new division of the waters. Punjab was now to get 4.22 million acre-feet, Haryana 3.5, and Rajahstan 8.6. It addition, Punjab was to be obliged to construct the Sutlej–Yamuna Link, a canal that would deliver Haryana's share of the water.

In April 1982, Indira Gandhi formally turned the first shovelful of soil for the canal. The next day the Sikh opposition denounced it. Not long after that, Gandhi herself was assassinated.

In the years that followed, Sikh farmers drove out many of the canal construction workers, and others were killed by separatist terrorists. The central government has, once again, dithered. A precarious peace returned to the Punjab in 1993, but the central government is well aware that anything it does – whatever actions it takes – will simply set the Sikhs off again. Haryana has on occasion tried to force a decision by turning off the water to Delhi (Haryana controls the headwaters of the Yamuna River, on which Delhi depends for its water), but the protests failed to stir the government into action, and water issues are now so snarled in the politics of ethnicity and ethnic nationalism that they cannot be untangled. The violent separatist movement is vigilantly watching for any sign, no matter how minute, that Sikh rights are being traduced. Everyone knows that a clumsy water decision could still set off a major insurrection.

In the middle of all this, the Indian capital of Delhi was itself facing dire shortages, and in 2002, as summer temperatures soared past forty degrees, there were riots in a number of suburbs. The city's water authorities forecast a looming, and imminent, crisis. The water table had gone down from 40 feet to 350, sucked dry by a rapidly increasing population that exceeded 15 million in 2002 for the first time, and by the unauthorized drilling of tube wells from thousands of homes. The numbers were stark: the city required 3 billion litres a year, but could only supply 2.5 billion. And at least 40 per cent of the water was lost to leaking pipes or outright theft every year. The city, they said, was in danger of drying up altogether by 2010.

Nevertheless, city politicians refused to raise the very low price of water, despite urging from the environmentalists. Sumita Dasgupta, a natural-resources specialist at the Centre for Science and Development in Delhi, is appalled at the waste he sees around him. "People in Delhi like to sleep while the tap is running in the morning. They like to wash their dishes in running water, and wash their cars and driveways with hoses kept running." He added, with some understatement, "It is a highly unsustainable lifestyle."[6]

~

At the other end of India there is another water dispute that has nothing to do with national frontiers, but still everything to do with "statism," ethnic differences, and politics, all snarled in the understanding that water is central to human life. War in the Cauvery Basin is not a real threat, but rioting, terrorism, and assassinations are real. It may not be war, but people are nevertheless dying over water allocations.

The Cauvery River is peninsular India's fourth-largest waterway. It rises in the Western Ghats Mountains of the state of Karnataka, which in olden times was the principality of Mysore. From there it flows eastward into the heartland of Tamil Nadu, once called Madras, across the verdant paddies and palm fields to the delta at Thanjavur, one of the most fertile and productive farming areas in India, and for that matter on the planet. Downstream, Tamil Nadu uses most of the water (489,000 of the total flow of 671,000 million cubic feet, more than 72 per cent), but doesn't control the head-waters, except for two small tributaries, the Bhavana and the Moyar.

The dispute, which is still before the Indian Supreme Court (when it is not being played out in fiery oratory and the occasional riot), is more than a hundred years old. In 1892, the British brokered an agreement between Mysore and Madras heavily loaded in favour of the downstream state – the British considered the port of Madras significant to Indian commerce. Madras farmers had in any case claimed inherited rights to Cauvery River water from centuries of use, citing as evidence the ancient and intricate irrigation works in the river's delta. The British saw to it that the treaty gave Madras power over any irrigation or diversion projects contemplated by upstream Mysore. Ironically, part of the reason for the arrangement was conservation. The British wanted to preserve the watershed, and forbade upstream Mysore from cutting down forests or grazing animals. Coffee production was restricted in the higher elevations for the same reason, to decrease the annual runoff and improve the retention properties of the soil. No one in Mysore liked the agreement. The princes seethed, but secretly. It was not yet time to contemplate a fight with the colonial power.

In 1924, the British pushed through another agreement, this one at least allowing Mysore to construct a number of dams and extend irrigation to 44,500 new hectares. The agreement was to expire in 1974, but there was trouble long before that. The reorganization of states after Indian independence in 1948 took some of the Cauvery Basin away from Madras (now rechristened Tamil Nadu), and awarded it to Mysore (now called Karnataka). The new, upstream state claimed that the redrawn boundaries had nullified the 1924 agreement, and that there was now no obligation to release any water at all to the downstream state. Tamil Nadu wanted to build a series of dams to better manage the Cauvery's flow in weak monsoon years, but Karnataka took it to court to prevent that development. The agreement, the brief said, had not allowed the river to be modified in such a way. There were petitions and counter-petitions to the central government and the

Supreme Court; there were political demonstrations, fasts by political leaders, and one nasty riot that killed a dozen people. In 1991,
the Cauvery Waters Disputes Tribunal, a creature of the central
government, passed an ordinance ordering Karnataka to release
water weekly from June to May; but a few days later the order was
overruled by another arm of the central government, with fateful
consequences. In December there was a riot in Bangalore, and the
homes of Tamils living there were torched. Tamil-owned farms in
the Cauvery Basin and in the southeast of the state were looted
and burned. Within days, the violence spread to Tamil Nadu,
where Karnatakan landowners were evicted and their homes
burned. Bandits closed the roads between the two states, and
troops moved to the border. More than a hundred thousand people
fled from Karnataka to Tamil Nadu.[7]

For the next few years there was plenty of rain and the tension
level subsided. But in 1995, the monsoons failed and the old animosities were reignited. Farmers lost their livelihoods and were
driven into the teeming slums of the cities, where they remained,
a potent force for any politician to manipulate. As the crops
withered, tempers became more inflamed. Threats of violence
developed into aggressive protests, then into riots. There were
more killings. The central government did nothing and the
Supreme Court said nothing. Tamil Nadu's position was simple.
Agriculture was the state's life, and without Cauvery water there
was no agriculture. In all equity, therefore, Karnataka should
release at least enough water to save the state's crops. This had,
after all, been ordered by the Supreme Court in 1991: How could
Karnataka place itself above the law? Karnataka's position was similarly uncomplicated. In 1995 Karnatakan reservoirs held only
23,200 million cubic feet, but the state's basic requirements
were for 24,000 million. Southern Karnataka and Bangalore were
already short of drinking water. Any release to Tamil Nadu would
imperil Karnataka's citizens and endanger its economic growth.

In January 1996, the Indian prime minister, P.V. Narasimha Roa, intervened. He demanded that Karnataka release an immediate 6,000 million cubic feet to Tamil Nadu to save the crops. He also set up an "expert council" to assess the crops of both states and to divide up the water as needed.

No one paid much attention. Crops continued to wither and tempers to flare. There was more violence in Mandya and Bangalore. Cars were wrecked, and at least one Tamil Nadu citizen was killed. Farmers patrolled the banks of the Cauvery to see that no water went downstream, and threats were made against officials controlling the sluices. A human chain was formed in Mysore, a kilometre-long barricade of grim-faced farmers. The politicians in Delhi and the state capitals argued about how much water should be released, but the growing militancy among the citizens was against any release at all.

In 1998 there was still no solution, and by 2002, politicians in the downstream state were threatening violence. "If the Cauvery problem is not solved," a Tamil Nadu member of the central Parliament wrote in August of that year, Tamil Nadu could go the way of Kashmir."

In Delhi, the prime minister appealed for calm. Water, he said, is a national issue. "Shared water goes far beyond the merits or demerits of the Cauvery dispute. Dozens of rivers cross state boundaries in India. Violent disagreements are, regrettably, becoming the norm. The government will have to act."

No one disagreed. But no one did anything either.

16 | THE CHINESE ARE USED TO THINKING BIG

They built the Great Wall, after all. So why not just reorganize the whole country, hydrologically speaking?

Cui Wei Yan, who lives in careful poverty in a single-room cave in the north of Shaanxi province with his elderly wife, Zhang Zhen Lian, their daughter and grandson, doesn't know it, but the bigwigs in Beijing are working away to solve his water problems – not through anything so mundane as laying pipe to the nearby village but through something grander, much grander. Through a project grand enough, in its way, to match even that *folie de grandeur*, the Great Wall of China. Cui's political masters plan to re-create agriculture and water-hungry industry in the arid north, among the rapidly advancing dunes on the fringes of the Gobi, and on the great plains around Beijing, where the water tables have been dropping more than a metre a year for the past decade. They plan to do this by essentially re-engineering the entire country, by taking massive amounts of water from the sodden south to the parched north, in a series of canals, conduits, reservoirs, and pipes, with a price tag somewhere around $58 billion.

It will be, by far, the largest civil-engineering project on the planet.

It may never get finished. And it might not work if it does.

~

Meanwhile Cui, who is seventy-eight years old and has lived in the same cave for all of his life, as his father did before him, and his grandfather before that, sets out every morning after sun-up with two buckets on a yoke across his shoulders to fetch the family's daily ration of water. He trudges up the steep hillside, heading for the cistern that the local authorities have set into the hillside a kilometre or so away. He drops his buckets into the cistern on a rope, and hauls out the water. It is not exactly clean – bits of grass and debris are floating in it – but then neither are his buckets. The operation takes him about forty-five minutes, longer, as he admits, than it used to. Back at the cave, he tips some of the water into the small tin basin the whole family uses for washing, and the rest into a rank barrel, from which cooking water will be drawn.

"We have a saying," said Cui, who is full of sayings and impish smiles, "that we don't wash our faces in the morning – we wash our eyes. We don't have enough water for the whole face."

Earlier, he had said of his way of life, "we bathe three times in our lives, when we are born, when we marry, and when we die."

That Cui's family lives in a cave is not so surprising. Hundreds of thousands of northern Chinese still live in caves set into the slopes that the Yellow River and its tributaries have carved from the hills in millennia past. It is the soil eroded from these deep gullies that is carried into the Yellow River, giving it a heavy dose of silt and, in consequence, its yellow colour. This is not as primitive as its sounds – the caves are practical dwellings, easily excavated from the three-hundred-metre-deep yellow loess soils of the hills, whose major geological fact, easily noticed even by outsiders, is their lack of rock or stone. You can make a home, therefore, with just a shovel and a lot of back-breaking labour, and many do, and it is a curious fact of rural life in the area that families often live directly beneath their fields. It is still common to see wisps of smoke from cooking fires puffing out between rows of crops, emerging through shafts driven into the loess. Still, though the caves are not so primitive, it is not exactly an opulent life. The family is very poor, and are vulnerable not just to the peculiar vagaries of Chinese political life but to climatic shifts and their resultant shortages. Nor are they unaware of their poverty. The village high on the hill above their cave has a television set in the commune centre, and Cui confessed to being envious of city dwellers he saw on the screen, who could apparently turn on a tap and get water whenever they wanted.

He was a peasant, he said. And despite decades of diligent Communist Party rhetoric about the nobility of the peasant life, he was not happy.

"You know what we call ourselves?" he asked after a long silence. "The bitter-life people."

The peasant farmers of Shandong Province on their new private plots were the first to go. In the early 1990s, the level of water in their wells began to drop. They weren't aware of it yet, for Beijing had said nothing, but the water tables were dropping all over northern and northeastern China. The peasant farmers knew only that there was less water than there had been. In 1994 some of the wells went dry, and the following year hundreds more. All over the region, farmers and their families took time from cultivation to dig. They pushed the wells down another metre, two more metres, and found water. In 1996, their wells dried up again. The water tables on the North China Plain had dropped more than four metres in three years, and no one knew where it would end. For several years rainfall had been meagre and rivers lower than usual, at a time when water demands were soaring. Extractions from river systems had historically been for agriculture, but China's astonishing economic boom, and its expanding industries and cities, were placing massive demands on finite water systems. China's own figures say that, between 1983 and 1990, the number of cities short of water tripled, to three hundred – almost half the cities in the country; those whose problem was described as "serious" rose from forty to one hundred. By the year 2000, the Chinese said in 1998, Beijing municipality would suffer a daily water shortfall of a half-million cubic metres, but it reached that dismal target before 1999 had arrived.

By 2002, the drought, and the shortages in the north, had gotten worse. In Shandong, villagers who had traditionally drawn water from a leaking reservoir rioted when workmen began to repair the leaks. In the northeastern city of Dalian, on Korea Bay, water was so scarce that it was being rationed for domestic users, and hundreds of bathhouses were closed. Even in the prosperous southern province of Guangdong, almost one hundred reservoirs had dried up, and rivers were reduced to trickles. Xi'an, the capital of Shaanxi province, has been in its long history the capital of China through thirteen dynasties, but this has come at some cost, especially to the environment. The forests of the heartland were cut to serve imperial architects, wetlands were drained for farming, rivers diverted and their water extracted, and the province went from water-rich to water-poor. The city was sited where it is because it was surrounded – at least on three sides – by rivers. Now only one of those rivers is flowing, and the ferrymen who used to take people across in hundreds of small voyages a day have been retired. There is no need for ferrymen when it is possible to walk across what used to be a major river.

A study by the Chinese Academy of Sciences in 2001 found that "groundwater depletion has seriously impacted the environment. Large tracts of land that overlie cones of depression have subsided, seawater has intruded into previously freshwater aquifers in coastal plains and groundwater quality has deteriorated due to salinization, seawater intrusion, and untreated urban and industrial wastewater discharges."[1]

Recognition that a crisis is at hand is fairly recent. It was only two decades ago that the region's water supply had first begun to seem fragile, and it is now only a few years since hydrologists and ecologists turned their attention to the problem. At Harvard, Henry Kendall, the Nobel Prize–winning physicist and eminent scholar, late in his life turned his formidable intelligence to the question of world food supplies, which he believed was the critical issue for the next few decades. He steered me to Lester Brown, whose controversial "wake-up call" entitled *Who Will Feed China?* (1995) had tracked the statistics and forecast that severe water shortages would force China to import food. His projections were later backed up by a study sponsored by the U.S. National Intelligence Council and performed by Harvard's Department of Earth and Planetary Sciences and by Sandia Laboratory, a . . . well, secretive isn't exactly the right word, since it has its own searchable Web site. Perhaps "well-connected" is the appropriate term. Sandia is famous for many things, among which are its simulation games. The Industrial Ecology Prosperity Game is a uniquely hard-nosed look at the effects of population increases, climate change, water shortages, and other ecological factors on the world economy.

Brown wrote with researcher Brian Halweil in the magazine *Worldwatch* in mid-1998:

A quarter century ago, with more and more of its water being pumped out for the country's multiplying needs, the Yellow River began to falter. In 1972, the water level fell so low that for the first time in China's long history it dried up before reaching the sea. It failed on 15 days that year, and intermittently over the next decade or so. Since 1985, it has run dry each year, with the dry period becoming progressively longer. In 1996, it was dry for 133 days. In 1997, a year exacerbated by drought, it failed to reach the sea for 226 days. For long stretches, it did not even reach Shandong Province, the last province it flows through en route to the sea. Shandong, the source of one-fifth of China's corn and one-seventh of its wheat, depends on the Yellow River for half of its irrigation water.

Although it is perhaps the most visible manifestation of water scarcity in China, the drying-up of the Yellow River is only one of many such signs. The Huai, a smaller river situated between the Yellow and Yangt'ze, was also drained dry in 1997, and failed to reach the sea for 90 days. Satellite photographs show hundreds of lakes disappearing and local streams going dry in recent years, as water tables fall and springs cease to flow. As water tables have fallen, millions of Chinese farmers are finding their wells pumped dry.

If there was no more water in the rivers, what then? they asked. The theoretical answer is simple enough: some of the "users" would have to quit using it. The economics were even simpler. The people who stop using it would have to be farmers. With a million or so litres of water, a farm would produce grain worth, say, $200. With the same amount of water, industry could produce goods that could be sold on the open market for twenty times that – and employ a good many more people.

In practice, much of the problem is mismanagement, or what Vacláv Smil, a researcher at the University of Manitoba, calls "Maoist stupidities." These include establishing water-thirsty industries in dry northern cities like Beijing, badly planned dams, poorly managed rivers, and attempts to grow unsuitable crops that require irrigation in the driest parts of the country. In northern Shaanxi, for example, an arid place where dunes are already beginning to roll over remnants of the Great Wall, they are still growing rice in the riverbeds. China has hardly experimented with more-efficient water-delivery systems; it badly needs to replace open irrigation ditches with sprinkler systems and drip agriculture. The first step should be to charge a fee – even a nominal fee – for water. "Right now, farmers basically get their water for free," Brian Halweil points out. "If there were even a marginal, symbolic charge, it would spur all sorts of conservation efforts." But even in China, as controlled as it is, removing subsidies is politically risky; the authorities don't want to reopen the social unrest that occurred during the Three Gorges forced relocations.[2]

The small farmers, the peasants, had been allocated their land in the restructuring of agriculture that followed Deng Xiaoping's reforms in the late 1980s, but with new ownership came new responsibilities and new worries – no longer would the collectives solve problems collectively. The farmers would be on their own. China was changing in other ways, too. The cities were growing at astounding rates, and prosperity was widespread. It was no longer unusual for the new urban elites to cruise past the farms in their gleaming automobiles, engaged on errands too mysterious to question.

"My father used to watch them go by," says Soo Ling, who moved to Shanghai as a teenager and now sells advertising space for a new-media company in Hong Kong. "He didn't hate them or resent them or anything. They went by in clouds of dust, and he'd just watch. He was puzzled by them, mostly. He didn't know who they were."

"But he went on farming?"

"It was all he knew. But . . ." she picked at her food, small savoury dumplings on a bed of well-watered lettuce, "it was more than that. To his generation, farming was what you did, you produced food. He knew nothing about ecology, but he was an ecologist." She shook her head. "No, that's too pretentious. But he understood the land, and loved it, and would never do anything to harm it, but it wasn't enough . . ."

"What do you mean?"

She recounted the story with which I was already familiar. Her father had been used to taking water when he needed it, from the reservoirs. The sluices would open, and the water would reliably flow. Then the authorities explained that the city needed the water: there was not enough for farming and for the city as well.

"My father knew what that meant. Beijing would get the water. It always would. Farmers would always lose."

"What happened?"

"The next year, the wells went dry again. Now my father is living in Beijing."

"Is he working?"

"Oh yes. He got a job, no trouble. He has a big apartment, and my brothers are living with him. He even has more money than he had before . . . a television! Actually, my brothers bought that. He hates the city. But what can you do?"

―

Why is China running out of water? The answer is the same as for the rest of the world: it isn't running out. It's only running out in places where it's needed most. It's an allocation, supply, and management problem.

In the summer of 1998, one of the wettest on record, it was certainly not running out in the humid south of China. The Yangt'ze, swollen by sustained downpours, reached its highest level since 1954 and killed more than 2,000 people in its rampages across the lower valley. That more people didn't die was a miracle; most of the world's major watersheds have only 30 or so people per square kilometre, whereas the Yangt'ze has more than 220.[3]

The industrial city of Wuhan escaped unscathed, though floodwaters had destroyed dikes upstream a day earlier, and it looked for a while as if more than seven million people would be overwhelmed by the water. Chinese sources agree that the flooding was exacerbated by deforestation efforts in the upper basin, and by dikes that prevented the water reaching its natural flood plain, ancient wetlands now drained for agriculture. The periodic Yangt'ze floods were one of the reasons given for why the Chinese government is spending more than $24 billion and resettling more than a million more-or-less resentful citizens to build the Three Gorges Dam. The other given reason is to supply water to the thirsty north.

Flooding has been more of a fact of life in the past than shortages. And there are still floods in the south. In 2002, another 650 people drowned in flooding in Changsha in Hunan province, as the Yangt'ze once again threatened to overflow its banks. An indication of how bad it could have been was that the Chinese media reported that 650 number by saying "only 650 people drowned." It took an army of soldiers and civilians, nearly 900,000 of them, to pile sandbags on the shores of the dangerously swollen •Dongting Lake, as the Yangt'ze flood crest raced towards it. If the lake, China's second-largest, had burst, water would have swept across a flat plan that was home to 10 million farmers. Two cities, Changsha itself and Wuhan (in the province of Hubei), which have a combined population of 13 million, would have been at risk. Water levels were three metres above the norm.

The floods illustrate China's central water dilemma. Three-quarters of the water is in the south, and three-quarters of the farming is in the north and northeast. The south includes the Yangt'ze, and about 700 million people. The much-drier north, inhabited by some 500 million people, includes the Yellow, Liao, Hai, and Huai rivers. All the northern rivers are stretched to the limit, and already the water supply for 200 million people is only assured through unsustainable mining of groundwater. The Yellow River, the "cradle of Chinese civilization," as we have seen, already dries up in bad years before it reaches the sea. However, development projects scheduled for its upper reaches will make matters considerably worse, including hydroelectric schemes and a canal to

Mongolia. The great west, like America's, is largely desert. But unlike America, there is no equivalent to California or the Pacific north, no Washington and no Oregon, only aridity. Towering mountain ranges cut off the wastelands from moisture-laden winds. The Altai Mountains to the north prevent rain clouds from reaching the Gurbantünggüt Desert, which fills the centre of the Junggar Basin in China's northwest corner. Similarly, the Himalayas in the south isolate the largest vegetation-free sand sea in China, the Taklimakan Desert in the country's far west. In western Inner Mongolia, in the Kansu Corridor and the Tarim Basin, less than ten centimetres of rain falls each year. There are large areas of true desert, where sometimes not a single drop of rain falls, year after year. The great Gobi Desert of Mongolia is one of the driest places on earth. And what rain does fall in China is highly variable, with little stability in supply, exaggerating the flood-drought cycle. Chinese farmers depend on the spring rain, but it is particularly erratic. Along the Yangt'ze, variability from year to year is around 45 per cent, and in north China it is more than 50 per cent. East of Beijing, the variability index can exceed 80 per cent.

～

Lester Brown's thesis about China's looming grain shortages was indignantly denied by the Chinese. But if you read Brown's argument carefully, it is far from an anti-China diatribe. On the contrary, he acknowledges China's substantial achievements in water management, and suggests that their problems are a factor of their successes, not their failures. Rising standards of income are leading inevitably to rising demands for water. He points out that the same thing has happened in other countries where development has been rapid, such as Japan, and there is no reason to suggest that China will be immune. At the same time, more and more agricultural land is being taken out of production, partly because of competing demands for water, but also because the expanding and ever-richer cities demand more space – for factories, housing, parking lots, and roads. It is politically necessary for the Chinese to create jobs and provide sanitary water supplies for the millions of citizens who live in the economically backward interior provinces, but giving in to their demands inevitably means increasing short-falls in the amount of water available for irrigation downstream, water used in producing China's food.

The economics of water use are compelling. In California, it is a given that industry makes more productive use of water than farmers do; the market value of goods produced by water use in industry is fifty times the market value of food produced with the same amount of water. By diverting water from irrigation to industry, the economy benefited, more jobs were created, and the tax base soared. Of course, food production went down, but food could always be imported. In China the same dynamics are at work, only more so: China needs to provide jobs for 15 million new workers every year. Brown's figures show that the demand for water by industry will increase from 52 billion cubic metres to 269 billion cubic metres by 2030. "In other words, non-agricultural uses that are now straining the system by drawing only 15 per cent of the supply would multiply nearly five times, while the agricultural needs now taking 85 per cent would have increased as well. Obviously, that can't happen."

When farmers lose groundwater or are cut off from irrigation sources, he says, they either survive on rainfall or they go out of the farming business. And food production declines.

China's problem, as we saw in Chapter 5, on pollution, is that supplies have been gravely compromised by the quality of the water that is still available. When almost 80 per cent of the rivers contain water unfit for human consumption, there is a serious danger that irrigated crops in the worst areas might poison the population instead of nurturing it.

China has not always been forthcoming about this or any water matter, but for some decades there have been strong signals that a crisis was anxiously predicted. In 1988, a new Water Law was passed, and the agencies responsible for water, soil conservation, and other regulations were consolidated in a bureaucratically superior ministry of water resources, which reports, as other senior ministries do, directly to the State Council. The ministry controls water-policy formulation; strategic planning, including flood control; water pollution and waste-water controls; economic regulation; and conflict arbitration. Sensibly, the sub-ministries were organized on a river-basin basis, each responsible for a major river system, such as the Yangt'ze, the Liao, the Huai, the Hai, and the Yellow river basins. Under them were the conservancy commissions, which looked after tributaries. The new system was designed to deal with what the announcement delicately called the "conflicts and shortfalls" of the older system, a bureaucratic nightmare in which no one knew who was in charge. That nothing had been done about the dikes and flood plains of the Yangt'ze by 1998 was not a reassuring sign that the new bureaucracy was any more flexible than the old, and late in that year the sub-ministry in charge of the river was reduced to innuendoes and criticisms of the local political bosses. River-basin management was all very well, but political satrapies hardly ever coincided neatly with basin boundaries, and it was difficult to get anything done.

But starting in around 1999, the Chinese stopped pretending the problems were not really problems at all but only the malicious acts of wreckers and saboteurs, and instead 'fessed up. One of the first signs was the sudden attention paid to the attempt to "save" the failing Baiyangdian freshwater lake system, the largest in north China, stretching across the flat plains that lie to the south of Beijing. In the 1950s, the lakes of Baiyangdian, in Hubei Province, covered more than 310 square miles. Today, local officials say only 186 square miles remain under water; Professor Liu Changming, of the Chinese Academy of Sciences, puts the figure closer to 44 square miles. What water remains is severely polluted. Nevertheless, there are concerted efforts in the national media and from Beijing to find remedies.[4]

By 2002, attitudes had radically changed. Early in the year, the central authorities issued a "water classification zone table," dividing the country into some 13,000 water-environment zones, *en passant* "classifying" 5,737 rivers and 980 lakes and reservoirs as zones for natural protection, drinking, industrial and agricultural use, sight-seeing, and entertainment. By mid-year, Beijing was sending out weekly, and sometimes daily, sharp bulletins that occasionally seemed to be verging on panic. For example, the State Environmental Protection Administration reported in 2002 that "the overall situation in China is grave." Pollution in all the seven major rivers was approaching catastrophe; acid rain was falling on 30 per cent of the country; the East China and Bohai seas were so polluted that fishing was threatened; China's deserts were expanding at about 3,000 square kilometres a year; there were major dust storms across northern cities that lasted forty-five days from March to May. Zhu Jianqiu, the agency's vice-minister, estimated that fully 90 per cent of usable natural grasslands in China, with a aggregate area of 135 million hectares, suffered varying degrees of degradation. Statistics issued in Beijing declared that water shortages were already causing $15 billion a year in lost industrial output. In May the same year, China's vice-premier, Wen Jiabao, declared that fully one-third of China's entire land mass – and bear in mind this is the fourth-largest country in the world – was facing what he called "severe land degradation" caused by water loss, soil erosion, overly saline soils, and accelerating desertification.[5] Wen was talking candidly to the Twelfth International Soil Conservation Organization Conference in Beijing. He blamed the problems on a whole shopping list of causes: careless management, inadequate infrastructure, irresponsible agricultural initiatives, and – a prime cause – deforestation of the country.

Earlier that month, travelling up from the Yellow River Falls (still impressive in June) towards China's northern frontier, we had passed a vast area of high hills deeply scored with erosional gullies. It was in one of those hills that Cui and his family lived in their one-room cave. Like much of China, all the hills were patterned with the parallel horizontal stripes of planting terraces; from the bottom of the gorges to the tips of the hills, and for as far as the eye could see – which was a very long way, as the road traversed many high elevations – the countryside had been mani-cured for hundreds and even thousands of years by the busy work of millions of peasant farmers, until it sometimes appeared that hardly a centimetre of the Chinese countryside had escaped this mammoth labour. But on thousands of hillsides, the only plantings that were now being done were of trees. The peasants had been told to plant food crops no longer. The constant irrigation of those hundreds of thousands of terraces had, in the end, exaggerated the natural erosion by denuding the countryside of its shade trees. Now, the policy of turning forest to farmland was being reversed. China was once 22 per cent forest – a figure that had been more or less constant through Chinese history. But, beginning in the 1950s, the trees were cut away and new fields constructed. By the year 2000, the percentage of the countryside covered by trees had been reduced to a paltry 6. That deforestation, it was now widely acknowledged, was a prime cause of southern flooding and of northern desertification.

Desertification in China, as elsewhere, is subject to the self-reinforcing aspects of feedback loops – cutting down trees elevates the local microclimate temperature, which calls for more water to irrigate thirstier crops, which depletes the water tables, which in turn elevates the temperatures further. The whole process is plain to see in the dusty landscape north of Yulin, in Shaanxi province, where I spent a morning talking to Dang Jian Qi, a professor of forestry at the local community college. Actually, he wasn't really a professor, but he had been planting trees, and managing plots of reforested land, for so long that the local government had sensibly signed him up to impart his knowledge to young students. We talked on a viewing platform hastily constructed for visiting bigwig Jiang Zemin. The view was instructive. To the north, the reclaimed forest tailed off into sand; a few kilometres beyond that, the dunes began. The Great Wall once stood proudly nearby; it has now crumbled into the sand and is overrun by dunes, so that only a few vestigial towers remain. On the other three sides the forest stretched to the horizon, neat rows of Jack pines, mostly, though there were test parcels of half a dozen conifer species. It looked like a success, and indeed it was – several thousand hectares of land had been reclaimed from the desert.

But it wasn't nearly enough, as the professor acknowledged. "When I was a boy," he said, "this was all verdant land, green as far as you could see. Now it is sand, and the sand is growing. We push it back here and there, but we need an effort such as it took to build the Wall itself to really succeed." Nevertheless, he said, the work was ongoing, and indeed our filming earlier that day had several times been interrupted by a low-flying military aircraft sec-onded to seed-laying duties, the plane criss-crossing the area just north of the Great Wall's ruins, putting down a dusty cloud of grass seed. And on every terraced hillside all through the northern province, seedlings had been planted. The intention, an official in Xi'an had said, was to double the amount of forest within a decade, from 6 per cent to 12.

Outside Yulin, however, they were growing rice in the riverbeds. To do so, they had imported hundreds of tons of impermeable clay and dumped it into the riverbed, so the water in the paddies wouldn't just seep away. The crops were moderate, but everyone acknowledged that they came at a steep price. Keeping water from replenishing the meagre groundwater helped a few farmers, but it was, they agreed, ecologically senseless.

Worse, the river's flow had diminished to a trickle. What had not diminished, however, was the amount of raw sewage pouring into it, or the amounts of industrial and agricultural waste chemicals. This was entirely typical of the Chinese landscape.

It was not surprising, therefore, that alarming reports occurred throughout 2002 of increasing number of red tides blooming off China's coasts, Most of the Chinese red tides consisted of two toxic organisms, *Alexandrium spp.* and *Prorocentrum dentatum*, both products of farm and household waste. The State Oceanographic Bureau, charged with monitoring the health of China's seas, is in the process of setting up monitoring stations along the coast in half a dozen provinces. Wu Jingyou, a bureau official, said in 2002 that their function would be to help fishermen cope with these toxic blooms, not to deal with the blooms themselves, which are produced by factors outside local bureaucratic control. In 2001, he said, shaking his head, there had been a record seventy-seven severe red tides, covering a total of more than 15,000 square kilometres. The East China Sea, one of the most polluted marine environments on the planet, reported twenty-seven red tides that year off the coasts of Zhejiang province alone. Some of them were a metre thick.

~

But now that China has turned its ponderous bureaucratic attention to the problem of pollution, successes have been lodged. One of the most dramatic has been the environs of Chengdu, the ancient capital of Sichuan province, on the upper reaches of the Yangt'ze. The city was built at the junction of the Fu and Nan rivers, themselves part of one of the major tributaries of the

Yangt'ze, the Minjiang. As recently as 1995, both rivers were grotesquely polluted; millions of litres of raw sewage poured into both of them each day, down open runnels winding through shantytowns lining their banks. Each had narrowed to a small channel; millions of tons of poisonous silt backed up the water into noxious ponds. The local government began dredging the river in 1993, widening and deepening the channels from 80 metres to 120. Shantytowns were razed and a hundred thousand inhabitants relocated and – not always true in China – rehoused. Upstream treatment plants intercepted household waste. The transformation was startling, and the city has subsequently won UN accolades for its achievement.

Curiously, Chengdu's success was partly inspired by an American ecologist, Betsy Damon, whose brainchild was an organization she called "Keepers of the Waters," aimed at "bringing together art and science to promote new ways of understanding water's place in the global ecosystem." Her biggest venture so far has been the Living Water Garden, a two-and-a-half hectare park at the Fu and Nan junction, which is at once a fully functioning water-treatment plant, a giant sculpture in the shape of a fish that contains a multitude of smaller sculptures, a living environmental education centre, a refuge for wildlife and plants, and a place to hang out and admire the views. It opened in 1998.

It was a curious collaboration. Damon is something of a performance artist, and one of her initiatives was to persuade the local television station to broadcast some of her "pollution dramatizations." Perhaps the most effective of these was to show a group of women standing in the river washing long strips of white silk that became dirtier and dirtier the longer they were washed. Nevertheless, the Director of the Fu Nan Rivers Comprehensive Revitalization Project, Zhang Ji Hai, incorporated her water garden, created with landscape designer Margie Ruddick, into a nineteen-kilometre-long park system called the Green Necklace. The Living Water Garden's treatment plant is a token only (it cleans a mere 250 cubic metres of water a day, not enough to affect the river-water quality), but that hardly matters. Its real purpose is inspirational, and paths throughout the "plant" allow the local people to observe the many birds, butterflies, and dragonflies that have taken up residence there and to watch how the once-dead river water becomes alive again.[6]

~

This is a small-scale project, but the Chinese are clearly not afraid of heroic engineering works, as the Great Wall and the Three Gorges Dam attest. If further proof were needed, look at the National Environment Protection Bureau's solution to the appalling pollution problem in the coal-mining town of Lanzhou in Gansou Province in central China. To clean up the pollution, the bureau never contemplated cleaning up industry first, or bringing in controls of any sort. No, the preferred solution was to expose the town to the prevailing winds by dismantling the three-hundred-metre Green Mountain to the east – giving new meaning to the phrase "letting in a breath of fresh air."

So it is not surprising that China's inclination is to bring works of heroic scale to water management also. This means, essentially, bringing water to the parched north from elsewhere – and that "elsewhere" could mean one of only two places: the flood-prone Yangt'ze or – a Russian paranoid's nightmare – Siberia.

In the late 1960s, when the Soviet Union and China almost came to a shooting war along their long frontier, one of the issues was water. The Chinese (at least this was the Russian view) had been staring at their maps too long, and had been hypnotized by the arid browns of the Gobi and the endless green of the Siberian tundra. The Chinese were famous for their long-term thinking and their indefatigable energy. Why would remaking a large part of their continental land mass deter them? Indeed, the tightly controlled Chinese media fanned these anxious flames by speculating more or less openly about the Soviet Union's eastern reaches, referring to them in code words easily understood in the Kremlin, where a similar language of obfuscation was readily practised. "Our northern resource" was a phrase that came into currency in the early 1970s; this, together with the sudden appearance of Chinese maps with the border mysteriously omitted, was more than enough for the Russians, who began issuing increasingly bellicose threats. But it was all for nothing. The notion made no real sense outside great-power rivalry. Why would China risk war for tundra of uncertain value, when it had plenty of water of its own, even though it was in demographically vexatious places? And so it has proved. Beijing's eyes are turned southward, now, not to the north, and the Siberian rivers continue to flow into the frozen Arctic, as they have done since the last ice age ended.

With water scarcity reaching the critical stage in sprawling showcase cities like Beijing and Tianjin (in the province of that name), the Chinese have dusted off one of Mao's more grandiose schemes, the South-to-North Project, which envisaged building three gigantic water-transportation systems from the water-rich Yangt'ze to the arid north. As updated by the Politburo's water planners, the project would involve a triple pipe, canal, and aqueduct network with a combined length of almost 3,500 kilometres. If ever completed, the scheme would dwarf not only California's own massive water-redistribution system, but even Colonel Gaddafi's aptly named Great Man-Made River in the northern Sahara, hitherto, at $32 billion, the planet's most costly civil-engineering project. (Some Chinese hydrologists, though not for publication, have gloomily compared the project not to California but to the Aral Sea water diversions in Central Asia, whose net effect, as we have seen, was to turn one of the planet's largest fresh-water lakes into a toxic desert, an object lesson in the unforeseen consequences of ecological engineering.) The official cost estimate of the South-to-North Project is a neat $58 billion, about twice the price of the Three Gorges Dam, China's most recent and most controversial mega-project, now nearing completion. Almost half a million people would have to be forcibly relocated. By the admission of the Chinese themselves, however, it will be decades, if ever, before this massive re-engineering of China's hydrological resources – in essence, a re-engineering of the whole country – could be completed.

Nevertheless, the project has about it an air of inevitability, in part because of its provenance in a speech by Mao (during a 1959 trip on the Yangt'ze River), and in part because it so neatly conforms to the pattern of Chinese "National Pride" projects like the Three Gorges dam, the Grand Canal, and the Great Leap Forward. But Brian Halweil, the Worldwatch staffer, is skeptical: "The three canals combined would, at best, be able to deliver about 60 million cubic metres of water to the north. The water the project can deliver is nowhere near the volume of deficit they are running. It's only about 5 per cent of the deficit."

The three routes of the diversion scheme all have their own particular challenges; each, in turn, would dwarf the Colorado-to-California water-diversions scheme, which after all is only a few hundred miles long and has to traverse only one real mountain range.

The western route of the Chinese project is the most speculative and uncertain. It would extract water from the upper reaches of the Yangt'ze and transport it into the Yellow River basin. Never mind that the 1,200-kilometre pipe and canal will take water from Tibetan regions, with all the political controversies that would entail; it will have to carve channels through several rugged mountain passes of 3,000 metres or more and traverse the most arid part of China, with consequent high evaporation rates. Estimated cost of this branch alone: $36 billion.

Early in 2003 the Xinhau news agency of China trumpeted a discovery that, if it proves out, may help make this branch of the project either unnecessary or less important. A survey of water resources in the Xinjiang region found a massive fossil reservoir beneath the Taklamakan Desert. Early engineering reports suggested this lake had about the same capacity as the 630-square-kilometre reservoir being filled behind the Three Gorges Dam. The reports suggested that there could be as much as 991 million cubic metres of water to be exploited annually, making the building of as many as five new dams and reservoirs unnecessary. Whether this was a long-term sustainable number was not mentioned.[7]

The central route calls for diverting water north from the Han River, a tributary of the Yangt'ze, by constructing a new canal to Beijing. But to make this canal flow in the right direction, the engineers will have to raise the Danjiangkou Dam and its reservoir about thirty metres, thereby forcing the relocation of more than a quarter of a million villagers in Hubei and Henan. Since even the elevated dam wouldn't hold nearly enough water for the planned diversion, as well as for meeting local needs, yet another new canal would have to be built, at ruinous cost, to bring water from the newly built Three Gorges Dam, thereby reducing that project's electricity-generating capacity by more than 5 per cent. This canal, too, would have to be punched through several intimidatingly high mountains.

The route that will follow the east coast is technically the simplest – after all, for much of its length it follows the same route carved out by the Grand Canal, another wonder of Imperial China, a waterway that once carried silk and rice from the south to Beijing. Nevertheless, even this route will require some dozen pumping stations to lift the water from the Yangt'ze mouth to the plains of Beijing. The real problem, though, will be the quality of the water that reaches the north. Those sections of the Grand Canal still in use are lined with thousands of rickety houseboats, whose inhabitants rely on the canal for waste disposal; in addition, hundreds of villages and towns along the way pour their effluent, untreated, into the canal itself; as a direct consequence, it is hazardous to use to canal's water for washing, never mind drinking. As though this wasn't enough, the new canal will also cross some of the country's most polluted river basins.

Plans to remedy all these problems have been announced, though to people on the ground they have so far been invisible. And even given the country's new fretfulness over pollution, some of the remedies, in the phrase of the *New York Times*'s Erik Eckholm, "stretch credibility. They include the forced closure of thousands of dirty factories that line the canal's route, many of which are owned by local governments that have evaded controls in the past. The government is also mobilizing to build sewage treatment plants in each of 119 counties along the canal, itself a mammoth undertaking in the hurried time frame."

Beijing wants all this to be done in time for the 2008 Olympics. Nobody believes this will happen, at least not to any more-than-shoddy standard. If it does, China will deserve its Olympic gold medal for the new sport of engineering.[8]

17 | SOLUTIONS AND MANIFESTOS

If you're short of water, especially clean water, the choices are conservation, technological invention, or the politics of violence.

Shto delat? Lenin asked once, in one of his most famous pamphlets. What is to be done?

On the crisis of contamination, the polluting of the world's waterways, the answer is pretty simple: spend money, spend money, spend money, in more or less that order. There is no mystery to ending pollution. You have either to prevent pollutants reaching the water at all, which means using any one of an array of available remedies, or you have to clean up the water once it has been dirtied, which is a policy problem – do a proper cost-accounting of the burdens of pollution, and make the people who dirty it clean it up.

This is all very well to say, of course. But even in the developed world, which has plenty of money to burn on frivolities, the protestations of polluters nevertheless find a ready hearing in the seats of governments. The oil patch pleads that the Kyoto Accord would bankrupt it; the major polluters, the car makers, the industrial farmers, the plastics industry, similarly plead that making them scrub the water clean after use would mean penury, and penury for them would mean recession or worse for the economy. Of course, these captains of industry have no historical memory, and so can't hear in their protestations the echoes of earlier special pleadings (that the ending of slavery would mean bankruptcy for all, that abolishing child labour would bankrupt healthy companies, and so forth). It is also surely a curiosity that these same captains of industry have so little regard for the entrepreneurial spirit that they can't seem to see that a) the world doesn't owe them a living, and b) invention so often flows from adversity. Everywhere you look, small companies and individuals are already inventing processes that would clean up the water, and some of these solutions are recounted below. What has so far kept them small is not that they are expensive, but the appallingly low cost of polluting.

In the developing world cleanup is much harder. The burdens of poverty and the wretched quality of available services mean that governments simply don't have the money to fix even the obvious problems. These governments are not without responsibility, of course – the population explosion is hardly the developed world's fault, except indirectly, through better health care. But the donor countries will have to help, with greatly accelerated aid programs aimed at specific problems, and not just with money pumped into the ever-welcoming coffers of the rich and corrupt. It is in our interests to do so. Not just because it is "right," or even because of the more cynical calculation that healthier countries are less likely to be a burden, but because we now know that the world is smaller that we thought. Atmospheric stations in Hawaii have measured pollution blowing in from Beijing; acid rain from Illinois has been found in Eastern Canada, and Canada's own acids in Scandinavia; pathogens from West Africa have been found in the Gulf of Mexico, having been carried there by the winds; the plume of toxic smoke that enveloped South Asia in 2002 poisons the air of countries that didn't cause it; the evil winds from the Aral Sea and from Chernobyl are found thousands of kilometres to the northeast; the sands of sub-Sahara can frequently be found in the swimming pools of the affluent bourgeoisie in the hills of Provence. There is no shortage of examples. We have only to look.

~

The crisis of supply is both easier to deal with, in that it doesn't necessarily take as much money, and harder, because the problems are so complex. In the end, there are really three ways of approaching a solution. The first is the most obvious, and that is to find fresh supplies, either by "making" more fresh water from the sea, or fetching it from elsewhere through massive diversions, as California does, as Ozal's Peace Pipeline from Turkey to Saudi Arabia would have done, as NAWAPA would have done in spades, as the Chinese are doing with their designs on the Yangt'ze. The second is to use less of it (through technological innovation, proper pricing, good management, and conservation). The third is to use the same amount of water but with fewer people (that is, head off the crisis by sharply limiting population growth).

Actually, there is a fourth solution: steal water from someone else.

Water survival strategy 1: If you need more water, get more water. That means you either import water from some place where there is a surplus, or you make more fresh water yourself.

Water diversions have been the solution of choice since human civilizations began, which means in ancient Egypt, Mesopotamia, and China. The technology may have been primitive, but the thinking of the early hydrological engineers such as Idrisi of ancient Egypt was, even by modern standards, remarkably sophisticated. Dams, canals, *qanats*, and water diversions were commonplace in most ancient civilizations, and in the ancient Sahara, for example, water-measuring techniques could calculate annual yields to the thimbleful. What has changed in modern times are the fabrication technologies.

Some years ago I had a curious conversation with a member of the International Water Resources Association.

"They're sort of like condoms," he said on the phone, "only bigger, a lot bigger. Huge. Like giant ribbed ticklers –"

The phone line wasn't very good. "Sorry, what was that?" I asked. "Ribbed what?"

"Ticklers. You know, like some women don't like the smooth condoms much, they prefer the, ah, well, ribbed ones because –"

"Yeah, okay," I said. I got the point. "But why are these bags ribbed?"

He sounded impatient. "Hydrodynamics. It's obvious. Why do fish have fins? If they weren't ribbed – if they weren't perfectly balanced, perfectly streamlined – they'd waste incredible amounts of energy twisting in the water."

"How big? How much bigger?"

He laughed. "A lot. Typically, 650 metres long and 150 wide at their widest. You know how long 650 metres is? Nearly seven football fields . . . These things are big. More than a football field wide. They've a draft of about 22 metres. They can carry a million three-quarters cubic metres of fresh water each time. That's a lot of water, a million and three-quarter tons of the stuff . . . But why don't you ask Jim Cran about them? He's the guy pushing the idea."

These "condoms" he was talking about were a modern spin on an older idea. Towards the end of the Second World War, smaller and more primitive versions, really just big rubber bags, were used for the high-risk activity of towing aviation fuel through the ocean to where it was needed. High risk, but perhaps still smaller risk than transporting the same amount of fuel in tankers during wartime: the tug that was needed to tow the bag left a much smaller sonar trace and was less likely to be thought worthy of attack. The bag itself was virtually invisible to sonar or, later, to radar, since its density wasn't much different from sea water itself. Of course, once detected, it would have been pretty easy to sabotage. One neat harpoon shot would have done it.

Later, a European consortium got the idea of using the technology to take fresh water to where it was needed, and tried an updated version. However, it wasn't "aquadynamic" – the closest marine model was a form of sea serpent. When it was towed through the water, it thrashed about, leaving a huge wake.

James Cran, who had come up with the same idea independently and at about the same time, called his version "Medusa bags." His model was a jellyfish. Jellyfish don't thrash. They sail serenely through the water, generally unaffected by storms. But the Water Resources guy was a little premature, it turns out. Bags the size he outlined haven't been built yet. They exist, though, in Cran's head.

"And in any case," he said, when I called him at the Medusa Corporation office in Calgary, a long way from the sea, "they're not ribbed, but strapped. Our original notion was simply a large polyester or nylon bag, a high-tensile woven polyester with plasticized PVC coating on both sides. We did a five-thousand-ton prototype and deployed it in Howe Sound, off Vancouver, where we promptly pulled the front off it with our tug, proving that we still had something to learn. We did a redesign. We placed fabric straps every few feet to distribute the pull longitudinally."

"Oh!" I said, "the ticklers!"

He didn't seem to hear me. "Then we did a ten-metre latex model – talk about your large condoms! – and found something interesting. When it was towed through the water, it was totally unaffected by wave motions. The motions simply continued on right through the bag and its contents – it was, after all, just water, like the ocean, with almost the same density. This was wonderful: Medusa would be the world's first cargo-supported vessel."

The idea had come to Cran after a visit to friends in San Francisco, during one of California's periodic droughts. "After dinner I went up to use the bathroom. 'Don't flush!' they told me.

There wasn't enough water. Not enough water in a city like San Francisco!"

Cran had been an engineer in Alberta's oil patch, but when the oil business fell on hard times, he turned his talents to designing software. "Spreadsheets and other products. In that way, I became a project planning specialist. I had the software, but not yet the projects. So I had to invent the technology to fit the projects that fitted my software. My first notion was a sort of hockey-puck-shaped bag, small, about 400 cubic metres, being towed very slowly, and from that I progressed to our current notions, bags of a minimum capacity of a hundred thousand tons, but more likely a million tons, or a million and a half. The science proves out. The specific gravity difference between fresh and salt water is enough to keep us on the surface; there is a 2.5-per-cent gravity difference, which gives us a two-foot head in pressurized bags, very streamlined, with a low rate of curvature." He gave the same dimensions I had heard earlier: 500 metres long, 150 wide, with a draft of 22 metres.

"We use ordinary tugs that proceed very slowly, at two knots – a speed, incidentally, that water engineers have found best for moving water through conduits, too. Our bags have a very large volume-to-area ratio, so the energy consumed per kilometre of movement is very low."

Cran took his idea to the Los Angeles Metropolitan Water District. "They pissed all over it," he said, unconsciously using a water metaphor. But more recently, the American West had once again been making anxious noises about running out of water. The droughts of the early 1990s notched up the rhetoric, and although in 1998 El Niño nearly washed the state away, there were still far-sighted water companies who wanted to secure new supplies. Alaska's governor, Walter Hickle, floated the scheme of tapping the Copper River and pumping it to southern California in an undersea pipeline. No one laughed, although the estimated $100-billion price tag made it – almost literally – a pipe dream. The water would cost $3,000 per acre-foot, which translates into

$2.43 per cubic metre. You can get desalinated water for that, and Cran could deliver for a fraction of that price.

Other water-delivery companies with similar ideas were already talking to potential buyers in the Middle East. "It's not well known," one of them told me, "that there is a corner of Oman, near Salaleh in the south, that gets washed by monsoons. What a shame to let that water get away. Why not use bag technology to tow it to where it is needed?"

Dismissed in California, Cran went to Israel, through contacts with the Montreal Bronfmans, and got to Shimon Peres, a man who loved to put things together. "I think he saw a chance to put Israel and Turkey together, so he was interested. Why shouldn't Israel, which needs water, get it from Turkey, which has lots?" Everyone knows the Levant – Israel, Jordan, parts of Lebanon, Gaza, Sinai – needs water, and Turkey has about as much as Scotland, much of it rising in the mountains, and very clean. That's where the Manavgat River scheme came from.

In 1989, Cran arranged for the Israeli Water Commission to visit Turkey. "There was magnificent water there – on the Manavgat, two huge lakes in the mountains, with essentially no exit except one through a rocky channel. Lower down, the town of Manavgat dumps its sewage, untreated, into the river, but until that point it is very clean. There is no farming upstream, and farming is a huge problem for fresh water." Tahal, the Israeli water agency, did a study of the Turkish situation, and pronounced it sound. So was Cran's bag technology.

Israel asked Turkey for a letter of intent – some proof it was serious about sending water. The letter never came. Somehow, Colonel Gaddafi of Libya heard about the project and tanked it. There were all kinds of Turkish engineering firms working on the Great Man-Made River scheme, and they would not be paid, he insisted, not a penny more, unless there was a guarantee that no water would be sent to Israel or anywhere without a comprehensive peace plan for the region. Turkey capitulated – almost.

Cran told the story: "'Okay,' they said, 'we won't send any water to Israel until then, but still no one can stop us building a terminal, a Manavgat Loading Terminal, for exporting water when the time comes.'" He snorted at this notion. This terminal, he said, was a joke. "I think at last glance, it had cost around $120 million. DSI, the Turkish state water organization and the largest constructor of dams and irrigation works in the world, was asked to build it. But terminal for what? No one told them. So they built the terminal for supertankers, despite the fact that no one had ever costed out water carried by supertanker, and there were no clients and no suppliers. Nor was anyone building a supertanker unloading terminal, in Israel or anywhere else. Worse, the Turkish terminal was designed to ship 180 million cubic metres a year (90 million treated and 90 million going for ballast). Israel, if it were to be a customer, wanted somewhere between 250 and 400 million cubic metres a year. So the terminal's potential was less than the smallest number the Israelis were looking at. The long-term demand in the area is around 800 to 1,000 million cubic metres. So, despite the cost, the terminal is just a toy."

Then came the famous handshake on the White House lawn, and Peres called on the Turkish foreign minister for that elusive letter of intent. Negotiations crept forward. The president of Turkey visited Jerusalem and announced that his country could deliver water via their terminal at $1 a cubic metre. "Everyone laughed. This was ridiculous. The Israelis had expected 25 cents, tops. Medusa can deliver the water for 20 to 25 cents a cubic metre – transport costs of around 10 to 15 cents, the rest for loading and unloading. And then Özal put forward his Peace Pipeline proposal. That would deliver water at $4 a cubic metre, even then the very top price for desal water. The whole thing was just a make-work scheme for Turkish engineers."

The idea surfaced again in 2002, however. Early in the year, the Israeli Cabinet agreed to buy some 50 million cubic metres of water a year from Turkey – only this time, Medusa bags were no

longer mentioned, but recommissioned, obsolete tankers. The deal
was that Israel would use the Manavgat terminal to fetch water,
and in return would help Turkey upgrade its army tanks and war-
planes. Less than three months later, that plan, too, was in trouble.
It would take longer than anyone thought to clean up and assem-
ble a fleet of tankers, and the price, once again, was too high – 85
cents a cubic metre – compared to the 10 cents expected from the
new desal plant at Ashkelon, when it came on stream. They should
have known: Turkey late in 2001 was sending some water to
northern Cyprus in tankers, but even that cost more than southern Cyprus paid to desalinate its own water.

At the start of 2003, the idea was not dead, but dormant. Israel
didn't want to disrupt its otherwise cordial relations with Turkey
by unilaterally cancelling the agreement, especially after a report-
edly moderate Islamist government was elected there in 2002.
Perhaps Turkey could deliver the equivalent of two years' supply
before payment was demanded, the water to be "stored" in Israel's
aquifers? That would give everyone plenty of time to look for
alternative sources if the Turks, for whatever reason, threatened to
screw down the faucet. It was all becoming too complicated.

On an April morning in 1998, a Japan Airlines 747 lifted off the
field at Osaka, heading for a primitive little airstrip that had never
seen a jumbo before: Cambridge Bay in the Canadian Northwest
Territories. The plane was empty, except for a few engineers and a
number of strangely large compressors. The payload on the way
back was to be chunks of Arctic ice. The plane had been hired by
a Japanese sake maker, and the ice – pure Arctic ice, pristine and
unsullied by the human propensity to pollute everything it touches
– was to be used as a marketing device for a new brand of liquor
called Arctic Pure. Never mind pricing of a dollar a cubic metre –
this would be more like a dollar a litre.

A month later, on the other side of the continent, a small tour company was set to make money from "the last of the unsullied rivers," the pristine and beautiful Feather River in Labrador, one of the few remaining places on earth that has never been logged, mined, used by humans, or polluted by factory or farm. That there was anything incongruous about helicoptering people in and letting them tour the river in gasoline-driven boats to experience the fabulous cleanliness, the company's promotional literature didn't say. There were also plans afoot to bottle the water, to take it to the restaurant tables of the Western world, on the no-doubt-correct theory that people were increasingly doubtful about the purity of European bottled waters. Both these schemes – the Arctic ice to Japan and the Feather River scheme – are neat metaphors for one Western ("Northern," in the new geography of the development industry) solution to the water problem: throw money at it.

Jim Cran's Medusa bags and Alaska's notion to pipe water to California are really just extensions of this approach. So are the plans occasionally mooted in St. John's, Newfoundland, and along the Alaskan coast to lasso icebergs and haul them south, chipping them up into tanker trucks before they melt (this was at least being thrifty: the water would melt into the sea in any case, so why not let it melt where it counts?). So are the extravagant notions to return Mississippi Delta water to the Texas High Plains, or the now-abandoned Russian projects to divert north-flowing Arctic rivers like the Ob and the Lena southward, either through the Urals into the Volga Basin or toward the Aral Sea. So, too, were the ever-more grandiose plans for "solving" North American water problems: the GRAND Canal scheme to divert water from James Bay into the Great Lakes, or the NAWAPA scheme to transfer water from British Columbia practically everywhere, even including the Great Lakes, the Mississippi, and California. So, too, is the Chinese leadership's expressed desire to re-engineer the country to take water to the hard-pressed north, the last of the grand projects that is actually being built.

It's not, after all, as though this approach didn't work. It works in California – it's long been a cliché that, in California, "water runs uphill to money."

> Every year the state moves 14 trillion gallons [53 trillion litres] of water, in directions mostly south, capturing it behind 1,200 dams on every river and stream of any size, before fluming it hundreds of miles, lifting it over some mountain ranges and pumping it under others, fitting the sere landscape with a caul of pipes, ditches, and siphons that irrigates an agricultural empire far greater but not unlike the one that bloomed in the deserts of Babylon and ancient Egypt. And if wealth has tamed the water, then the water has made California wealthier still. If it were a nation, the state's $1 trillion economy ranks it seventh among national economies, and in addition to providing 55 per cent of America's fruit, nuts and vegetables, California has managed to become the sixth-largest agricultural exporter in the world by intensively farming a region that receives less than 20 inches [50 centimetres] of rain a year.[1]

River-diversion projects will still be built – sometimes because there simply is no choice. The Okavango is a case in point; you have to get water somewhere. But the day of the massive diversion projects has effectively ended. There are few places left in America or Western Europe for damming river water. The barely ticking-over Central Arizona Project was among the last of the diversion schemes contemplated by U.S. water agencies and, late in 1998, the Army Corps of Engineers had diverted its formidable attention to East Coast harbours, hoping to invent ways they could accommodate the coming generation of super-super container ships, some of which would need a deepwater harbour with a draft of up to fifteen metres at low tide. That there are few places anywhere that are still worth damming is one of the signs that the water crisis is real:

demand hasn't stopped rising, and will double in the next few decades, but new supplies are increasingly hard to find.

~

Is there really a demand for water towed around in humungous plastic bags? Cran insists there is.

There are three main markets for delivered fresh water: the agricultural market, which generally can't afford water prices of more than 10 cents a cubic metre; the municipal market, which can afford 50 cents a cubic metre or higher (in California they generally pay around $350 per acre-foot, around 30 cents a ton); and the homeowner market, the bottled-water crowd, for whom the cost of water is negligible – the container and delivery is what you're selling. Cran's market is the municipal one, and his main competition is the desalination industry.

"I figure we need a one-third price advantage over desal to take account of the horrendous political and bureaucratic problems of import-export. Our figures show we can do this. Of course, we can generally only deliver to coastal cities. But that's okay. So does desal. And there are real exchanges possible. If we deliver to Tel Aviv, it means Tel Aviv won't be diverting Jordan River water from agriculture to the city, and farming will have more water.

"Look at it this way. There are three places in the world where there is massive movement of water through piping systems: in California, in Israel, through the National Water Carrier, and in Libya's Great Man-Made River. All of these take water from the interior to the coast. This is ridiculous! Instead of California pumping water over mountain ranges and across deserts to take it to L.A., why not leave it where it is and use Medusa water for the coast? One of our customers could be, for instance, the city of Las Vegas. They want Colorado River water now guaranteed to Los Angeles. Why don't they buy our water, "give" it to Los Angeles, and simply keep the Colorado River? And the legal and political quagmire! I could ship water from the Sacramento Delta to Los Angeles, but what a red-tape nightmare! One of Jimmy Carter's last acts was to declare all the rivers of north California and Oregon 'wild and scenic,' which meant not only that they couldn't be tampered with but they couldn't even be studied."

But there are many other places in the world where water bags could be useful. Singapore, for example, has no water of its own and is entirely dependent on the goodwill of Malaysia. Yet Singapore once seceded from Malaysia, and half its water comes in a single conduit – an acceptable situation when relations are good, but as Malaysia slid into recession late in 1998, the resentment against the more ebullient Singaporeans once again surfaced, with water always a potent weapon at hand. In 1998, Singapore commissioned the first of three planned desal plants, but a hedge wouldn't be a bad idea. Water bags from Indonesian rivers could lessen the political hazard of dependency on a single source. And in the Cauvery dispute in India, water bags could bring water down the coast to Tamil Nadu (Madras) from the more water-rich north, freeing up water for inland farmers.

There have been other notions for carrying water through the ocean. One of those was studied by the Intertanker group and involved using superannuated oil tankers for transporting water. There are hundreds of single-hulled tankers rusting away in ports all over the world which are suddenly illegal for transporting oil; when the oil market collapsed in the late 1980s, the carriers tried to find markets for water, but failed. Soon also, dozens of newer supertankers will reach the mandatory retirement age. Why shouldn't they go into the water-carrier business, for which safety standards are, for obvious reasons, not so strict? But the cost of the cleanup is estimated at $5 million per ship, and may in the end prove to be impossible. Benzene is not something you want in your drinking water.

And then there was the other tugboat solution, the towing of icebergs to where the water is needed. Greenland, Labrador, Alaska, British Columbia, and Antarctica all have plenty of fresh water, and the icebergs are in any case breaking off and drifting out to sea. Why not capture them and take them to where they would be more useful? This concept has been widely explored. One of the proponents was a Saudi prince, Mohammed Al-Faisal, who sponsored a study on the feasibility of towing Antarctic icebergs to Mecca. In the end, his group towed a block of ice around San Francisco Bay for a while, and proved to their own satisfaction that no iceberg would ever cross the equator.

In any case, how to "mine" the iceberg when it gets to its destination? They are so damn unstable! As Jim Cran put it, "You'd get people on the thing strip-mining it, and just the mining cost, at around $9 a ton, is way too much compared with desalination. Medusa is a chip off the same block, so to speak, but we don't have to go so far, and we don't have to worry about either instability or melting. We can choose the shortest distance for the water to travel, and the draught is controlled. No icebergs could ever get into the Persian Gulf. They're too deep."

This is not to say that water planners have discarded the notion. Both South Africa and Namibia have studied the notion of towing – or steering – icebergs from Antarctica. There have even been proposals for an iceberg-processing plant at Saldanha Bay on South Africa's west coast, should the things ever make it safely across the southern ocean. The latest study, by J.M. Jordaan, was enthusiastic, but recognized that "a four month journey over five thousand nautical miles having to be traversed, under extreme weather conditions and sea states, with the associated losses due to meltdown, appears a formidable obstacle to be overcome."[2] Even if they did make it safely, near-shore icebergs would be vulnerable to any pol-

lutants in the offshore water, and treatment of supposedly pristine
iceberg water would be required.

In the early part of the new millennium, literally hundreds of
companies were studying alternative distribution strategies. One
of the heavyweights was Luxembourg-based World Water S.A., a
partnership of a Saudi investment company (hitherto best known
for being the world's largest Toyota distributor); the NYK Shipping
line, (formerly Nippon Yusen Kaisha, and now the world's largest
ocean transporter); Nordic Water Supply of Norway (which has
actually used Medusa-style bags to carry water from Turkey to
Cyprus); and Alaska Water Exports, since renamed Aqueous
Corp., headed by Ric Davidge, a former Reagan-administration
deputy secretary and one-time director of the Alaska Hydrologic
Survey. World Water took as its mandate "to economically provide
high quality water to any area in the world"; much of its research
budget was being poured into transportation alternatives.

Despite all these efforts at moving water in bulk, getting fresh
water from the sea – essentially making new fresh water through
desalinating sea water – is everyone's best technological hope for
solving the looming crisis. After all, 97 per cent of the water on
earth is sea water, and 96.5 per cent of sea water is really fresh
water in disguise, the other 3.5 per cent is the dissolved solids that
make it unusable for humans. Get rid of those solids – and you can
both drink it and irrigate your plants with it.

The Israeli economist Amikam Nachmani, in a paper called
"Water Jitters in the Middle East," is typical of the political
observers of the water problem in calling desalination "the only
realistic hope" for dealing with freshwater shortages.

The fact remains that Middle East water needs have increased
considerably of late. The reconstruction of Kuwait, Lebanon,

and Iraq, the influx of immigrants into Israel, and so on, all point to the need for a Marshall Plan for Middle East water. Is desalination a practical solution? Israel's official view, as presented in January 1992 to the Moscow multilateral water summit of the Arab–Israeli peace conference, sees desalination as the only long-term remedy for water-poor areas like the Middle East.

A desalination project for 10,000 people costs the equivalent of one military tank; for 100,000 people, the price is roughly that of a jet fighter. Investing in desalination of brackish water or sea water, or recycling sewage [for agriculture], is cheaper than attempts to settle disputes over available water sources, most of them already overused.

Nachmani quoted the political scientist Frank Fisher, who said that "100 million cubic metres of water, which is the bone of contention between Israel and its neighbours, is not worth a war: a day of war costs $100 million, whereas desalination of 100 million cubic metres [of water] also costs $100 million." Similarly, Nachmani maintained, "the diversion of water from one place to another is much more expensive than the development of new technology for cheaper desalination. In addition, desalination plants create a sense of security by virtue of the state's ownership of the resources and installations it considers vital (although Kuwaiti and Saudi desalination plants were targets of Iraqi attacks during the [first] Gulf War)."[3]

No doubt this is why Israel was rumoured to have tabled a "demand" for $10 billion for water matters at the Camp David meetings; it was to be part of the price of the late-lamented "Peace Process," and a basic part of bringing stability to the region.

In April 2000, with Israeli experts already using doomsday rhetoric about the future of water availability, the Israeli government finally approved, in principle, a series of large-scale desalination plants to be located at Ashkelon, Ashdod, Hadera, and other places, capable of producing somewhere around 200 million cubic metres of fresh water from the Mediterranean each year, about a tenth of the country's consumption. The first of those, the 100-million-cubic-metre plant at Ashkelon, was not, however, expected to be finished until late 2003. Ironically, the plan was resisted by the country's farmers, the biggest water hogs of all, who mounted a fierce opposition campaign in the Knesset; they were afraid desal water would be too expensive, and raise the cost of water across the system. Nevertheless, the plan was pushed through, and water planners in late 2002 were saying airily that, by 2007 or so, "everyone will have as much as they want."

There are already more than 7,500 desalination plants in operation worldwide, some of them tiny. Two-thirds of them are in the Middle East, 26 per cent in Saudi Arabia alone. The world's largest plant, in Saudi Arabia, produces 485 million litres a day of desalted water; the Saudis and other oil-rich countries are using desalinated water for most of their domestic consumption. Israel's water controller, Dan Zaslavsky, has estimated it would cost about $2.5 billion to desalinate enough water to remove the water irritant from regional politics, a much higher figure than the $100 million from Frank Fisher, but still considerably less than a war, even a minor war, when a sophisticated airplane can cost not much less than that.

Only some 12 per cent of the world's desal capacity is produced in the Americas, and until the Yuma plant was conjured into being with federal subsidies to deliver clean water to Mexico, most of the plants in the Americas were located in the Caribbean and Florida. The Californians have historically rejected desal; they got their water so cheaply (again through federal subsidies) that it hardly seemed worthwhile. However, as drought and groundwater overpumping heighten concern over water availability, desal plants are springing up all over the state.

In August 1999, as we saw in Chapter 14 on North America, Tampa Bay Water in Florida announced it had hired a consortium to build a new reverse-osmosis desal plant that would produce 25 million gallons of fresh water a day, at a declared cost of 45 cents a ton. At second glance, the economics didn't seem quite as rosy – over the life of the plant, the water would cost more than $2 a ton, but this is still better than most other sources. The economics were possible because Tampa Bay is less naturally saline than the Gulf of Mexico, and because the plant would be built next door to a major power plant that would supply it with cheap electricity. Nevertheless, the price was such that it attracted a delegation from Singapore, which was planning a 36-million-gallons-a-day plant – at an average cost of nearly $8 a ton. And then, at the Ashkelon plant, the Israelis were set to produce desalted water at a fraction of that – rivalling the cheap water that California supplies to its profligate farmers.

The worldwide desal capacity is still pretty small. Just enough, as the British Columbian scientist and desal engineer Bowie Keefer puts it, to serve as the flow of one very small river. "But if you push energy efficiency to its ultimate, though, the possibilities are significant. To produce enough desal water to make up the equivalent of a decent-sized river would cost billions of dollars. But," he said, echoing Nachmani and Zaslavsky, "when you look at the cost of military hardware, it's not prohibitive."

Like others in the field, Keefer thinks desal is the water world's best hope: "Look at it this way. If you plunked the whole of California down in say, Mauritania, where there's no water, and said 'Okay, keep the California lifestyles going, only with desal water instead of the Colorado,' you could. You could keep a pretty advanced economy going. The irony, though, is that food on the world market is too cheap. That new Mauritanian California would be producing food about twice as expensive as the current market prices. The farmers couldn't sell it. It's still cheaper to grow grain in Manitoba. But it does show you the potential.

"Of course, it was a dream under the Nixon and Reagan administration in the U.S. They had these Soviet-style notions of plunking down a nuclear plant in the middle of nowhere and causing the desert to bloom – that was when nuclear power was supposed to be too cheap to meter, remember that?"

Even without nuclear, the potential is still there, he said. "The potential is for desert reclamation, worldwide and affordable." Desalination is a proven technology. Its main problem has been cost.

The most beguiling – and massive in scope – desalination solution is the Red–Dead Canal, originally proposed by the American Walter Clay Loudermilk and latched on to by the Jordanians. Jordan's engineers maintain that the canal could provide enough fresh water through desalination to solve Jordan's problems and enable Israel to pay what the Jordanians call its "water debts" to the Palestinians, Jordan, and Syria combined. Their analysis has been received skeptically by the Israelis, but has never been rejected outright. Israel has floated an alternative, the Dead–Med Canal, which would, at least from the Israeli perspective, have several advantages: it would be entirely within Israeli control, and it would be shorter and therefore cheaper.

The Red–Dead plan would be, in effect, to extend the Jordan River south to the Gulf of Aqaba on the Red Sea. Except, of course, that the water would flow the other way, northward, for the 275 kilometres from the gulf into the Dead Sea. The water would be pumped from the gulf, and along the Israeli–Jordanian boundary through the Arava Valley. It would have to flow uphill about 200 metres, somewhere around Mount Edom, but from there it would begin a 660-metre descent to the Dead Sea, the lowest place on earth.

From here, the planners wax ever more lyrical and enthusiastic, until their reports begin to sound like a feasibility study for some kind of Disneyfied theme park called Waterland. About 990 million cubic metres a year would hurtle down the valley in a series of great zigzags, creating boating lakes, swimming pools, and hydroelectric power as they went – enough power to drive the world's biggest desalination plant. Some 40 per cent of the descending gulf water would be rendered potable, and the leftover brine would decant into the Dead Sea, lifting its levels back to 1960s values and upping its tourist potential. The Dead Sea is much more saline than the ocean, and so too would be the brine left over from desalination. The tourists would float just as effortlessly as before.

Sure, it would cost billions, but once built, it would pay for itself – the steep drop would generate sufficient energy to desalinate enough water to make the project self-sustaining. Well, that's the theory. Who pays for getting it up the initial 200 metres is seldom discussed. The economic models are ambiguous, and the ecological consequences would no doubt be severe. There doesn't seem to be much political enthusiasm, either.

Still, the proposal generates more interest than some of the alternatives. These include a water pipe from the Nile, siphoned under Suez, a notion that Egypt, already wrestling with water deficits, contemplated for political reasons (watering Sinai would be a potent political act) but can realistically not afford. Turkey has proposed its Peace Pipeline to Saudi Arabia, and no doubt a branch line could reach out to the West Bank. And so on and so on.

In the end, though, it always seems to come back to desal – with or without a Red–Dead Canal.

The process of desalination, the removal of dissolved minerals (including but not limited to salt) from water, is not excessively complicated. Nor does its source water necessarily have to be the sea, though it usually is, desal plants taking ocean water through offshore intakes and pipelines, or from wells on the beach. But desal plants can also use brackish groundwater, or reclaimed (recycled) municipal water. Brackish water is not usually as salty as sea water, and is therefore cheaper to desalt, although for safety's sake, waste-water desal plants usually also incorporate anti-pollution devices and biocides (usually chlorine).

The water-treatment industry is fiercely competitive, and a number of different techniques have been developed, including reverse osmosis (RO), distillation, electrodialysis, and vacuum freezing. Only RO and distillation have so far been thought commercially feasible.

The basic principle of RO is the same as Keefer's smallest device, the Nutcracker: it pumps the salty water at high pressure through permeable membranes or filters. There are other stages, such as pre-treatment, to separate out larger particles that would clog the membranes. And sometimes the water is passed through several stages of membranes before being declared "clean." The quality of the water produced depends on the pressure, on how salty the water is, and on the efficiency of the membranes. Since the membranes themselves are the heart of the process, what they are made of is usually kept a closely guarded commercial secret.

Proponents of RO like to point out that their plants take less energy to operate than distillation plants, have fewer problems with corrosion, and have higher recovery rates – about 45 per cent – for sea water. Also, RO can remove unwanted contaminants, such as pesticides and bacteria; and RO plants take up less space than distillation plants for the same amount of water production.

Distillation also has advantages. Its plants don't have to shut down for cleaning or replacement of equipment as often as RO plants do. You don't have to pre-treat the water, because there are no filters to clog, and distillation plants don't generate as much waste.

Distillation, familiar enough from school science kits (or to anyone who has attempted a little homemade moonshine), is also simple enough: the water is heated, evaporated to separate out the dissolved minerals, and cooled back to water. As with RO, there are several competing technologies. The most common methods of distillation include multi-stage flash distillation (MSF), multiple-effect distillation (MED), and vapour compression (VC). In MSF the water is heated and the pressure is lowered, so the water "flashes" into steam. In MED the water passes through a number of evaporators in series, and vapour from one series is subsequently used to evaporate water in the next. The VC process involves evaporating the feed water, compressing the vapour, then using the heated compressed vapour as a heat source to evaporate additional water. Some distillation plants are a hybrid of more than one desalination technology.[4]

What's left over from either desal process is about one-third fresh water (up to half in very efficient plants) and two-thirds highly saline brine. The fresh water, though, is very clean, ranging from 1 to 50 parts per million of dissolved salts from distillation, and 10 to 500 parts per million for RO plants (drinking-water standards are usually around 500 milligrams per litre, equivalent to 500 PPM). Because it's usually purer than most drinking water, it can be mixed with less-pure water before distribution.

In May 1992, a desal plant owned by the city of Santa Barbara was analyzed by consultants for the California Water Commission. It produced 6.7 million gallons per day (MGD) of clean water, and generated 8.2 MGD of waste brine with a salinity approximately 1.8 times that of sea water. An additional 1.7 MGD of brine was generated from filter backwash. They figured this was about 1.7 to 5.1 cubic yards per day of solids, equivalent to one to two truck-loads per week.[5]

～～～

Energy is the big enemy of desalination, and a major barrier to its widespread implementation. Do you really want to purify water by adding greenhouse gases to the atmosphere? Distillation is a notorious energy hog, but even RO uses vast amounts of energy. "Desal is a tricky world," Jim Cran says. "There are some slippery economics out there." Because you can't lend money at interest under Islamic law, the cost factors for desal plants in Muslim countries can be misleading and financing costs ignored. So, often, are energy costs. The Saudi plants are sitting, literally, on a sea of oil. Sometimes they simply use the gas flumes from oil production to desalt their water, and energy costs are therefore written off as nil. As Cran put it, "Some of the U.K. water companies, privatized by Thatcher and hungry for world markets, have engineering divisions offering desalinated water, and over long whisky lunches they'll talk about $1 a cubic metre for cleaned-up sea water. More honest people will admit that, when energy and other costs are all taken account of, the real cost is somewhere around $2 to $2.20. For the Saudis, the world's major desalinators, the cost is closer to $4."

For Santa Barbara's desal plant to produce 7,500 acre-feet (9.25 million cubic metres) of fresh water would take 50 million kilowatts. To bring the same amount of water from the Colorado River to the Los Angeles Metropolitan Water District, pumping it over mountain ranges and along hundreds of miles of aqueducts, takes only 15 to 26 million kilowatts.

Price estimates of water produced by desalination plants in California range from $1,000 to $4,000 per acre-foot (81 cents to $3.24 per cubic metre). By contrast, in 1991 the Metropolitan Water District (MWD) paid $27 per acre-foot for water from the Colorado River and $195 for water from the California Water Project. Non-interruptible untreated water for domestic uses in San Diego is purchased from the MWD for $269 per acre-foot; treated water costs an additional $53. The least-expensive new supplies, other than desalination, would cost $600 to $700 per acre-foot. In Santa Barbara the development of new groundwater wells in the mountains would cost approximately $600 to $700 per acre-foot. Enlarging the Cachuma Reservoir, even if it proved feasible, would bring in new water at about $950 per acre-foot. During the recent drought, the city purchased water from the

State Water Project on a temporary, emergency basis at a cost of $2,300 per acre-foot.

Desal is not without its controversy. One of the most vigorous debates concerns using nuclear reactors to power desal plants. In May 2002 scientists near Chennai (formerly Madras), in India, were finalizing plans to hook a new desal plant into an existing nuclear power plant; the plan was to produce 5,400 cubic metres of fresh water a day. Other countries, too, are looking to nuclear energy to deal with the power-hunger of the desalination process. Among them are the United States, Indonesia, and South Korea. The Koreans, for their part, have designed a 300-megawatt reactor for desalination and electric generation. Tunisia is talking to France, and Morocco to China, about nuclear plants to desalt sea water. A consortium of Canadian and Russian scientists were, in 2002, pooling resources to build floating nuclear desal units that could be barged to the world's dry spots on demand – a prospect (Canadian reactors on Russian vessels) that was sending environmentalists into paroxysms of indignation tinged with paranoia. The system's proponents point out that a coal-fired plant produces 964 tons of carbon dioxide for every gigawatt hour of electricity, whereas a nuclear plant produces only 7.8 tons.

Desalination techniques are getting better, and therefore cheaper. In 2002, the tenders for Singapore's first water-desalination plant finally came in, all with bids lower than the water supplied by the Public Utilities Board. The bids, to produce 30 million gallons a day, or 136,000 cubic metres, ranged from $0.78 to $1.40. The PUB charges consumers $1.17 per cubic metre, not including a 30-percent conservation tax. The plant was designed to produce about a tenth of Singapore's current demand. Interestingly, plans to build a total of six more such plants, producing as many as 100 million gallons, have been put off because the recycling of grey water (recycled household water, called NEWater in its Singaporean euphemism) has provided "new" water much more cheaply. The cost of building a reclamation plant is estimated at one-half that of a desalination plant.[6]

The research, meanwhile, is ongoing. The Middle East Desalination Research Centre, created in 1996 in Muscat in the gulf state of Oman, is where much of the more interesting work is being done. In April 1998 its director, Eric Jankel, put out tenders for projects that included the development of "small, home-use RO facilities" and a number of other innovative small-scale desalination techniques (along with, it's fair to say, a series of worthy but not-so-sexy projects, such as "scale-prevention tech-niques for thermal desalination").

A British firm, Light Works Limited, was already involved in work for the centre on what it called seawater greenhouses, essen-tially using the ocean as a heat sink to produce fresh water "by humidifying warm air using sea water, and then recovering pure water from the humid air using a condenser, also cooled with sea water." Running this process inside a humidity-controlled greenhouse had the added benefit of cooling and humidifying what the study called "the growing environment." They could not only get fresh water from the sea but grow plants in the very envi-ronment that produced the water in the first place.

And a team from San Diego was working on solar-powered desal plants, using off-the-shelf technologies like solar-dish concentra-tors and a solar-powered variable electric motor drive. They hoped to be able to produce small, portable systems for widespread distri-bution in Middle Eastern and North African countries, offering a "near-term solution to water shortages in remote locations."

The Dutch company Aqua Vita Water Systems has already produced just such a mobile water-disinfecting-and-desalination system for on-the-spot desalination of sea water into fresh drinking water, using an electrochemical process, which is easy to operate and requires scant technical knowledge. It is capable of producing 300 litres of fresh water with 1.68 kilowatts of power. In early 2003, they were on the point of producing a self-supporting system that would use available wind power for its energy needs.

Nor are they alone. A hydrologist called Robert Parke, for example, has proposed dozens of desal solutions, ranging from the mundane to the esoteric, including inverting transparent polymer domes over decommissioned tankers to turn them into mobile solar-powered desalinators.[7] At last count, my mailbox contained seventy-four other projects and proposals, from companies and consortia all over the world, all of them promising the cheapest, cleanest desal water on the planet, at prices irresistible to consumers and sure to make billions for their proponents.

It is entirely possible that some of these inventions will do everything their developers claim.

⁓

And there is one more possible thing to do with sea water – use it directly, without desalinating it.

Obviously, such water wouldn't be used for drinking or for irrigation, but virtually every water authority "wastes" potable water by using it for such mundane chores as flushing toilets. Sea water could do these chores easily. Retrofitting current systems would be difficult or impossible, but new conduit systems that discharge through treatment plants directly to the sea, and that use materials resistant to salt corrosion, would be plausibly cost-effective. Such a scheme has already been implemented in Hong Kong, resulting in savings of from 18 to 20 per cent of freshwater consumption.[8]

Water survival strategy 2: *If you can't get more water, well then, clean up what you are currently polluting (the second of the two overlapping water crises) or, simply, use less water in the aggregate. That is, reduce your demand. This can be done in three ways: by conservation; by pricing mechanisms; or by making existing water consumption more efficient, through a combination of a new water ethic and skilful use of imaginative technologies.*

Eugene Stakhiv, the Army Corps of Engineers scientist, is no more a believer in a water apocalypse than the Spaniard Ramón Llamas; he believes passionately that the doomsayers are devaluing the notion of human ingenuity. Just look at the record, he says, how often prophets of doom have been proved wrong by some new invention: the hydraulic pump, or more productive seed varieties, or more advanced farming techniques, or a revolution in technology.

These inventions can come from unlikely sources. Take the chemical giant Monsanto, for instance, once notorious for providing the U.S. Army with Agent Orange and now reborn, at least in its public relations, as a biotech company with a stated commitment to sustainable development. Hendrik Verfaillie, head of the company's agricultural division, told a Toronto conference in 1998 that solutions to the world's food problems lie in genetically engineered crops that produce more food per hectare with less labour, less water, and less fertilizer. His company estimates that world food demand will increase by a factor of three in the next fifty years. This would require another 38 million square kilometres of croplands. "If that were the case, we'd have to burn down the rain forest," he said, with considerable hyperbole. "We would have to eliminate all the wetlands and tax the environment in a way that would be totally unacceptable. Another possibility is to increase productivity by a factor of three. And one way to do that is through biotechnology."[9]

Still, it never does to be overly credulous about corporate spokespeople and their public statements. There are other facets to biotechnology's newly kindly eye. Biotech companies have been diligently patenting seeds, which means that farmers, many of them poor peasants, who want to grow even traditional varieties, might henceforth have to pay a royalty. Worse, many of the re-engineered seeds have a self-destruct mechanism built in so they can no longer replicate, forcing farmers back to the biotech ware-houses for the next season's crop seeds.

When you start to look around, however, it's hard to disagree with Stakhiv's high opinion of human ingenuity. For example, water is now being "stored" in deep aquifers: winter river flows are being pumped into the chalk aquifers underlying London, a response that has reversed the downward trend of the water table; Arizona, not yet needing its full Colorado River allotment, is storing the surplus in deep aquifers there. This replenishment has twin advantages: it doesn't divert water from its natural basin systems; and it minimizes evaporation, making the storage more efficient than surface reservoirs, especially in high-heat arid regions. And in Senegal, a Canadian–Senegalese joint venture is restoring water to the country's so-called fossil valleys, irrigating and repopulating land along three thousand kilometres of dried-up riverbeds, refilling the valleys with water from the Senegal River, without at the same time endangering native fisheries, agriculture, or navigation.

Some of the ecological science can be arcane and sophisticated. Such, for example, are metapopulation theory and patch-dynamics, the science of manipulating the populations of plants and insects within territorial ranges to maximize growth. The technique mainly has been used to kick-start the regeneration of blighted lands such as old municipal dumps; but in more arid countries, the same theories are being introduced as a way of increasing the moisture retention of degraded earth.

Even rain-makers (more accurately described as "rain-boosters") are having some success, mostly because a South African scientist had noticed an odd phenomenon: emissions from a paper mill seemed to be making it rain (or rather, rain harder). Graeme Mather, the scientist, analyzed the emissions and found two hygroscopic salts, potassium chloride and sodium chloride. Hygroscopic substances have a strong chemical affinity for water, and were apparently making water droplets bigger and heavier, precipitating them as rain. He seeded other clouds with the same particles, released from flares under the wings of small aircraft, with positive – and measurable – results.

And in September 2002, the Russian Emergency Situations ministry claimed credit for precipitating a light drizzle over Moscow, a rainfall that temporarily cleared the acrid fog then blanketing the capital from forest fires in the hinterland. Viktor Beltsov, the ministry's spokesman, said the rain was caused by an ionizer atop one of the ministry's buildings. The device was a metal cage crisscrossed by tungsten filaments that emitted a powerful vertical flow of oxygen ions that stirred the air and raised humidity. (Perhaps tactfully, the Russian Academy of Sciences declined comment.)

More tellingly, a team of Israeli scientists reported in 2002 that some salty residues from ocean spray make their way into the upper atmosphere and create rain. Often they scrub the air of pollution and precipitate it to earth as acid rain. The observations were made during a study using satellite data to look at clouds forming over South Asia. Over land, rising pollution caused towering storm clouds over India and Bangladesh; over the ocean, the same pollution caused much lower clouds and produced rain much more easily. Ocean spray was identified as the crucial difference.

But for me, a small and largely forgotten example of technological inventiveness, and one of the most beguiling because it was so wrong-headed, is the work of the Russian Feodor Zibold in the early part of the twentieth century.

Zibold was a "natural scientist," as they used to be called, a philosopher of nature, a man consumed with a childlike curiosity about how things really worked. There were people like him all over Europe a hundred years ago, men of independent means and independent minds, all imbued with the kind of sunny optimism that goes with exploring where none other had been before. Charles Darwin is the most famous exemplar of this agreeable species. Zibold came later, but he was operating on the fringes of Europe, in the last tottering days of Czarist Russia, and could be forgiven his dilatoriness. And unlike Darwin, he was often spectacularly wrong, which only adds to his charm.

One day late in 1906, when he was taking the waters in the Crimea, he came across a local legend that the ancient Greeks, who had built an important provincial capital at Theodosia (now Feodosia, Ukraine), had mastered the morning dew: they had become so proficient at collecting and dispensing it that they had supplied the whole city with its fresh water, using neither well nor brook.

Zibold was captivated by this notion and determined to recover the secret. Once he started to look around, the evidence was everywhere. Lying about on the ground were stony tumuli, undoubtedly remnants of ancient dew collectors, and surrounding them were clay pipes, which had clearly been used for conducting the water to storage cisterns. Filled with enthusiasm and energy, he bullied the local agricultural community into helping him build his own massive dew collector, a stone reservoir twenty metres across, in its centre a pyramid of stones and pebbles six metres tall. Halfway through construction, they ran out of money, and it wasn't until 1912 that the marvel was finished. And, indeed, it worked – a yield of 350 litres per day was reported, not exactly city-slaking news, and not much water per ton of rock, but something.

In subsequent years the yield dropped, for no apparent reason, and by 1917 the local farmers had more on their minds than dew.

They were more intent on surviving the Bolsheviks' maniacal enthusiasm for collective farms, which in fact worked a lot less well than Zibold's dew farm, and the famous dew collector was abandoned. The remnants are still there, however, and curators of the shabby little museum in town will happily show them to you and explain how Zibold's "condensers" were in fact ancient Scythian tombs, and his clay pipes dated back no further than the Middle Ages, when they had been used to bring water into town from a nearby spring.

And the point of the story? Zibold's work has inspired dozens of emulators, among them the Grenoble researchers Daniel Beysens, Irina Milmouk, and Vadim Nikolayev, who have developed workable dew collectors for use in areas where actual rainfall is scarce. In a paper they presented in 1998 to an international Conference on Fog and Fog Collection, they showed that their small collectors can produce "rainfall in the order of .1 to .5 millimetres per day, and are still open to improvement, the ultimate goal being continuous formation of dew, day and night. . . . Optimizing the parameters [might] lead to a new generation of condensers able to provide clean water wherever ordinary means cannot be used."

The fog conference is itself ample evidence of humankind's intellectual fecundity. It was held, appropriately enough, in the damp climate of a Vancouver summer. More than 140 scientific papers were presented, ranging from the mind-bogglingly arcane ("Fog Chemical Climatology over the Po Valley Basin") through the poetic ("Fog from Space") to the intensely practical ("Fog Water Collection for Agricultural Uses in the Darjeeling–Himalaya, India"; "Design, Construction, and Operation of a System of a Fog Water Collector"; "Evaluation of Fog-Harvesting Potential in Namibia"; "Fog Collection as a Water Source for Small Rural Communities in Chiapas, Mexico"; and the Grenoble paper, which was called "Dew Recovery: Old Dreams and Actual Results"). Since fog, as one delegate put it, "is simply a cloud on the ground," harvesting drinking water from fog was a major theme of the conference. The consensus? A collector can generate up to ten litres of water per metre of fog-collecting mesh per day.

Even in arid regions, the yields are impressive. A pilot project in the Soutpansberg Mountains of northern South Africa yielded 1.13 litres of water per square metre of mesh per day, and the test project produced enough fresh water to supply the needs of a local school.

In the Namib Desert, one of the oldest on earth, there is a dazzling array of creatures with idiosyncratic adaptations to the extreme heat and dryness: a lizard that hops from foot to foot to diffuse heat absorption; a beetle that curls into a ball to roll down dunes to conserve energy; a spider that spins a small cone-shaped web to track and condense dew; and a fog-collector beetle. This little beetle stands on its forefeet, with its back to the fog, to condense it directly into runnels that lead to its mouth. Another beetle excavates tiny furrows to direct condensation to where it is needed. Namib geckos, for their part, have learned to lick their own eyeballs, on which water has condensed. There is even a plant that exudes salt that extracts moisture from the air and feeds it back to the plant itself.

Scientists at the Gobabeb Research Station, appropriately situated between Namibia's sand and rock deserts, have spent years studying and mimicking these ingenious inventions, refining tiny emitters that carry droplets of water directly to the roots of plants. They are also working with mesh "fog traps," whose yield is already between a litre and six litres of water a day (depending on altitude and distance from the sea), and have developed solar distillers to sweeten saline waters.

Israel, which is mostly arid (with an annual rainfall of less than two hundred millimetres), has always been in the forefront of water-conservation techniques. The Israelis have been famously efficient at deploying drip irrigation, low-pressure spray irrigation, wind-trap funnels for moisture control, cloud-seeding, and the rest of the panoply of arid-region remedies. The Jacob Blaustein Laboratory has invented a hothouse cultivation technique – closed-cycle hydroculture, in which evaporation is recycled – that wastes only thimblefuls of water. Drip irrigation has always been associated with wealthy farmers and high-value crops, because it has been capital-intensive – farmers need several kilometres of hose and hundreds of emitters per hectare of land. Typical users were the makers of high-end wines in California, who have learned to control the delivery of nutrient-enriched water to the nearest teaspoonful, and have even managed to use controlled water stress, which forces the plant into desirable growth habits by modest amounts of water deprivation at the right time, in order to concentrate the juices. But "a new spectrum of drip systems keyed to different income levels and farm sizes, beginning with a $5 bucket kit for home gardens, now exists and can form the backbone of a second green revolution, this one aimed specifically at poor farmers in sub-Saharan Africa, Asia and Latin America."[10]

These systems fit the expectations of poor farmers whose plots are, typically, tiny, often a hectare or less: they are affordable, they give rapid payback (increasing yields two- or even threefold in a short time span), they are divisible into small units, and they are parsimonious in water use, essential where water is in short supply. India and Nepal are already using these systems, and, on the fringes of the Sinai and other deserts, Bedouin, formerly nomads, are employing drip irrigation on oasis crops. A test plot in India increased yields from 500 kilograms per hectare for conventional methods to 670 kilograms (a 34-per-cent yield gain), using 55 per cent less water. A test by the Swiss aid agency Swiss Development Co-operation, in the Indian state of Maharashtra in 1999, found average water savings of 55 per cent, labour savings of 58 per cent, and reduction of expenditures on fertilizers and pesticides averaging 16 per cent. "With the drip systems, some farmers obtained cash profits for the first time, while others saw substantial increases – 50 per cent to more than 300 per cent – in their net profits."[11]

Sandra Postel and her colleagues, with World Bank backing, are proposing a major new international initiative to spread low-cost drip irrigation through private micro-enterprises to a million new hectares, with the aim of reducing the hunger and increasing the incomes of 150 million of the world's poorest rural people over the next fifteen years. Their estimates are that such an initiative could boost annual net income by $3 billion and inject three times that amount into the poorest of the developing world's economies.

Low-pressure spraying, which can be used for feed-grain crops, consumes 30 per cent less water than conventional high spraying, and 60 per cent less than conventional furrow irrigation. Laser levelling of fields has reduced irrigation water consumption considerably – in parts of California's Central Valley by as much as one-third. In the Sacramento Valley, rice farmers have done even better, with protective screens on river diversions and the development of less-water-intensive crop varieties. (They point out, rather smugly, that rice fields also serve as seasonal wetlands for migrating waterfowl.)[12] It's not just farmers that are having successes: American industry, which began diligently recycling its water as prices went up, has seen its consumption drop by almost 35 per cent since 1950, according to a survey released by the United States Geological Survey in 1998. Overall American water use dropped by 9 per cent between 1980 and 1995, according to the same study.

In August 1995 the Israelis,[13] bureaucratically incarnate in the Interim Secretariat of the Convention for Combating Desertification (INCD), got together with their Jordanian and Palestinian counterparts and came up with a fifteen-year plan to conserve water and combat desertification, budgeted at around $400 million. Their approach, flexible and inventive, was to turn the ecological disadvantages of drylands to economic advantage. Among other things, they analyzed the development of closed irrigation systems; the integration of flood-dependent production systems that build on, rather than try to suppress, flood pulses; the

introduction of water-treatment and solar-energy systems that
exploit radiation and heat from the sun; the cultivation of desert
crops; forestation and better management of grazing lands; the
development of aquaculture to exploit both sun and saline water
for the cultivation of fish, seafood, and algae; and the promotion
of ecotourism, based on bird migration and the rehabilitation of
endangered desert mammals. There was even a breeding program
for camels and a camel-study course, which among other things
would use camels for transportation, draft, and tourism ("Camel
races are popular in many Arab societies . . .").

Central to the proposal was the construction of an International
Centre for Combating Desertification in the Negev Desert in
Israel, a home for scientists and students from, at least initially,
Jordan, Palestine, and Israel, "though the participation of associ-
ates from other Mediterranean countries will be promoted and
given high priority." This institute would study desert crops,
finding better arid-land and salt-resistant crops, as well as markets
for them. It would intensify Israel's already-steady commitment to
closed-systems agriculture – the development of greenhouse technologies for arid conditions.

There are aquifers containing substantial amounts of water of
varying salinity levels underneath much of the Middle East and
North Africa, from Morocco in the west to Saudi Arabia in the
east. In the Negev, salinity levels range between three thousand
and six thousand parts per million of total dissolved solids, though
learning how to use saline water for irrigation is a challenge. Major
crops that are today irrigated with saline water include cotton,
wheat, corn, table tomatoes, and melons. A number of secondary
crops are grown with saline water as well. Research currently
focuses on Bermuda grass, potatoes, grapes, and olives. The institute would breed and select crops for salt tolerance.

To conserve water, there would be studies of "savannization,"
the sparse planting of trees, currently considered an effective way

to prevent land degradation in semi-arid ranges, as well as the more traditional forestation techniques. There was even an art course: Desert Landscaping and Gardening with Drought and Salt-Resistant Plants, with a sub-course, Dune Stabilization.

~

In neighbouring Jordan, a Canadian development agency is helping fund a study in the arid Muwaggar region to help Jordan capture and make the most efficient use of rainwater for food and feed production. The project focuses on rangeland in marginal arid regions where traditional farming is impossible and conventional irrigation methods are prohibitively expensive. The goals are to improve rangeland, enhance the cultivation of cereal crops, and combat desertification, which is rapidly accelerating, owing to the impact of human activities such as overgrazing, improper ploughing, and the careless movement of motor vehicles. The project was conceived by Mohammad Shatanawi, director of the Water and Environment Research and Study Centre at the University of Jordan.

The Muwaggar area was chosen because groundwater is deep and scarce, rainfall is unpredictable, and flash floods are common. Michel Rahbeh, a hydrologist and research assistant, says the soil surface is so hard that most of the water runs off when it rains, yet the local soils are highly fragile and subject to desertification.

The traditional way to harvest water in Jordan is to collect and store it in cisterns for human and animal consumption. "The aim of this project is to engineer these old practices on a larger scale for agricultural production," Professor Shatanawi says. "This is about applying practical know-how to improve the utilization of natural resources." The project's sponsors are trying to "copy and rediscover the wisdom that has been accumulated in semi-arid areas over millennia and to adapt it carefully to the current situation with modern tools. We must be very careful here, as it is easy to do more harm than good in a sensitive environment."[14]

On the outskirts of the central South African town of Bloemfontein, when I was growing up some forty years ago, there was a "sewage farm," evaporation ponds for the town's waste water combined with irrigation channels to nearby cornfields. As schoolchildren, we were dutifully bused out to see how it worked, part of a civics course that took us also to the municipal water works. I can't say that it did us much good: we were too squeamish to look too closely, and the boys spent most of their time during the visit threatening to push each other in. In later years the "farm" was abolished. The city spread out beyond it, and the population increase, combined with more-modern methods of chemical treatment in underground silos, meant the end of the older system.

This change was fairly typical of what was happening elsewhere in the world. The irrigation of crops with waste water, even treated waste water, fell out of favour. There were concerns about the spread of disease from consuming products irrigated with raw water from sewage systems. The "sewage farms" came to be seen as unsightly and unsanitary.

It took Israeli ingenuity, pushed by their water shortages, to bring them back into favour. In the mid-1990s, Israel was already reusing 70 per cent of its sewer water, safely irrigating twenty thousand hectares of land. By century's end, reclaimed waste water will supply more than 16 per cent of Israel's total water needs.[15]

This is what Sandra Postel calls the "elegant solution" of the agro-sanitary approach – partial treatment in waste-water lagoons and reservoirs, followed by irrigation, to solve simultaneously problems of pollution, water scarcity, and crop production in dry lands.[16]

(Nevertheless, it's fair to say that the Israelis, who have been getting A-pluses so far in this chapter, have had their failures, too.

A water-saving drive by the Infrastructure ministry, launched to great fanfare in 2001, was a flop. Part of the drive was the suspension of daytime watering in public parks in the summer months, and an ad campaign urging householders to cut water use. "We expected savings of 15 percent of all consumer use, which is about 700 million cubic metres a year," remarked Dr. Yosef Dreizin, the Water Commission's director of planning. "But instead of 100 million cubic metres, we got only 50."[17]

In the American West as well, waste-water reuse is coming back into favour. Tucson, Phoenix, and Los Angeles all recycle portions of their waste water. St. Petersburg, Florida, has closed the cycle altogether, reusing all its waste water and discharging nothing to the rivers or ocean. The waste water not only irrigates city parks and residential lawns, but has drastically cut the need to buy fertilizer. In El Paso, Texas, city engineers are pumping treated waste water into a deep aquifer, "mining" it several years later and several kilometres "downstream," and using it to water parks and public spaces. As Postel puts it, "a major push by development agencies, governments at all levels, and private engineers to combine low-cost sewage treatment with irrigation could go a long way towards solving the vexing triad of pollution, scarcity and health problems now plaguing so much of the world."[18]

Perhaps the most interesting technological intervention has been in Namibia itself; Namibian determination and German technology have turned Windhoek into the only city on the planet to fully recycle all of its waste water, including sewage, into drinking water, a "source" of roughly 3.5 million cubic metres a year – almost 20 per cent of the city's total consumption. And yes, it is safe. This is perhaps the most intensively tested and monitored water in he world.

Nevertheless, safety *is* an issue, and careless recycling has its hazards. In the United States, sewage sludge, even after treatment, has on occasion been found to contain pathogens and toxic chemicals, including PCBs, DDT, dioxins, and salmonella, and even lead, mercury, polio and hepatitis viruses, parasitic worms, asbestos, and radioactive waste. Some industry spokespeople have voiced concerns that the "beneficial uses" of sludge (which means using it as fertilizer) could introduce viruses into the food chain. Still, faced with few options for disposing of urban wastes, large cities are selling their "beneficial" sludge at a loss. New York City, for example, has been paid more than $100 million to "fertilize" a ranch in Sierra Blanca, Texas, with 400 tons a day of possibly toxic waste.[19]

Technology, or human inventiveness, is one answer to the challenge of conservation. But on its own, technology is not enough. Much of the current water wastefulness is political in its origins. Pricing and public policy – economics, broadly defined – are critical to any solution.

In the early 1990s, a UNICEF report, roundly condemned in the United States as naive, interventionist, and somehow "collectivist" (the current euphemism for communism now that communism has gone), declared that, if the "unserved poor" were the focus of water and sanitation schemes, 80 per cent of the world's population would have had safe drinking water by the year 2000. The report's point, largely ignored in the debate that followed, was that almost all the rich countries' aid money (somewhere around 80 per cent) was still going to wasteful mega-projects. This was not supposed to be happening. Virtually all donor countries – prodded no doubt by a reformed World Bank under Jim Wolfensohn – had promised to shift investments away from high-cost mega-projects, which seldom seemed to work in developing countries, to low-cost "village technologies," which almost always had an immediate effect. Nevertheless, the NGO WaterAid produced a report by researcher Maggie Black that found some interesting exceptions to the mega-project bias.

In Tegucigalpa, Honduras, a program based on the creation of barrio water boards to install and run slum services has reduced the community's water expenditures from 40 per cent to 4 per cent of household income.

In Karachi, Pakistan, the installation of what its organizers called a "donkey system, not a racehorse system" of sewerage brought sanitary sewers to more than six hundred thousand slum dwellers who had been written off by the politicians. It worked by piping household toilets and drains underground into "laneway" holding tanks, which were subsequently pumped out at regular intervals. Not your top-of-the-line developed-country system by any means, but far better than what had been before, which was essentially nothing. And it was installed for a mere $100 per household, instead of the $1,000 a full-service system would have cost. The tanks and sewers were paid for and managed by groups of households organized by "lane communities." Akhter Hameed Khan, the community organizer who got the project underway, has said that "the most important step was to liberate the people from the demotivating myths of government promises."[20]

The "pour-flush" toilet, flushed manually with two litres of water, is being installed in dozens of Asian cities, saving somewhere between six and eighteen litres per flush. UNESCO has developed a "VIP latrine," a hygienic version of the old pit latrine, that is solidly constructed, odourless, and easy to clean.

A new South African company has invented a "household sewage treatment" process. A former Communist Party functionary and architect, John Bizzell, has teamed up with a South African science group to develop a small plant that takes the effluent from a standard septic tank and converts it to water that complies with the Department of Water Affairs and Forestry (DWAF) Special Standard for non-potable water, among the highest in the world.

Bizzell is enthusiastic about its properties: the water can then be discharged to ground or recycled for flushing and irrigation. In

South Africa, in a regular suburban house, 45 per cent of water is used for flushing and a further 20 per cent for irrigation of the garden. This means big savings in the cost of water to a homeowner (water is currently expensive in South Africa, due to cross-subsidization to provide free water for the poor) and a considerable saving in the precious commodity itself.

"In a nutshell, the works include a thousand-litre tank filled with a media of three-inch-diameter and one-inch-thick perforated PVC discs. These maximize surface area, so that in a thousand-litre tank there is almost as much surface area as a football field. Biomass builds up on this surface area to a thickness of three to five millimetres. The effluent is introduced at the bottom of the tank and mixed with millions of tiny air bubbles. Any suspended solids cling to the surface tension of the bubbles which zigzag to and fro across the tank as they slowly rise through the biomass, which devours the suspended solid particles as they pass. The effluent eventually spills out as a clear liquid at the top into a disinfecting chamber, where it is treated by ozone to kill *E. coli* and other pathogens. The resultant product is theoretically potable but would not taste like tap water. It is then recirculated through the flushing system or discharged to the garden for irrigation and evapotranspiration."

In South Africa, single-house units retail for about R10,000, or less than $1,000. The running costs are minimal – equivalent of a forty-watt bulb burning continuously. The beauty of the system is that, being modular, it can be indefinitely added to in order to serve a street, a neighbourhood, or a township – or a highrise tower. It could be especially effective in remote places whose municipalities cannot afford to put in traditional trunk sewers, rising mains, pumping stations, and sewage plants.

Bizzell admits that efficacy dips with reduced temperatures; nevertheless, the system has been tested in the Drakensberg mountains of KwaZulu–Natal province, where the thermometer regularly dips to minus-five Celsius. But since the tank is hermetically sealed, with neither gas effluents nor odours, it could even be installed inside a basement in North American or European latitudes.

The same Dutch company that has produced the "Redosal" portable desalinator, Aqua Vita Water Systems, has also come up with a production model, using membrane-electrolysis technology, that can purify up to two hundred thousand litres of contaminated water an hour.

In 2002, the company donated a working model, capable of producing over five thousand litres of clean water per hour, to the NGO "Flying Doctors." The system was designed to provide medical workers with a mobile water-purification system to produce potable water generated from any available water in any remote area of the world, independent of stationary power sources. The system weighs only 126 kilos, and is easily portable. It is capable of cleaning and disinfecting water from rivers, lakes, sewers, or other waste-water systems at a cost of around five cents per ten thousand litres.

And then there is Bowie Keefer . . . You can't, after all, get through a water book without ever quoting Samuel Taylor Coleridge and his "Rime of the Ancient Mariner," one of the more famous and most mangled quotations in the water dictionary: "Water, water, everywhere, nor any drop to drink." The old fellow, a gloomy traveller who was also fond of metaphors involving storm petrels and that famous harbinger of doom, the albatross, was lamenting the notion that, adrift as he was on a stricken ship, he could see nothing but water from horizon to horizon, but could never drink any of it. The literature of shipwreck has often used the device of water-deprivation to melodramatic effect, and it was also a staple of early Hollywood, when there were many scenes in which water-fat actors lolled about gasping for something to drink, and threatening cannibalism on their fellow travellers.

Bowie Keefer is one of the people who has made all this obsolete: he developed a device that enables the stranded mariner to, indeed, drink sea water without harm.

Keefer was formerly a research fellow with the British Columbia Research Council, and subsequently spent some time with the Canadian Forces in the Sinai, where he became interested in desalination of water. In the 1980s, he was experimenting with variants of filtration techniques, working with reverse osmosis. He came up with a way to recover some of the energy stored in the residual brine from a first filtration, to further clarify water. His smallest device – small enough to fit into a pocket – uses human energy to produce about a litre of fresh water per hour, more than enough to fend off the Ancient Mariner's gloom. The next size up can produce around five litres an hour if operated manually, or 150 litres a day driven by twelve-volt power, enough to keep a decent-sized yacht comfortable.

Jim Cran, who had worked with Keefer earlier, is one of his admirers. "With this thing, a guy in a life raft can dip a pipe into the sea, squeeze the handle of the device, and after fifteen minutes or so get a glass of fresh, potable water." Keefer knows his device has saved lives. A few years ago an elderly couple, set adrift in a life raft when their yacht was holed by a rambunctious pod of whales, was kept alive for fifty-five days by water produced with a Keefer device.

Keefer bought the rights to the device from the B.C. Research Council, but has subsequently sold it to a Minnesota pump company, Recovery Engineering (now Katadyn). "They're doing well with it, I hear," he said wryly.

Of course, it's not just stranded yachtsmen who stand to benefit, nor even American sailors, whose lifeboats are now equipped with the Keefer-inspired devices. Small versions, powered by human muscle, could prevent disease in a thousand different situations, from the Kenyan woman dipping water from a slimy pond to a Saharan nomad faced with a remote well that has succumbed to salination or pollution. It will, after all, clean as well as desalinate water.

And so on and so on . . . Out of sight, invisible to the macro-economists and off-books to the World Bank and the International Monetary Fund, initiatives are being undertaken and small triumphs gained. Generally, somewhere in the background is an NGO presence, prodding, advising, helping, and, in many cases, pitching in to work. Just up the Great Rift from the now-vanished cultivators of Tanzania, the local women have reinvented contour irrigation runnels, helped this time by a Swedish NGO representative with a simple hand-held levelling device made of rubber tubing and water. The Dogon of Mali are using stone terraces and dug "water sinks" to hold the water that, in the rainy season, used to run off across the hard-pan soil and vanish. Runoff agriculture, rainwater harvesting, the use of tough, imported grasses as erosion controls – all these features are improving food yields and preserving water supplies in many places around the world. Small-scale dams, solar pumps with few moving parts, easy-to-use technologies like bicycle-driven electric generators all help populations survive in water-marginal places. So do micro-loans to village co-operatives – the old hippie cliché of empowering the people. The prime cliché of the NGO movement seems to work: Don't give 'em food – teach 'em to farm; don't give 'em a tractor, but an understanding of farming; learn from and adapt traditional techniques: *qanat*s in Morocco, cisterns in Jordan, contour furrowing in Kenya and Tanzania.

Not so long ago I met an NGO representative in the Dogon village of Sangha, on the lip of the Bandiagara Cliffs. He was an American, from Cleveland, and several years back he had helped install solar-powered pumps in this and other villages. "We learn from our mistakes," he told me. "The first one, we just put in. A year later it had been wrecked. Then we began to understand the power of the sun in local lore, and we explained the technology in a way compatible with the spirituality of the local people. Then they installed the next one themselves." He pointed. "There it is. Meticulously maintained. These people are poor and technologically ignorant. But our biggest mistake was to assume we knew better than they did."

The next morning, just before I set off with Guiré, the Dogon who was showing me the cliff villages of his people, the NGO guy came around to see us off. "Here's the real thing I learned," he said. "We have to teach our engineers and our politicians ecology. They have to understand the completeness of cycles. Engineers can be like children. They'll sometimes wreck things by building things, without thinking clearly enough. They make things work in a marvellous way. But an engineer's vision is narrow. They don't really understand what they're doing. So of course they do harm."

As we walked along the escarpment, peering down at the tiny dwellings clinging to the stone cliffs, I asked Guiré what he thought of the American. He just laughed. "By ecology he is talking about the world as it is," he said. "And by engineers he means people who, by solving problems, create other problems." He laughed again. "He is a good person. But sometimes the best thing to do is nothing at all."

~

"The key to water is market economics," Maurice Strong was saying. "Too-cheap water subsidizes inefficiency. And there are too many perverse subsidies: public funds are being used for anti-public purposes." Strong, the secretary-general of the Rio Conference on the Environment in 1991, a board member of Serageldin's World Water Commission, and an éminence grise in dozens of earth-friendly environmental groups (including his own Earth Council), was talking about a study the council had commissioned called Subsidizing Unsustainable Development.

I told him of Sandra Postel's notion that conservation was humankind's last oasis, that the easiest way to find water was to stop losing it. Inefficient water-delivery systems were hugely wasteful of water. In the Philippines and Thailand, more than half the "available" water simply disappears through leaks and rotting infrastructure, not to mention theft. Even in Jordan and Yemen, the annual water supply could be increased by one-quarter if the disappearing water was made to reappear.

"Yes," he said, "we have to bring supply and consumption into balance. And the most efficient way of doing that is pricing. Most farmers still use the old irrigation furrow method of watering their crops, although it is the least-efficient way of doing it. Surge-irrigation techniques, laser levelling of fields, all those systems can reduce water consumption dramatically. They are expensive. There is no incentive to introduce them unless water becomes expensive too."

Even in Israel, I remembered, this was a fact of economic life. Israeli farmers still receive substantial water subsidies, which have encouraged profligate water consumption and the growing of export crops with a huge water appetite, such as cut flowers and tropical fruit. And some studies have shown irrigation to be an ambiguous benefit. In some regions, dry-land farming (rainfall-dependent cropping) might have yielded equal tonnages; and it has been suggested by some Israeli agronomists that perhaps the Blooming of the Negev should never have been attempted. A recent increase of about 12 per cent in the price of water caused a 10-per-cent drop in consumption, and a consequent drop in agricultural production. This decline seems to have had zero effect on the Israeli GDP; the freed-up water was gobbled up by industry, which has a greater economic multiplier than farming.

In Australia's Murray–Darling River basin, planners shifted from a command-and-control approach to water (useful in times of abundance) to a market-oriented property-rights system that allowed "owners" of water to trade their rights with users prepared to pay a negotiated price. To no one's surprise, the new approach brought about a more rational and equitable allocation of the resource.

"In the Ogallala area," I said, "they've been using new low-energy sprinklers. Water use is down in the High Plains of Texas by 44 per cent. Some of this decline is by taking land out of production altogether. But not all of it."

Strong nodded. He knew all that – after all, he owned, and still owns, part of a Colorado aquifer himself, and was once asked to chair the Western Resource Council's water committee, though not an American citizen. He rummaged in his briefcase and pulled out a sheaf of papers, clippings, and notes. "It's the same as with energy," he said. "If gasoline was priced to take account of its real cost, including the cost of remedying pollution, it would lead immediately to reduced consumption. So with water. What, for instance, about a water-depletion tax? Why should the taxpayers subsidize the unsustainable withdrawal of groundwater?" He rummaged some more. "Ah, here it is," he said. " 'Subsidizing Unsustainable Development.' Done for the Earth Council."

I took the paper home. Water was Chapter 2 of the study. There were the usual graphs of water-scarce countries and freshwater withdrawals, the usual rundown on humankind's profligacy, without which no environmental paper would be complete. "It is time for a reality check," the authors say.

> Aquifers are being drained, rivers are drying up, more than a billion people don't have access to safe water, and vast tracts of irrigated land are being lost to salinity; over all, water is being lost in flood proportions and used inefficiently and for low-value purposes. And what are governments doing? Subsidizing this ecological vandalism, natural resource waste and economic perversity by selling water well below actual supply cost, much less [than] market value. The message of a subsidy is clear: don't worry about conservation or higher efficiency or recycling.[21]

The paper then runs through the litany of abuses I had already become familiar with through Marc Reisner, Sandra Postel, and Peter Gleick: U.S. subsidies to agricultural irrigators amounted to nearly $500 per acre, sometimes on marginal land worth much less than that (the total annual subsidy is probably greater than $2 billion); three-quarters of the benefits of the Animas La Plata Dam in Colorado go to farmers, who are paying only 3 per cent of the cost; 70 per cent of the farmers' profits in California's Central Valley – which is supposed to be the richest farmland in the world – came directly through taxpayer subsidization. And more numbers: in Australia, water charges are largely nominal in the Murray–Darling River basin, the major agricultural region; the foregone budget revenue from illegal connections to water mains in developing countries is around $5 billion; budgetary savings from fixing leaking pipes would be around $4 billion a year – subsidies and potential savings in the drinking-water sector alone would be around $44 billion; cost recovery from farmers in developing countries is no better than 10 to 20 per cent; "farmers in perpetually parched Tunisia pay no more than a seventh of the cost of the irrigation water they receive . . ." The figures were familiar, and familiarly depressing.

The "real" costs of removing subsidies have been the subject of often fierce debate, with estimates ranging from 0 to 10 per cent of GNP. In 1997, more than two thousand economists, among them six Nobel Prize winners, signed a declaration that "there are many potential policies to reduce greenhouse gas emissions for which the total benefits outweigh the total costs." They urged national governments to introduce policies that would be worth implementing whether or not global warming was a threat – the "no regrets" policies. Another suggestion was to extend the notion of internationally tradeable budgets for emissions. Each country would be "allowed" to produce so much greenhouse gas. If it "wanted" to produce more, it could buy the rights from countries that produce less. Strong, when he was chairman of the now-defunct electric utility Ontario Hydro, instituted an exchange of this sort with Costa Rica. He "bought" the rights to a section of Costa Rica's rain forest (thus preventing it from being harvested), which was supposed to offset emissions from his utility. An emissions exchange has already been set up in Chicago.[22]

What would happen if subsidies were removed? Wouldn't farm prices go up? Isn't the reason for the subsidies cheaper food prices in the first place? Wouldn't many farmers go broke? Wouldn't the poor be penalized most – the poor can't pay high prices for anything, and if water were treated as an economic good, priced to regulate demand, wouldn't the poor have to go without?

The answers given by André de Moor and Peter Calamai in their Earth Council report tally with theoretical studies elsewhere. No government has yet had the guts to remove water subsidies altogether, but where it has been done regionally or locally, there has been considerable price elasticity – a 10-per-cent price increase typically yields a drop in demand of 15 to 20 per cent. Metropolitan Boston saw demand drop by 30 per cent when it increased prices and mandated water-saving plumbing fixtures. In Canada, where taxpayers typically subsidize water-delivery systems to the tune of about $3 billion a year, residential water use under a flat-rate system averages 450 litres per person per day, and a mere 270 in communities with a full user-pay system. A study in Egypt had already shown that a forced drop in water use would affect the GDP only marginally. A similar study in Morocco, where 92 per cent of water is used by farmers, with only a 10-per-cent recovery, showed that dropping most subsidies would cut water use by one-third – and affect GDP by only 1 per cent. And the researchers predicted that the drop could be transformed into a surplus with more trade liberalization and by factoring in the ecological gains made by using less water. In reducing soil salination and water-logging, more land would be kept in production for longer.

In fact, de Moor and Calamai say, the notion that the urban poor would be damaged most by an end to subsidies turns out to be incorrect. In fact, subsidies mostly benefit the rich, since they are most likely to be connected to a public supply in the first place. Outside the developed world, the urban poor often have to rely on small private water-sellers, and typically pay between $2 and $3 per cubic metre, twelve times higher than the price of public city water. The same is true even of farmers. It is the wealthier farmers who tend to occupy the best land and have access to wells and pumps. The most extravagant example of welfare for the wealthy is in the Central Valley in California, where millionaire owners of agribusinesses are subsidized by taxpayers all across the nation (while at the same time they rail against "big government" and "government interference").

The answer, the report suggests, is demand management instead of the supply-oriented system now in use. Don't charge according to what people can pay, but according to what the water is worth, taking into full account the development costs of delivery systems. Water has an economic value in all its competing uses, and sound water pricing will achieve more sustainable patterns of water use and generate the new resources necessary to expand services. For farmers, too, prices should be adjusted to encourage sustainable practices. If you stopped subsidizing water to plant water-thirsty crops alien to deserts or to irrigate pasture to raise cows for beef (at a 50,000 ratio – one kilo of beef from 50,000 kilos of water), if you stopped doing the wasteful things for which the American West is famous, you would have water to spare. If you want the same notion in the jargon of economists, treating water as an economic good should primarily serve the purpose of financial sustainability through cost-recovery.

Technology has its role to play here, too. Demand can be reduced through improved technology, in the cities through such devices as low-flow faucets and toilets, and in agriculture through better surveillance and control devices.

The newly privatized water companies in Western Europe, particularly in France and Britain, are forcing through some of these changes. They have, after all, to make a profit for their share-holders. Privatization has not been without cost: there have been outraged stories in the British tabloids about "fat-cat water pluto-crats" paying themselves multi-million-pound salaries while threatening to disconnect the urban poor who cannot pay the newly inflated rates. After privatization, the money consumers paid for water went up 106 per cent, while corporate profits increased some 690 per cent, and there was considerable evidence that the companies were not investing sufficiently in an aging infrastructure, with the possibility that areas of Britain would be left without sufficient water in a prolonged drought. But even from a social-democrat point of view, the increased price of water is the right change (though not the fat-cat salaries or the profits). It is important to signal water's value, and to price it so that waste hurts. The political trick is to balance conservation incentives against the irreducible needs of the urban poor.

"The point," says Strong, "is that water is one of those issues that no one country can solve in isolation. They have to be solved transnationally. We have to build institutions that transcend national governments."

But what about the privatization initiatives? Very few issues in the water world draw more passionate opposition that the massively expanding private water behemoths. The anti-corporatist lobby groups are vociferous in their opposition to turning any water-delivery or purification systems over to private enterprise. In the aftermath of high-profile water problems, such as the tragedy in Walkerton, Ontario (in which *E. coli* seeped into the municipal water supply and killed a handful of citizens), the muted grumbling turned into an uproar. Bafflingly, however, the main argument seemed to be that, since water was part of the global commons, it should therefore not "belong" to anyone, which rather missed the point that it was the delivery of the commodity, not the commodity itself, that was at issue. The other argument revolved around pricing, and there were dozens of stories circulating that private water companies had boosted prices beyond the reach of the poor, and had even cut off some of the truly indigent altogether. Most of these stories proved, on examination, to be false. (In Buenos Aires, for example, the French water conglomerate Suez has modernized treatment plants that were on the verge of collapse, and has been highly profitable – despite lower prices to consumers.) In this all-or-nothing debate, the notion that public authorities could suitably regulate private companies seems hardly to have emerged.

According to a *New York Times* report early in 2002, "the case for privatization germinated decades ago after the World Bank unsuccessfully tried to fix the public water supply system in Manila. Despite five repair attempts over the years, water loss was as high as 64 per cent. Fundamentally we realized that without a change in incentives – some very logical, sensible things – this was not working, a World Bank spokesman said."

By 2002, private water management was a $200-billion business worldwide, and ran water systems for nearly 10 per cent of the world's populations, prompting *Fortune* to call it "one of the world's great business opportunities." Hans Peter Portner, a fund manager at Banque Pictet in Geneva, who handles the bank's Global Water Fund, predicts that privatized water systems will expand to serve about 17 per cent of the world's population by 2015.

But the opposition to privatization is so vociferous, and the issue so sensitive, that even the water behemoths themselves are taking on a Green tinge, and begin to sound like anti-market activists. In August 2002, Gérard Mestrallet, the head of Suez, was quoted as saying, "Water is too essential to life to be a commodity.

Public-private partnerships are appropriate, but it is absolutely irresponsible to privatize in developing countries."

Much of the opposition to privatization has been misguided
and misinformed. But not all of it. A few cases in point:

- The city of Atlanta, Georgia, to take an affluent-world
  example, had contracted a few years ago to turn its water management over to United Water, a subsidiary of the French
  conglomerate Suez, itself part of Lyonnaise des Eaux. But in
  May 2002, the state's Environmental Protection Agency was
  forced to issue boil-water orders, even to the affluent North
  Buckhead neighbourhood; the tap water wasn't safe – it was a
  rusty colour and contained bits of unnamed debris.
- Newly privatized water in the Empangeni region of South
  Africa raised prices so high that thousands of local residents
  were forced to dip their buckets into the local rivers. They
  couldn't afford the metered water, and drank the polluted river
  water instead. The result – a sudden spike in cholera cases, first
  to ten thousand, and then to five times that number.
- In San Isidro de Lules, Argentina, Vivendi International was
  forced to abandon its water-management contract after furious
  protests followed a rapid doubling in the prices of water to
  householders. To the local residents, the raises had been foreseeable – Vivendi had bid so intensively for the contract that it
  had virtually forced itself into negative cash flow, and would not
  have survived without raising prices. (Vivendi, of course, had its
  own view of this affair. In response to a critic's remark that
  Vivendi had failed to recognize people's emotional attachment
  to water, which was after all "a gift from god," Olivier
  Barbaroux, president of Vivendi's water business, quipped, that,
  "Yes, but [God] forgot to lay the pipes.")[23]

The real point about privatization, though, is not whether there
are problems, or whether for-profit corporations should control an

essential resource, or whether public scrutiny of private companies is feasible. The real point, in much of the world, has not been whether the rich would benefit from private water supplies but whether there would be any water deliveries at all: strapped Third World governments were in no position to repair their crumbling infrastructure or to extend it to the relentlessly expanding slums in the new megacities. I have already made the point that even studies by NGOs suspicious of private enterprise had found baseless the notion that ending water subsidies would boost prices beyond the reach of the poor. And then a World Water Forum in The Hague early in 2000 declared forthrightly – and correctly – that, without privatization, water matters would get steadily worse. It was, the meeting said, privatization or nothing, and the ideological debate was therefore pointless at this stage.

Water survival strategy 3: *By definition, water use will go down if there are fewer people. But is this likely to happen? Was Thomas Malthus, the doomsayer who forecast worldwide famines because populations were growing faster than the food supply, right after all?*

Some years ago, in the Kenyan town of Narok, I remember chatting to a storekeeper about his family. He had ten children "so far," five with one wife and five with another.

Had the first wife died?

"No, she was used up. So I put her aside and got another."

In 1997, Kenya had 6.7 live births per woman, and the population was growing at 3.3 per cent, the world's fastest, according to the Kenya National Museum in Nairobi. Without a change in the birth rate, even if the economy grows at a respectable rate, Kenya is likely to become poorer. Water poorer, too.

The museum has a permanent display on the population problem, which it calls the country's most serious – more serious by far even than AIDS. There are cases showing the various birth-control methods, and exhibits explaining the procreation process.

Busloads of eager schoolchildren, in their British-legacy school uniforms, tour through the exhibits every day. The museum regards this education task as far more important than the history and wildlife displays in other exhibition rooms.

I told the curator of the population exhibit what I had found in Johannesburg. There, the black population has historically been growing at a much faster rate than the white. Popular expectations were that it would continue to do so, and that population growth would continue to worry the politicians. But, to the experts' surprise, the black birth rate has fallen dramatically in the past few years, to a point where it is now just at replacement level.

"Yes," the curator said, "this is because all the girls are going to school. Teach the girls, and the problem is halfway to being solved. It can happen here too." She looked out at the chattering, eager schoolchildren. "Too bad," she said. "What they really come here to see is Ahmed." (Ahmed is the giant elephant that became a national institution in the 1970s; Jomo Kenyatta had even protected him through a special presidential decree, and had him "preserved" even after death.)

I looked outside to where Ahmed stood, stuffed and looking desiccated, trunk reaching for the sky, tusks the colour of old tea. She was wrong. The boys wanted to see Ahmed, true, but the girls were clustered around the birth-control cases. Their small black faces, round and eager, were rapt. Boys will be boys, but these girls would control the future. This is accepted as a truism all over Africa and the developing world: educate the girls, and the women will have fewer babies.

Population projections are a familiar source of gloomy prognostications, not only from Malthus but in recent times from such people as Paul Ehrlich, author of *The Population Bomb*. Given the startling expansion of human populations, particularly in developing countries, it seems easy to agree. Ehrlich himself has gone even further and has become an unashamed advocate of enforced birth control; and another Green doomsayer, Garrett Hardin, has been quoted as declaring that "the freedom to breed is intolerable," a pristine example of eco-fascist thinking.[24] But in 1996, United Nations population projections showed some unexpected changes. They forecast that, in 2050, the world population would be smaller by nearly half a billion than earlier forecasts had shown, a decrease attributed to what it called "significant reductions in birth rates throughout the world." The population curve was flattening out – without coercion. According to figures subsequently calculated by Population Action International (PAI), an NGO whose raison d'être was slowing world population growth, this decline had some interesting ecological consequences: it meant, PAI said, that "there will be between 400 million and 1.5 billion fewer people living in water-short countries in the year 2050 than previously projected." This was the good news. The bad news was still bad: "Even under this improved scenario, renewable freshwater scarcity will continue to remain a problem for millions of people around the world." PAI projects that, by 2050, the percentage of the world's population living in water-stressed countries will increase by anywhere from three- to fivefold.[25] A Consultative Group on International Agricultural Research (CGIAR) report in 1998 quoted Nobel laureate Norman Borlaug as believing that the "average yields of all major food crops must increase by 50 per cent by 2025 if food needs are to be met." The report also quoted the FAO as saying that two-thirds of the growth in agricultural production must come from lands already in production.[26]

Still, the lower birth rate means that certain countries – Sri Lanka or El Salvador, for instance – will delay their falling into "water stress" by at least a decade, ten years in which conservation measures could be imposed. In India, the change would be even more dramatic. Earlier population forecasts had shown India becoming water-stressed as early as 2015. "Under the new UN projections, however, it is conceivable that India will not cross the water-stress benchmark until 2035. . . . More important, if its total fertility rate were to fall in accordance with the low projection,

India's population could actually begin to decline towards the middle of next century, bringing the country back out of water stress within a decade."

China is a special case, as we have seen, mostly because China is so huge. But there are some short- and medium-term solutions even there, including water diversions from the Yangt'ze Basin, less water-intensive crops, and conservation, the last oasis – the adoption of local-emitter (drip) irrigation for water-intensive crops, more efficient sprinkler systems as opposed to furrow irrigation, and higher pricing for water in the cities. As China's cumbersome industrial facilities modernize, they will use less water: modern factories in the United States and Europe use less than one-third of the water for the same industrial output as Chinese factories do, and China's headlong modernization will incorporate new techniques. Chinese planners can separate industrial and household wastewater streams and use recycled household water for irrigation. They might opt for chemical or heat-treated sewage rather than employing water-borne methods. All this will buy time. But 1.5 billion relatively affluent Chinese will use a lot more water than a billion not-so-affluent Chinese. And that water is just not available.

Thus the Malthusians, the gloomy prognosticators of doom, are back for one more kick at the can.

Water survival strategy 4: *Steal water from others.*

In ancient Jewish, Christian, and Islamic traditions, the ultimate source of the waters of life lie beneath that politically potent piece of real estate called Jerusalem – a metaphor for the recognition that the solution to the problems of water is ultimately political. Who owns water? Who processes it? Who controls it? Who wants to steal it? Who can? These are the ultimate water questions.

In the present century, as water use presses inexorably up against water availability, the lamentable political fact has been recognized that some 47 per cent of the earth's area is made up of river basins shared by more than one country – more than 60 per cent in Africa and Latin America – and that the only accepted body of international law on the subject (the UN convention on the Law of the Non-Navigational Uses of International Watercourses, adopted in 1997) had still not been implemented as the calendar tipped into 1999.

There have been many efforts to codify international water law. The UN's offshoot, the Food and Agriculture Organization (FAO), published an index in 1978 that listed more than two thousand treaties and "international instruments" dealing in one way or another with international watercourses and aquifers, some of them dating back to the first or second centuries. Most of these agreements were bilateral, and dealt with shared boundaries or rivers that flowed from one country into another. Most of the disputes in water history (and almost all the cases discussed earlier in this book) are generated by the perceived notion of one state that action by another has caused some (often-unspecified) harm. It has been the task of the codifiers of law to try to cobble together a list of general principles from all these many arguments.

There are three prime movers among these codifiers: the Institut de droit international (IDI), the International Law Association (ILA), and the International Law Commission (ILC). The IDI is a non-official organization of self-perpetuating and self-defined internationally worthy jurists founded in 1873, which has gained some important acceptance: its conclusions have often been cited in diplomatic negotiations. The ILA, by a curious coincidence also founded in 1873 – it must have been catching – is another professional body, numbering around a thousand, drawn from a wide variety of countries; it, too, has no official backing, but its recommendations on international law are also taken seriously. The final body, the ILC, is a creature of the United Nations; its thirty-four members are elected directly by the General Assembly and work on projects mandated by the UN.

All three of these bodies have developed legal principles on water law, each with slightly different shades of meaning. In 1961, the Institut de Droit adopted what it called the Salzburg Resolution on the Use of International Non-Maritime Waters, and eighteen years later followed that with a further declaration, the 1979 Athens Resolution on the Pollution of Rivers and Lakes and International Law. The Salzburg Resolution set up the principle of "equitable utilization," by which it essentially meant that sovereign rights over international watercourses are limited by the "right of use" of other states sharing the same water. Disputes should be "settled on the basis of equity, taking into consideration the respective needs of the States, as well as any other circumstances relevant to a particular case." Their approach requires maximum flexibility, much discussion, and the setting up of bodies of experts to foster co-operation.

The ILA's Helsinki Rules on the Uses of the Water of International Rivers also supported the equitable-use notion. States are entitled to a "reasonable and equitable share in the beneficial uses of the waters of an international drainage basin." The commonly accepted notion of "first in time, first in right" and the "use it or lose it" principle so enshrined in the American West were called into question. The Helsinki Rules made it clear that an existing use may have to give way to a new use in order to come up with an equitable distribution. There was provision for compensation, but "first come, first served" would no longer apply. This notion hasn't gone down so well in, say, Egypt, which might under these rules have to modify or even abandon its millennia-old irrigation practices if, for instance, Ethiopia wants to build new hydroelectric projects on the Upper Nile. That Egypt would get compensation – unspecified – has been brushed aside by the Egyptians, who flatly refuse to acknowledge any such principle.

The ILC, for its part, said that, while equitable utilization was all very well, a state had no right to any use if such use harmed another riparian state. "A watercourse state may not justify a use that causes appreciable harm to another watercourse state on the ground that the use is 'equitable' in the absence of agreement between the states concerned." In Article 7 of the draft on the Law of the Non-Navigational Uses of International Watercourses, the commission was even more explicit: "Prima facie at least, utilization of an international watercourse is not equitable if it causes other states harm."

Stephen McCaffrey, professor of law at the University of the Pacific's McGeorge Law School in Sacramento, California (from whose essay "Water, Politics, and International Law" this quick survey was drawn, though responsibility for errors in my summary of his work is entirely my own),[27] is dubious about this whole approach. It may work when the harmed state is weaker than the other, he says, but, "in other cases, the ILC's solution is at least questionable. It would appear to encourage a 'race to the river' and to reward the winner. That is, a state that develops first would thereby establish an entitlement to the amount of water used or to the particular use made of the watercourse, and other riparian states would not be permitted to trench upon that amount or adversely affect that use without the permission of the former state." As McCaffrey points out, the ILC reversed its approach in 1994, and its final draft formed the basis of the UN convention. Thus, equitable utilization has become the underlying principle of whatever international law now exists.

The inherent difficulties in codifying practice into law hasn't stopped international efforts; virtually every water conference now includes a session on "Water and Law." As the December 2000 issue of *Water International* put it, lawyers and politicians are wrestling with four key principles:

- Legal entitlement (what is the scope of the resource and who is entitled to it?)
- Framework for allocation (where all needs cannot be met, who is entitled to what quantity or quality of the resource?)

- Institutional mechanisms, including governance issues (who is responsible for implementing, or overseeing the implementation of, the laws?)
- Compliance verification, dispute avoidance, and resolution (how are rights and obligations enforced?)

McCaffrey points out somewhat cynically that, since geography generally has favoured mature development of watercourses by downstream states (Iraq in relation to Turkey, Egypt in relation to Ethiopia and Sudan), the upstream states generally prefer "equitable use," since they can thereby lever some advantage for themselves. Downstream states generally favour an intolerance of "harm," since it allows them, generally, to go on doing what they have been doing anyway.

In transnational water disputes, which is the most dangerous? When the upstream nation is more powerful than the downstream, and therefore more cavalier about taking into account downstream needs? When the downstream nation is more powerful, in which case the upstream nation risks retaliation for any careless handling of the supply? Or when both countries are water-stressed and more or less equal in power?

The pessimists will say all three are dangerous. Egypt, a powerful downstream riparian, has several times threatened to go to war over Nile water; only the fact that both Sudan and Ethiopia have been wracked by civil war and are too poor to develop "their" water resources has so far prevented conflict. In the Euphrates basin, Turkey is militarily more potent than Syria, but that hasn't stopped the Syrians from threatening violence. And there are endless examples of powers that are similar in military might, but have threatened war: along the Mekong River, along the Paraná, and other places. In the Senegal Valley of West Africa, water shortages contributed to recent violent skirmishes between Mauritania and Senegal, complicated by the ethnic conflict between the black Africans and the paler-skinned Moors who control Mauritania.

There are those who think the threats of conflict overblown. The Canadian security analyst Thomas "Tad" Homer-Dixon, a name that pops up as a footnote in numberless academic papers, is one of the skeptics. Homer-Dixon's research found virtually no examples of state violence associated with renewable resources like fish, forests, or water, but many associated with non-renewables like oil or iron. He pooh-poohs the alarmist fears, though he acknowledges that water supplies are needed for all aspects of national activity, including the production and use of military power, and rich countries are as dependent on water as are poor countries. Moreover, about 40 per cent of the world's population lives in the 250 river basins shared by more than one country. But the story is more complicated than it first appears. Wars over river water between upstream and downstream neighbours are likely only in a narrow set of circumstances. The downstream country must be highly dependent on the water for its national well-being; the upstream country must be able to restrict the river's flow; there must be a history of antagonism between the two countries; and, most important, the downstream country must be militarily much stronger than the upstream country. He found only one case that fit all his criteria: Egypt and the Nile.

Not everyone agrees with this analysis, thinking it overly optimistic. Frederick Frey, a political scientist with the University of Pennsylvania, argues that water is different from other renewable resources such as fish or wood. "Water has four primary characteristics of political importance: extreme importance, scarcity, maldistribution, and being shared. These make internecine conflict over water more likely than similar conflicts over other resources.

Moreover, tendencies towards water conflicts are exacerbated by rampant population growth and water-wasteful economic development. A national and international 'power shortage,' in the sense of an inability to control these two trends, makes the problem even more alarming."[28]

There is another way of looking at the notion of water conflicts, which Homer-Dixon acknowledges and urges on the world's policy-makers. Water shortages may not lead to war, but they most certainly lead to food shortages, increased poverty, and to the spread of disease. They increase the migrations of peoples, further straining the massive megaslums of the developing world. Standards of living deteriorate, social unrest and violence increase, leading, as the doomsayer Robert Kaplan put it, to "the coming anarchy." Bangladesh may never go to war with India – even before the recent settlement, the Bangladeshis were too poor to do much more than grumble – but the stress caused by water shortages led to massive migrations of people, upsetting the ethnic balance of several Bangladeshi and Indian states, and leading to the rise of terrorist and nascent revolutionary movements.[29]

By other definitions, then – water wars.

~

The first question I asked when I started this book – Is water in crisis? – was simple to pose but elusive of answer, its meanings on the one hand hidden in polemic and propaganda, and on the other rooted in the difficulties of defining any ecology (with their infinite linkages, causes shading off into consequences, and sudden upwellings of pattern from apparent chaos). There are many people who seem to take an almost-pornographic interest in the imminence of ecological collapse; and many, by contrast, who stubbornly refuse to look it in the snout, though it be about to bite them. I listened to the tree huggers and the apostles of free-marketism (the romance of apocalypse on the one hand, a fatal narrowness of vision on the other). I listened to people comfortable with the vocabulary of crisis (Lester Brown of the Worldwatch Institute) and to those who eschew it (Peter Gleick and the ever-thoughtful Sandra Postel). But in the end, for me, the central figure in the water universe was a man I came to think of as almost outside ideology, the Russian Igor Shiklomanov.

This is no amateur hydrologist. Shiklomanov has been diligently collecting water information for decades. He was beavering away long before the ruble collapsed, long before Boris Yeltsin, long before Gorbachev and perestroika, far back into the stultifying days of Soviet bureaucracy, now dismissed by the code phrase "the period of stagnation." Shiklomanov is to water data what a pig is to truffles. His data banks of water, of snow and ice, are used and trusted by researchers all over the globe. He is careful, meticulous, scrupulous in pointing out shortcomings in his own work. I've heard him talk a dozen times, I've read a fair proportion of his formidable output, and I have yet to catch him in hyperbole or exaggeration. He never cries doom. He never uses the word *crisis*. He never speaks of precipices over which we might tumble – he hardly uses metaphors at all. He never scolds or harangues. He simply collects numbers, which he presents straightforwardly, in a genial, avuncular manner, without any embellishment. Because of this, I found him much more disturbing – even frightening – than all the many predictions of Apocalypse Soon. For what his columns of numbers and his charts and graphs show is the steadily narrowing gap between clean-water supply and water demand, between water for drinking and water for food and water for sanitation and water for industry. His figures show more clearly than any polemic that water is bound up with matters of money and management and politics, and that solutions to the looming crisis in fresh water (my phrase, not his) are only to be found in the collective political will (my phrase again, definitely not his – the word *collective* has many vibrations to a Russian ear, most of them bad). In Maurice Strong's phrase, "not world government, no, but a world system."

Still, whenever my reading on water threatened to overwhelm me, either with gloom (yet another river fatally polluted, another groundwater table irretrievably and inexcusably exhausted, another million or so hectares poisoned beyond repair, another screed on the dying oceans) or with tedium (yet another academic research paper rivetingly entitled something like "The Hydrological Characteristics of Schist Formations in the Permian Underlayers of the Modder River Catchment Area"), I thought of Eugene Stakhiv and his confidence in the ability of humans to invent themselves out of trouble. And then I thought about Feodor Zibold and his dew collector, and felt a little better. For what Zibold stands for, in all his fecklessness and stubbornness, is the enduring fact of human ingenuity. Out of nothing – out of folk-legend and entirely erroneous guesses – Zibold created something. Water wars might be caused by human folly, but they might still be prevented by human inventiveness. What Zibold – a curious stand-in for humankind – tells us is that we are not without weapons in these wars we are waging against our own worst nature.

CHAPTER NOTES

~~~~~~

I | WATER IN PERIL
Is the crisis looming, or has it already loomed?

This chapter draws in part from Leif Ohlsson's *Hydropolitics*, an excellent compilation of global water problems by a variety of experts, and from Peter Gleick. And one quotation is from Sandra Postel's *Last Oasis*, one of the indispensable books for anyone interested in global water issues. The on-the-ground reporting in Kenya and Niger is from personal observations. Selected academic papers used as sources include: "Techniques for Assessing Changing Water Resources in the Twenty-first Century," by A. Rango, USDA Hydrological Laboratory, Beltsville, Md.; "The Consequences of Water Scarcity: Measures of Human Well Being" by Peter Gleick. Almost all the good work on Australia seems to have been produced by John J. Pigram, director of the Centre for Water Policy Research at the University of New England, in Armidale, N.S.W., Among his papers: "Water Reform: Is There a Better Way?" and "Water Reform and Sustainability: The Australian Experience."

1   Article in *The Australian* from the World Water Development Agency, Tokyo, March 4, 2003.
2   Unless otherwise noted, all references to "tons" refer to metric tons, and all references to dollars are to U.S. dollars.
3   Figures from Leif Ohlsson's introduction in *Hydropolitics*, p. 3-28.
4   George Rothschild, director-general of the International Rice Research Institute in the Philippines.

2 | THE NATURAL DISPENSATION
How much water is there, who has it, and who doesn't?

There are numerous references and texts on the hydrological sciences. I have relied on the updated *Online Encyclopaedia Britannica*, whose data have proved reliable, as well as the studies of Igor Shiklomanov's State Hydrological Institute, in St. Petersburg, and Peter Gleick's excellent compilation called *Water in Crisis*. Academic papers include: "Groundwater and Its Uses in the World," by I.S. Zektser; and "Over Exploitation of Groundwater," by M. Ramón Llamas, University of Madrid. The "second update" of Population Action International's pamphlet "Sustaining Water, Easing Scarcity," by researchers Tom Gardner-Outlaw and Robert Engelman, was essential. Technical papers include "Water Availability and Scarcity in Central America and the Caribbean Region," by the hydrologist A.V. Izmailova, St. Petersburg; "Outlines on the Water Crisis in Latin America," Alda da C. Rebouças, São Paulo; "The Comprehensive Assessment of the Freshwater Resources of the World – Policy options for an integrated sustainable water future," by Johan Kuylenstierna et al.; "Global Renewable Water Resources," by Igor Shiklomanov; "World Water Resources," by Shiklomanov, summary of his monograph for the UN by the same name; "Principles for Assessment and Prediction of Water Use and Water Availability in the World," by N.V. Penkova and Igor Shiklomanov; "The World Atlas of Snow and Ice," by V.M. Kotlyakov and N.M. Zverkova; "Digital Atlas of World Water Balance," by Daene C. McKinney et al., University of Texas (also available at the Web site www.ce.utexas.edu/prof/maidment/atlas/atlas.htm); "Water Supply, Sanitation, and Environmental Sustainability, the Financing Challenge," by Ismail Serageldin, World Bank publication; "Water Quality for Human Needs: Criteria, standards, and needs," by Richard D. Robarts; "Main Water Users and Assessments of Water Withdrawals," by Mathieu Bousquet et al.

[1]   Michael Parrish, *New York Times*, Sept. 6, 2002.

[2]   Larsen B study by UIS National Ice and Snow Data Center,
      1998; Larsen A study by David Vaughan and Christopher Doake
      of the British Antarctic Survey, recounted in *Nature*, Feb. 1998.
      Some details on glacier movement from *Britannica Online*.

[3]   *Encyclopaedia Britannica*, Vol. 14, p. 1155.

[4]   Ibid.

5 Ibid.
6 *National Geographic*, March 1998.
7 For a good discussion of the hydrosphere, see *Encyclopaedia Britannica*, Vol. 20, pp. 715 *ff.*
8 *Britannica Online*, "Hydrologic Cycle," <http://www.eb.com:180/cgi-bin/g?DocF=micro/ 283/98.html>.
9 *Britannica Online*, "The Hydrosphere: Distribution and Quantity of the Earth's Waters," <http://www.eb.com:180/cgi-bin/g?DocF=macro/ 5002/99/0.html>.
10 Adapted from *Britannica Online*, "The Hydrosphere: The Hydrologic Cycle," <http://www.eb.com:180/cgibin/g?DocF=macro/ 5002/99/8.html>.
11 From an editorial in *Worldwatch*, July/Aug. 1983, p. 3.
12 *Britannica Online*: "The Earth Sciences: Hydrologic sciences: Study of the waters close to the land surface: Evaluation of the catchment water basin: Groundwater."
13 From Atef Hamdy, Mahmoud Abu-Zeid, and C. Lacirignola (Mediterranean Agronomic Institute, Bari, Italy), "Water Crisis in the Mediterranean: Agricultural water demand management, "Water International 1995.
14 From various sources, including *National Geographic Atlas of the World*, sixth edition.
15 National Wildlife Federation fact sheet, untitled, 1998.
16 John J. Pigram, "Water Reform and Sustainability."

3 | WATER IN HISTORY
How humans have always discovered, diverted, accumulated, regulated, hoarded, and misused water.

The historical anecdotes and quotations were collected from numerous sources. Particularly useful on biblical and Koranic citations was Joyce Shira Starr, whose book *Covenant over Middle Eastern Waters* contained excellent historical chapters. Roman engineering is well summarized in Trevor Hodge's book, *Roman Aqueducts and Water Supply*, published in 1989. Academic papers I found useful included: "Water Conservation Techniques and Approaches," by D. Prinz, at the University of Karlsruhe, which includes an excellent summary of ancient techniques, such as Tunisian *meskats*, qanats, and the collection of fog and dew. Two engaging books about attitudes to water in ancient times are *The Legendary Lore of the Holy Wells of England*, by R.C. Hope, and *The Folklore of Wells*, cited below.

1   Details from *A Thousand Years of East Africa*, by John Sutton
    (Nairobi: British Institute in Eastern Africa, 1992).
2   *Larousse Encyclopedia of Mythology* (London, Hamlyn, 1969), p. 11.
3   Camille Talkeu Tounounga, *Liquid of the Gods: Dogon Creation Myths*
    (UNESCO Courier, May 1993).
4   *Folklore of Wells*, Sir Rustom Pestonji Masani, Norwood, Pa.,
    Norwood Editions, 1974.
5   James Gray, *Man Against the Desert* (Saskatoon, Western Producer
    Book Service, 1967).
6   *Encyclopaedia Britannica*, Groundwater, Vol. 8, p. 433.
7   *Fresh Outlook*, April 2001.

4 | CLIMATE, WEATHER, AND WATER
Are we changing the first, and will changes in the other two necessarily follow?

Passages from Timbuktu, the Namib desert, and other places were from personal observations. The post-Kyoto proceedings of the Eminent Persons Group on Climate Change were obviously useful, as was the ongoing debate in both technical and popular journals about CO_2 production and its effects on climate. The Israeli government information office is a cornucopia of interesting material on combating desertification. A useful paper was Maria C. Donoso's "Assessment of Climate Variability Impact on Water Resources in the Humid Tropics of the Americas." Others include "Country-Specific Market Impacts of Climate Change," by Dr. Robert Mendelsohn et al., at Yale; "Meeting the Challenge of Population, Environment and Resources: The Costs of Inaction," by Henry Kendall et al., Union of Concerned Scientists; "Global Climate Warming Effects on Water Resources," by V. Y. Georgiyevsky; "Induced Climate Change Impacts on Water Resources," by Eugene Stakhiv, U.S. Army Corps of Engineers; "Water, Climate Change, and Health," by Paul R. Epstein, Harvard.

1   Dr. Reed P. Shearer et al., California Institute of Technology, in
    Science, July 1998.
2   Ranjan S. Muttiah, Temple, Texas, et al., *Water International*,
    Sept. 2002, p. 407.

3   For a discussion of the Sahara, see Tor Engelman et al., *The Desert Realm: Lands of majesty and mystery* (Washington, D.C., National Geographic Society, 1982), p. 112.

4   *Britannica Online*, <http://www.eb.com:180/cgi-bin/g?DocF=macro/5000/74/155.html> "The Biosphere and Concepts of Ecology: Major Ecosystems of the World: Terrestrial ecosystems: Deserts: Environment."

5   *National Geographic*, Nov. 1979.

6   Israeli Government Information Office fact sheet (untitled). For some of Israel's recommended solutions, see Chapter 17, "Solutions and Manifestos."

7   Ibid.

8   *Geophysical Research Letters*, July, 15 2000.

9   Magnus Magnusson and Hermann Pálsson, *The Vineland Sagas: The Norse discovery of America*, London, Penguin, 1965.

10   *The Economist* Survey of Development and the Environment March 21, 1998.

11   Reported in *The Economist*.

12   Data from "Water, Climate Change, and Health," a paper by Paul R. Epstein, associate director of the Center for Health and the Global Environment at Harvard Medical School.

13   Reported by Traci Watson, *USA Today*, Oct. 29, 1998.

14   *The Economist*, Nov. 7, 1998.

15   Eugene Stakhiv, U.S. Army Corps of Engineers, "Induced Climate Change Impacts on Water Resources," June 1998.

16   International Development Research Center home page.

5 | UNNATURAL SELECTION
Contamination, degradation, pollution, and other human gifts to the hydrosphere.

The sources are too scattered and numerous to list all of them. Some data along the Volga River, in East Europe, among the U.S.–Canadian Great Lakes, and other places are from personal observations and interviews. Particularly useful were Linda Nash's survey called "Water Quality and Health," in Peter Gleick's *Water in Crisis*, and reports from the NGO WaterAid, particularly a thorough report by researcher Maggie Black, cited below. Other sources include a case study called "Water Supply of Middle Sized

Latin American Cities Endangered by Uncontrolled Groundwater Exploitation in Urban Areas," by J. Bundschuh et al., at Salta University in Buenos Aires. There are excellent "pollution maps" in the National Geographic's revised sixth edition of the *Atlas of the World* (1995), from which some of the pollution data was drawn.

1   M. Meybeck, "Surface Water Quality: Global assessment and perspective," 1998.

2   Quoted by Chichester, *The Resurrection of the Rhine*. The quotations from Gottfried Schmidt are from the same source.

3   *The Times*, Oct. 26, 1998.

4   www.expatica.com, a Web site of Expatica Belgium, via Environment News.

5   U.S. Geological Survey Web site <epa.gov/nerlesd1/ chemistry/pharma/>, quoted in *Fresh Outlook*, Nov. 2002.

6   *Fresh Outlook*, Nov. 2002.

7   National Wildlife Federation Fact Sheet, undated.

8   Details from Linda Nash, "Water Quality and Health," in Peter Gleick, *Water in Crisis*, p. 25.

9   Quoted in Nash, in Gleick, p. 31.

10   F.B. Smith, *The People's Health, 1830–1910*, London, Croom Helm, 1979, cited by Maggie Black, "Mega-Slums," *WaterAid*, 1998.

11   Details of the megacities from many sources, but especially Maggie Black's excellent WaterAid report, 1998.

12   Black, "Mega-Slums."

13   Nash, "Water Quality and Health," p. 27.

14   Peter Popham, the *Johannesburg Star*, Oct. 12, 2000.

6 | THE ARAL SEA
An object lesson in the principle of unforeseen consequences.

Descriptions of the region are from personal observations and visits. The best journalistic accounting I have seen was Don Hinrichsen, in *People & the Planet* magazine. Also useful was Arun P. Elhance's "Conflict and Cooperation over Water in the Aral Sea Basin," in *Studies in Conflict and Terrorism*, Vol. 20, pp. 207-218.

1   The details of this Action Program were presented to a UNESCO conference in Paris in June 1998. The paper was called "The Future

of the Aral Sea Basin – How can independent states survive in water scarcity?"

2   Iskandar Abdullaev, *Water International*, June 2002, p. 266 *ff.*

3   Environmental Security and Shared Water Resources in Post-Soviet Central Asia, by David R. Smith, in *Post-Soviet Geography*, June 1995.

4   Arum P. Elhance, "Conflict and Cooperation over Water in the Aral Sea Basin, *Studies in Conflict and Terrorism*, 20: 207-18.

5   *People & the Planet*, Vol. 4, No. 2, 1995, reporting by Don Hinrichsen.

7 | TO GIVE A DAM
Dams are clean, safe, and store water for use in bad years, so why have they suddenly become anathema?

Marc Reisner's *The Cadillac Desert*, which is almost compulsively readable, contains much fascinating data, especially about U.S. dams, but also about the siltation problem faced by all dams. Janet Abramovitz's study, cited below, is thorough and thoughtful, unhyped but chilling. The Internet resounds with polemics on dams. Especially useful is the RiverNet Web site. There's a good summary of the Three Gorges project from an official Chinese perspective in C. Yangbo's paper, "The Three Gorges Project: Key project for transferring water from south to north," written in Yichang, China, in 1998.

1   *Encyclopaedia Britannica*, Vol. 1, p. 611.

2   Marc Reisner, *The Cadillac Desert*, p. 469.

3   Peter Theroux, "The Imperiled Nile Delta," *National Geographic*, Jan. 1997.

4   Janet Abramovitz, "Imperiled Waters Impoverished Future," Worldwatch paper, 1996.

5   "Athabasca Chipewayan First Nation Inquiry: W.A.C. Bennett Dam and Damage to Indian Reserve 201 Claim," Indian Claims Commission Proceedings, No. 10 (1998): pp. 117 *ff.*

6   Dianne Murray, "Dams and Dying Fisheries," Hudson Bay Research Centre, Carleton University, Ottawa.

7   Reisner, *The Cadillac Desert*, p. 472.

8   Abramovitz, "Imperiled Waters," p. 15 *ff.*

9   Philip Fradkin, *A River No More*, p. xiv, 199.

10 Interview by BBC Delhi correspondent Sunita Thackur.

8 | THE PROBLEM WITH IRRIGATION
Irrigated lands are shrinking, and irrigation is joining dams on an ecologist's hit list.

Sandra Postel has said many sensible things about irrigation in various publications, especially her *Last Oasis* and *Pillar of Sand*, and her survey, "Water & Agriculture," in Peter Gleick's *Water in Crisis*.

1   Much of the data in this section come from Postel's excellent survey, "Water and Agriculture," in Gleick.
2   Eigeland, *The Desert Realm*, p. 65.
3   Figures from the U.S. Salinity Laboratory, Riverside, California.
4   Tor Eigeland et al., *The Desert Realm*, p. 167.
5   Postel, "Water & Agriculture," p. 58.

9 | SHRINKING AQUIFERS
If the water mines ever run out, what then?

Notes and observations from personal visits to North Africa and the American southwest form the basis for this chapter. The Libyans have produced a number of interesting studies of the Great Man-Made River project, among them a useful technical paper by the hydrologist Saad Al-Ghariani, professor of water sciences at Al-Fateh University in Tripoli. Much useful material was gleaned from technical papers, such as "The Aquifers of the Grand Basins: A vital resource for development and for the struggle against desertification in the arid and semi arid zones," by Jean-Marc Louvet, Paris (Observatoire du Sahara et du Sahel); "Socio-Economic Aspects of the Demand for Domestic Water in Sub-Saharan Africa: What lessons for water management?" by Étienne J. Maïga et al.; and "The Dwindling Water Resources of Lake Chad: An alarming trend of the twenty-first century," by Abubakar B. Jauro, Lake Chad Basin Commission, N'Djamena, Chad. Marc Reisner's *The Cadillac Desert*, again, was an indispensable source on the Ogallala Aquifer. So was Michael's Glantz's *Forecasting by Analogy*, 1989, whose summary on the Ogallala was based on a study by Donald A Wilhite, the Center for Agricultural Meteorology and Climatology, University of Nebraska at Lincoln. Other references are cited in the text.

1   Sandra Postel, *Pillar of Sand: Can the Irrigation Miracle Last?*
      W.W. Norton & Co., New York, 1999.
2   From an editorial in *Worldwatch*, July/Aug. 1998, p. 3.

3   *Britannica Online*: Groundwater.

4   Ibid.

5   *Britannica Online*: The Earth Sciences: Study of the Waters Close to the Land Surface.

6   *Britannica Online*: Sahara: Physical Features.

7   Edna Jaques, *Drifting Soil* (chapbook). Moose Jaw, Sask., The Times Company Limited, 1935 (second edition), quoted in *Man Against the Desert*, by James H. Gray, Saskatoon, Western Producer Book Service, 1967.

8   Details from ibid.

9   For a much more detailed and highly readable account of the Ogallala problem, see Reisner, *The Cadillac Desert*.

10   This and following paragraphs from Sandra Postel, *Worldwatch*, Sept./Oct. 1999, p. 36.

11   *New York Times*, January 26, 2003.

10 | THE MIDDLE EAST
If the water burden really is a zero-sum game, how do we get past the arithmetic?

Much of this chapter was produced after on-the-ground observations and interviews with many of the participants. Endless amounts of material is available on Middle East water policies and politics, both in print form and online. The various Water Plans cited are all easily available from many sources. Helena Lindholm's study in Leif Ohlsson's *Hydropolitics* provides an intelligent political overview. Papers include "The Water Sources of the West Bank of the Jordan Valley and Their Utilization," by Aharon Yaffe, Jerusalem; "Water Resources Scarcity in Yemen: A time bomb under socio-economic development," by Jac A.M. van der Gun; "A Sober Approach to the Water Crisis in the Middle East," by Jad Isaac, Applied Research Institute, Jerusalem; "Troubled Water of Eden," by Daniel Hillel for *People & the Planet* (a reworking of his excellent book, *The Rivers of Eden: The Struggle for Water and the Quest for Peace in the Middle East*, published by the Oxford University Press in 1994); "Roots of the Water Conflict in the Middle East," by Jad Isaac and Leonardo Hosh (paper delivered at the University of Waterloo, Canada 1992); *The Politics of Scarcity: Water in the Middle East*, by Joyce Starr and Daniel Stoll (Westview Special Studies series on the Middle East, 1988).

1   Thomas Naff, quoted by Lindholm in "Water and the Arab–Israeli
    Conflict," in Ohlsson, *Hydropolitics*, p. 76.
2   Quoted by Stephen McCaffrey, "Water, Politics, and International
    Law," in Peter Gleick, *Water in Crisis*, p. 92.
3   Nir Becker and Naomi Zeitouni, "A Market Solution for the
    Israeli–Palestinian Water Dispute," in *Water International*, Dec. 1998.

11 | THE TIGRIS–EUPHRATES SYSTEM

*Shoot an arrow of peace into the air, and get a quiverful of suspicions
and paranoia in return.*

Turkey has an ever-proliferating library of studies (some of them useful,
many just propaganda) on the GAP project, from which much of this chapter
is drawn.

1   Quoted by Amikam Nachmani in "Water Jitters in the Middle East,"
    Studies in Conflict & Terrorism, Vol. 20, No. 1 (1997), p. 67.
2   Turkish figures from that country's State Hydraulic Works show
    water availability at 2,110 cubic metres per year per capita for Iraq,
    1,830 for Turkey, 1,420 for Syria, 300 for Israel, 250 for Jordan, and
    100 for Palestine.
3   İlter Turan, "Turkey and the Middle East: Problems and Solutions,"
    Water International, 1993.
4   Mikhail Wakil, "Analysis of Future Water Needs for Different Sectors
    in Syria," in *Water International*, Vol. 18, No. 1, 1993.
5   Quoted by Stephen McCaffrey, "Water, Politics, and International
    Law," in Peter Gleick, *Water in Crisis*, p. 101.
6   McCaffrey, ibid, p. 93.

12 | THE NILE

*With Egypt adding another million people every nine months, demand is
already in critical conflict with supply.*

I travelled virtually every kilometre of the Nile, and the descriptions come
from those personal visits to Egypt, Sudan, Ethiopia, Uganda, and Kenya.
Jan Hultin's "The Nile: Source of life, source of conflict," in Leif Ohlsson's
Hydropolitics, is a good start for anyone interested in the subject. Many of the
ancient legends cited were collected by the indefatigable if atrocious Henry
Morton Stanley, who, at least on these, has proved reliable. Other sources

include: "Water Resources Assessment in South Sinai, Egypt," by A.M. Amer et al.; and "National Environment Measures and Their Impact on Drainage Water Quality in Egypt," by M.A. Abdel-Khalik et al., Cairo.

1   Henry Morton Stanley, *In Darkest Africa*, London, Samson Kow, Marston, Searle, and Rivington, 1890, p. 448.

2   Ibid.

3   Ibid., p. 487.

4   Sandra Postel, *Last Oasis*, quoting a study by Dale Whittington and Elizabeth McClelland of the University of North Carolina.

5   Harry Thurston, *Island of the Blessed*, Doubleday Canada, 2003, pp. 344 *ff*.

6   Reuters News Service, Dec. 17, 2000.

13 | SOUTHERN AFRICA

At the other end of Africa, politically induced famine overlapped a catastrophic natural drought, and millions were at risk of starvation. But some countries were getting water policy more or less right.

The Botswana and the "on-the-ground" passages in this chapter are from personal observations and interviews. The South African paper, the *Mail & Guardian*, through its Web site, has been keeping close watch on the Southern African water situation, and several quotations are from that source. Material has also been drawn from an excellent study called *Water Management in Africa and the Middle East*, by Kathy Eales, Simon Forster, and Lusekelo Du Mhango, quoted by Population Action International, and from a letter to the author from South African architect John Bizzell.

1   Adapted from *Into Africa*, by Marq de Villiers & Sheila Hirtle (Key Porter Books, Canada, and Weidenfeld & Nicolson, 1997).

2   South African Environmental Project Web site, Nov. 1998.

3   *Mail & Guardian*, Aug. 1997.

4   Ibid., Nov. 1996.

5   Eales, Forster and Du Mhango, *Water Management in Africa and the Middle East*.

6   *Mail & Guardian*, Nov. 1996.

7   Ibid., Dec. 1996.

8   Frank Herbert, *Dune*, Appendix 1: The Ecology of Dune (New York, Ace Books, 1990).

9   Interview, 2002.

14 | THE UNITED STATES AND ITS NEIGHBOURS
Whose water is it, anyway, and how are they using (and abusing) it?

Again, Marc Reisner's *The Cadillac Desert* is must-reading for anyone interested in American water politics. Sandra Postel, in *Last Oasis*, and in her PBS mini-series, is particularly good on the lower Colorado River. Other sources include *Water Markets: Priming the Invisible Pump*, written by Terry Anderson and Pamela Snyder for the Cato Institute; *A Life of Its Own*, by Robert Gottlieb, excellent on the California water cartels; a *Wall Street Journal* piece by Mitchel Benson on water law (Oct. 1, 1997); and various pieces by U.S. Water News Online. The Canadian Environmental Law Association, based in Toronto, is an indispensable source for anyone interested in Canadian water politics, and the way the Canadian authorities are fumbling the transition to privately run water utilities and the handling of water under the NAFTA rules.

1   Quoted by Sandra Postel, *Last Oasis*, p. ix.

2   U.S. Water News Online, June 1977.

3   Clark S. Knowlton, "International Water Law Along the Mexican–American Border," paper given to a symposium of the American Association for the Advancement of Science, El Paso, Texas, April 1968.

4   From the Cascadia Times Web site, "Water in the West." The Cascadia Times's mission is "to develop a common understanding of critical environmental issues in the western United States and Canada." It was founded in 1995.

5   Reisner, *The Cadillac Desert*, p. 7.

6   Ibid., 484.

7   *Los Angeles Times*, Dec. 18, 1998.

8   *The Economist*, Jan. 24, 1998.

9   *New York Times* online, February 2003.

10 Details from a story by Jean Heller, *St. Petersburg Times*, online in the ITT Industries Handbook to Global Water Issues.

11 Scott Allen, *Boston Globe*.

12 Tom Carpenter, *Canadian Geographic*, Nov./Dec. 2002.

13 These happy thoughts about terrorist attacks thanks to David Isenberg, who makes his living as a security consultant in the U.S. I'm grateful to Robert David Steele, CEO of OSS (Open Source Solutions), an intelligence-gathering company, for steering me to Isenberg.

14 *Globe and Mail*, May 23, 1998.

15 Dave Biggar, of the O'Leary Fish and Wildlife Association, quoted in the *Halifax Chronicle-Herald*, July 25, 2002.

16 *Fresh Outlook*, Aug. 2002, p. 9.

17 Dan Braniff, in a letter to the author.

18 Southam News Service, Aquaculture News, April 1998.

19 J.C. Day, Kristan M. Boudreau, and Nancy C. Hackett, "Emerging Institutions for the Bilateral Management of the Columbia River Basin," *American Review of Canadian Studies*, Summer 1996.

20 *The Economist*, March 29, 1997.

15 | SOUTH ASIA AND THE INDIAN SUBCONTINENT
Some water issues are intractable, and some that look intractable have, to everyone's astonishment, been amicably solved.

All three of the Indian water conflicts have been usefully summarized in Elizabeth Corell and Asok Swain, "India: The domestic and international politics of water scarcity," in Leif Ohlsson's *Hydropolitics*. Much of the background on the Cauvery River problem came from personal interviews with people in the region. Other sources include the papers "Integrated water quality management plan for Bhadri Lake, India," by Sekhar M. Chandra; "Environmental Impact of Ground Water Utilization and Development: The Indian perspective," by Dr. D.C. Singhal; and "Intersectoral Competition for Land and Water Between Users and Uses in Tamil Nadu," by S. Suresh.

1 *The Hindu*, Sept. 25, 2002.

2 James J. Novak, *Bangladesh: Reflection on the Water*, quoted in Ranabir Samaddar, "Flowing Waters and Nationalist Metaphors: The Dispute between India and Bangladesh over the Ganges," *Studies in Conflict and Terrorism*, Vol. 20, No. 2, 1997.

3 Quoted by Tara Patel in *New Scientist*, Nov. 30, 1996.

4 Figures from Corell and Swain, "India," in Ohlsson, *Hydropolitics*, which contains a meticulous summary of the politics of the region.

5 Stephen McCaffrey, *Water, Politics, and International Law*, 94.

6 Rama Lakshmi, *International Herald-Tribune*, Aug. 26, 2002.

7 Corell and Swain, "India," in Ohlsson, *Hydropolitics*.

16 | THE CHINESE ARE USED TO THINKING BIG
They built the Great Wall, after all. So why not just reorganize the whole country, hydrologically speaking?

Coverage of China's water woes was given a huge impetus in summer 1998, when the Yangt'ze, once again, catastrophically overflowed, leading to endless charges and countercharges about the role in the disaster played by the Three Gorges Dam project, now under construction. For the first time, the summer saw the emergence of an activist Green movement inside China, and more recently the Chinese authorities have themselves gone public with a panoply of studies on the country's manifest environmental problems – in fact, China, once so secretive, is now more candid than most countries about its environmental woes. Much of the information in this chapter came from personal interviews in China in 2002.

1   Liu Changming et al., *Water International*, June 2001, p. 265 *ff.*
2   Brian Halweil, Virtual Information Center.
3   Ed Ayes, *Worldwatch*, Nov./Dec. 1998.
4   *Daily Telegraph*, London, June 15, 2001.
5   *China Daily*, May 28, 2002.
6   Anne H. Mavor, Positive Futures Network.
7   AP report from Beijing, Feb. 9, 2003.
8   Erik Eckholm, *New York Times*, Aug. 27, 2002.

17 | SOLUTIONS AND MANIFESTOS
If you're short of water, especially clean water, the choices are conservation, technological invention, or the politics of violence.

The water world seems to be split between those who believe that there are technical (that is, engineering and scientific) solutions to the looming crisis in fresh water, and those who believe the problem is more one of politics and management. Sandra Postel leans to the latter. So does Maurice Strong, who was secretary-general of the Rio conference on the environment in the early '90s, and who has been active in water politics ever since, partly through the Earth Council, which he founded, through his influence on the World Bank and more recently in his advisory role to the government of China. Of course, the two positions overlap, as I have tried to show: the best conservation methods come from fine-tuned engineering. It's why people like

Maurice Strong have argued so vigorously for the international sharing of scientific data. Other sources include papers such as "Transboundary Fresh Water Disputes and Conflict Resolution: Planning an integrated approach," by Edy Kaufman et al., in *Water International*; "The Economic Value of Water," by Kindler Janusz, Warsaw; and "Water & Food," by W.E. Klohn et al., of FAO (includes useful sections on, for example, how much water it takes to produce food). Stephen McCaffrey's survey of international water law in Peter Gleick's *Water in Crisis* was a source for my own summary (though he is not to be blamed for conclusions I may have extracted).

1   Wade Graham, "A Hundred Rivers Run Through It,"
    Harper's Magazine, June 1998.
2   Paper presented to an International Association of Hydraulic
    Research meeting in Namibia.
3   Amikam Nachmani, *Water Jitters in the Middle East*, 67-93.
4   Details from the California Water Commission.
5   Woodward-Clyde Consultants, EIR for the City of Santa Barbara and
    Ionics, Inc.'s Temporary Emergency Desalination Project, March 1991.
6   From the *Straits Times*, Singapore, June 1, 2002, p. 1.
7   Robert Parke in an e-mail to the author.
8   D. Rajyalakshmi in *Asian Water* no. 15, 1999.
9   David Lees, "Food by Design," *Financial Post* magazine, Oct. 1998.
10 Sandra Postel et al., in *Water International*, March 2001, pp. 3 *ff*.
11 Ibid, p. 8.
12 Richard Golb, executive director, Northern California Water
    Association.
13 Professor Avishai Braverman, Mordechai Cohen, et al.
14 Newspaper report from Leila Deeb, Amman.
15 Postel, *Last Oasis*, 129. Postel's book contains a good survey of
    wastewater treatment.
16 Shaul Streit, "On-Land Treatment and Disposal of Municipal Sewage:
    Agro-sanitary integration" (a paper for a World Bank seminar), 1992,
    quoted by Sandra Postel, *Last Oasis*, 129.
17 *Jerusalem Report*, July 29, 2002.
18 Postel, *Last Oasis*, 135.
19 Sheldon Rampton, *Harper's Magazine*, November 1998.
20 Araf Hassan, *Innovative Sewerage in a Karachi Squatter Settlement*,
    Orangi Pilot Project, 1986.

21   André de Moor and Peter Calamai, *Subsidizing Unsustainable Development: Undermining the Earth with public funds.* Earth Council 1997.

22   *The Economist,* Dec. 20, 1997.

23   *New York Times,* Aug. 26, 2002.

24   *The Economist,* Dec. 20, 1997.

25   Tom Gardner-Outlaw and Robert Engelman, "Sustaining Water, Easing Scarcity: A second update," for Population Action International, Washington, D.C., 1997.

26   Third System Review of the Consultative Group on International Agricultural Research, October 1998.

27   McCaffrey, *Water, Politics, and International Law,* p. 99.

28   Frederick W. Frey, "The Political Context of Conflict and Cooperation Over International River Basins," *Water International* No. 18, 1993.

29   See particularly "Environmental Scarcity and Global Security," by Thomas Homer-Dixon, Foreign Policy Association Headline Series, No. 300.

BIBLIOGRAPHY

Some of the essential sources were mentioned in my Introduction and in the Endnotes. But of course there are many, many others.

Books and Pamphlets

Abramovitz, Janet N.: *Imperiled Waters, Impoverished Future: The Decline of freshwater ecosystems*. Worldwatch Institute, Washington, D.C.: 1996.

Anderson, Terry L. and Snyder, Pamela. *Water Markets: Priming the invisible pump*. Washington, D.C.: The Cato Institute, 1979.

Basson, M.S., *Overview of Water Resources Availability and Utilisation in South Africa*. Pretoria: Department of Water Affairs and Forestry, 1997.

Brown, Lester R. *Who Will Feed China? Wake-up call for a small planet*. New York: W.W. Norton & Co., 1995.

CGIAR System Review Secretariat, *Third System Review*, The International Research Partnership for Food Security and Sustainable Agriculture, Oct. 8, 1998.

Clarke, Robin, *Water: The International Crisis*. MIT Press, 1993.

Cossi, Olga, *Water Wars: The Fight to control and conserve nature's most precious resource*, New Discovery Books, New York, 1993.

Dixon Hardy, P, *The Holy Wells Of Ireland*, Hardy & Walker, Dublin, 1840.

Encyclopaedia Britannica (both print and on-line editions).

Eigeland, Tor et al., *The Desert Realm: Lands of majesty and mystery*, National Geographic Society, Washington, D.C., 1982.

El-Ashry, Mohamed T., & Gibbons, Diana, editors, *Water and Arid Lands of the Western United States*, 1988.

Engelman, Robert & LeRoy, Pamela, *Sustaining Water: Population and the future of renewable water supplies*. Population Action International, Washington, D.C., 1993.

Engelman, Robert, *Why Population Matters*, International Edition, Population Action International, Washington, D.C., 1997.

Farid, Claire, Jackson, John & Clark, Karen, *The Fate of the Great Lakes: Sustaining or draining the sweetwater seas?* Canadian Environmental Law Association and the Great Lakes United Buffalo State College, Buffalo, N.Y., 1997.

Fradkin, Philip L., *A River No More: The Colorado River and the West*, University of California Press, Berkeley and Los Angeles, 1996.

Gardner-Outlaw, Tom & Engelman, Robert, *Sustaining Water, Easing Scarcity: A second update*. Population Action International, Washington, D.C., 1997.

Gleick, Peter H., *The World's Water: The Biennial report on fresh water resources 1998–1999*. Island Press, Washington, D.C., 1998.

Gleick, Peter, editor, *Water In Crisis: A guide to the world's fresh water resources*. Oxford University Press, New York, 1993.

Glantz, Michael, *Forecasting by Analogy*. NCAR. (Includes summary on the Ogallala based on a study by Donald A Wilhite, the Center for Agricultural Meteorology and Climatology, University of Nebraska at Lincoln), 1989.

Gottlieb, Robert, *A Life of Its Own: The Politics and power of water*. Harcourt Brace Jovanovich, New York, 1988.

Graves, Robert, editor, *New Larousse Encyclopedia of Mythology*. Paul Hamlyn, London, 1959.

Gray, James H., *Men Against the Desert*. Prairie Books, Western Producer Book Service, Saskatoon, 1967.

Herbert, Frank, *Dune*. Ace Books, New York, 1990.

Hillel, Daniel, *The Rivers of Eden: The Struggle for water and the quest for peace in the middle east*. Oxford University Press, 1994.

Homer Dixon, Thomas F., *Environmental Scarcity and Global Security*. Headline Series, Foreign Policy Association, Inc., New York, 1993.

Hope, R.C. *The Legendary Lore of the Holy Wells of England*, Elliot Stock, London, 1893.

Kendall, Henry, et al., *Bioengineering of Crops: Report of the World Bank panel on transgenic crops*. The World Bank, Washington, D.C., 1997.

Kendall, Henry W. et al., *Meeting The Challenges Of Population, Environment, and Resources: The Cost of inaction*. A Report of the Senior Scientists' Panel, Union of Concerned Scientists. The World Bank, Washington, D.C., 1995.

Loude, Jean-Yves & Lièvre, Viviane, *Le Roi d'Afrique et la reine mer.* Quoted in *Balafon*, the in-flight magazine of Air Afrique Vol. 126.

Lowi, Miriam R., *The Politics of Water: The Jordan River and the Riparian States*. McGill Studies in International Development (No. 35), 1984.

Magnusson, Magnus & Pálsson, Hermann, *The Vineland Sagas: The Norse discovery of America*. Penguin, London, 1965.

Masani, Sir Rustom Pestonji, *The Folklore of Wells*, Norwood Editions, Norwood, Penn., 1974.

National Geographic Society, *Water: The Power, promise, and turmoil of North America's fresh water*, Special Issue [Vol. 184 No. 5A] National Geographic, Washington, D.C., 1993.

Ohlsson, Leif, editor, *Hydropolitics: Conflicts over water as a development constraint*. Zed Books, London.

Postel, Sandra, *Dividing The Waters: Food security, ecosystem health, and the new politics of scarcity*. Worldwatch Institute, Washington, D.C., 1996.

Postel, Sandra, *Last Oasis: Facing water scarcity*. W.W. Norton & Company, New York, 1997.

Quiller-Couch, M. & L. *Ancient Holy Wells Of Cornwall*, Charles. J. Clark, London, 1894.

Reisner, Marc, *The Cadillac Desert: The American West and its disappearing water*. Penguin Books, New York, 1986.

Starr, Joyce Shira, *Covenant Over Middle Eastern Waters: Key to world survival*. Henry Holt & Co, New York, 1995.

Sutton, John, *A Thousand Years Of East Africa*. British Institute in Eastern Africa, Nairobi, Kenya, 1992.

Walton, K., *The Arid Zones*, Hutchinson, London, 1969.

Worster, Donald, *Rivers of Empire: Water, aridity, and the growth of the American West*, Oxford University Press, New York, 1985.

Young, Gordon J., Dooge, James C.I., and Rodda, John C., *Global Water Resource Issues*. Cambridge University Press, 1994.

Academic & Technical Papers

(Papers unsourced in this list can be found in the Proceedings, International Conference on Water, entitled "Water: The Looming Crisis," Paris, June 1998, published by UNESCO).

Abdel-Khalik, M.A. et al. (Cairo), "National Environment Measures and Their Impact on Drainage Water Quality in Egypt."

Amer, A.M. et al., "Water Resources Assessment in South Sinai, Egypt."

Bousquet, Mathieu et al., "Main Water Users and Assessments of Water Withdrawals."

Bundschuh, J. et al. (Salta University, Buenos Aires), "Water Supply of Middle Sized Latin American Cities Endangered by Uncontrolled Groundwater Exploitation in Urban Areas."

Chandra, Sekhar M. (Warangal, India), "Integrated Water Quality Management Plan for Bhadri Lake, India."

Donoso, Maria C., "Assessment of Climate Variability Impact on Water Resources in the Humid Tropics of the Americas."

Dukhovny, Viktor (Scientific Information Committee of the five-state Interstate Water Commission), "The Future of the Aral Sea Basin – How can independent states survive in water scarcity?"

Eales, Kathy, Forster, Simon and Du Mhango, Lusekelo, "Water Management in Africa and the Middle East."

Elhance, Arun P., "Conflict and Cooperation over Water in the Aral Sea Basin," in *Studies in Conflict and Terrorism*, Vol. 20, pp. 207-18.

Epstein, Paul R. (Harvard), "Water, Climate Change, and Health."

Falkenmark, Malin, "The Massive Water-scarcity Now Threatening Africa – Why isn't it being addressed?" *Ambio*, Vol. 18, No. 2.

Falkenmark, Malin, "Middle East Hydropolitics: Water scarcity and conflicts in the Middle East," *Ambio*, Vol. 18, No. 6.

Falkenmark, Malin, "Vulnerability Generated by Water Scarcity," *Ambio* Vol. 18, No. 6.

Falkenmark, Malin, "Global Water Issues Confronting Humanity," *Journal of Peace Research*, Vol. 27, No. 2.

Frey, Frederick W., "The Political Context of Conflict and Cooperation Over International River Basins," in *Water International* vol. 18, 1993.

Gardner-Outlaw, Tom and Engelman, Robert, "Sustaining Water, Easing Scarcity: A second update," for Population Action International, 1997.

Georgiyevsky, V. Yu., "Global Climate Warming Effects on Water Resources."

Gleick, Peter, "The Consequences of Water Scarcity: Measures of human well being."

Hasan, Arif, "Innovative Sewerage in a Karachi Squatter Settlement," Orangi Pilot Project, 1986.

Hillel, Daniel, "Troubled Water of Eden, for People and the Planet" (a reworking of his excellent book, The Rivers of Eden: The Struggle for Water and the Quest for Peace in the Middle East, published by the Oxford University Press in 1994).

Homer-Dixon, Thomas, "Environmental Scarcity and Global Security," in Foreign Policy Association Headline Series, #300.

Isaac, Jad (Applied Research Institute, Jerusalem), "A Sober Approach to the Water Crisis in the Middle East."

Isaac, Jad and Hosh, Leonardo, "Roots of the Water Conflict in the Middle East" (paper delivered at the University of Waterloo, Canada, 1992).

Izmailova, A.V. (St. Petersburg), "Water Availability and Scarcity in Central America and the Caribbean Region."

Janusz, Kindler (Warsaw), "The Economic Value of Water."

Jauro, Abubakar B., "The Dwindling Water Resources of Lake Chad: An alarming trend of the twenty-first century." (Lake Chad Basin Commission, N'Djamena).

Kaufman, Edy et al., "Transboundary Fresh Water Disputes and Conflict Resolution: Planning an integrated approach," in *Water International*.

Kendall, Henry et al., "Meeting the Challenge of Population, Environment, and Resources: The Costs of inaction" (Union of Concerned Scientists).

Klohn, W.E. et al., "Water and Food," for FAO.

Kotlyakov, V.M. and Zverkova, N.M., "The World Atlas of Snow and Ice."

Kuylenstierna, Johan et al., The Comprehensive Assessment of the Freshwater Resources of the World – Policy options for an integrated sustainable water future."

Llamas, M. Ramón, University of Madrid, "Over-exploitation of Groundwater."

Louvet, Jean-Marc (Paris), "The Aquifers of the Grand Basins: A vital resource for development and for the struggle against desertification in the arid and semi arid zones," for Observatoire du Sahara et du Sahel.

Maïga, Étienne J. et al., "Socio-economic Aspects of the Demand for Domestic Water in sub-Saharan Africa: What lessons for water management?"

McKinney, Daene C. et al. (University of Texas), "Digital Atlas of World Water Balance" (also available at the Web site www.ce.utexas.edu/prof/maidment/atlas/atlas.htm).

Mendelsohn, Dr. Robert et al. (Yale), "Country-Specific Market Impacts of Climate Change."

Meybeck, M. (Marie M. Curie University, Paris), "Surface Water Quality – Global assessment and perspective."

Murray, Dianne, "Dams and Dying Fisheries," Hudson Bay Research Centre, Carleton University, Ottawa, 1998.

Neu, H.J.A. "Runoff Regulation for Hydropower and its Effects on the Ocean Environment." *Canadian Journal of Civil Engineering*, Vol. 2, pp. 583-91.

Neu, H.J.A. "Man-Made Storage of Water Resources – A liability to the ocean environment?" Part 1, *Marine Pollution Bulletin*, Vol. 13, No. 1, pp. 7-12; Part 2, *Marine Pollution Bulletin*, Vol. 13, No. 2, pp. 44-47.

Pigram, John J. (Centre for Water Policy Research at the University of New England, in Armidale, N.S.W.) "Water Reform: Is there a better way?"

Pigram, John J., "Water Reform And Sustainability: The Australian experience." Paper for the Centre for Water Policy Research, Ninth World Water Congress, Montreal, Sept. 1997.

Prinz, D. (University of Karlsruhe), "Water Conservation Techniques and Approaches" (includes an excellent summary of ancient techniques, such as Tunisian *meskats*, *qanats*, and the collection of fog and dew).

Rango, A. (USDA Hydrological Laboratory, Beltsville, Md.), "Techniques for Assessing Changing Water Resources in the Twenty-first Century."

Rebouças, Alda da C. (São Paulo), "Outlines on the Water Crisis in Latin America."

Robarts, Richard D. (Saskatoon), "Water Quality for Human Needs: Criteria, Standards, and Needs."

Roberts, Bruce R., "Water Management in Desert Environments: A comparative analysis." *Lecture Notes in Earth Sciences*, Vol. 48, 1993.

Robinson, Nicholas, editor, "Agenda 21: Earth's Action Plan" [annotated], The Commission on Environmental Law of IUCN – The World Conservation Union, Oceana Publications Inc., New York, 1993.

Serageldin, Ismail, "Water Supply, Sanitation, and Environmental Sustainability: The Financing challenge." Directions in Development, World Bank, Washington, D.C.

Singh, Udai P. & Helweg, Otto J., editors, "Supplying Water and Saving the Environment for Six Billion People." Proceedings of Selected Sessions from the 1990 ASCE Convention, 1990.

Shiklomanov, Igor, and Penkova, N.V., "Principles for Assessment and
    Prediction of Water Use and Water Availability in the World."
Shiklomanov, Igor, "Global Renewable Water Resources."
Shiklomanov, Igor, "World Water Resources" (summary of his monograph
    for the UN by the same name).
Singhal, Dr. D.C. (Roorkee, India), "Environmental Impact of
    Groundwater Utilization and Development: The Indian perspective."
Smith, David R., "Environmental Security and Shared Water Resources
    in Post-Soviet Central Asia," in *Post-Soviet Geography*, June 1995.
Stakhiv, Eugene, "Induced Climate Change Impacts on Water Resources"
    (U.S. Army Corps of Engineers).
Starr, Joyce and Stoll, Daniel, "The Politics of Scarcity: Water in the
    Middle East" (Westview Special Studies series on the Middle
    East, 1988).
Suresh, S. (Bangalore), "Intersectoral Competition for Land and Water
    between Users and Uses in Tamil Nadu."
Union of Concerned Scientists, "World Scientists' Call For Action At the
    Kyoto Climate Summit," Cambridge, Mass., 1997.
van der Gun, Jac A.M., "Water Resources Scarcity in Yemen: A time
    bomb under socio-economic development."
Water Technology Board Tenth Anniversary Symposium, "Sustaining Our
    Water Resources." National Academy Press, Washington, D.C., 1993.
Yaffe, Aharon (Jerusalem), "The Water Sources of the West Bank of the
    Jordan Valley and Their Utilization."
Yangbo, C. (Yichang, China), "The Three Gorges Project: Key project
    for transferring water from south to north," 1998.
Zaretskaya, I.P., "Water Availability and Use in the Danube Basin," and
    "State of the Art: Expected water availability and water use in the
    Danube Basin."
Zektser, I.S., "Groundwater and its Uses in the World."

INDEX